COPING WITH CRISIS AND HANDICAP. Proceedings of a national symposium, ed. by Aubrey Milunsky. Plenum, 358p ill bibl index 81-2565. 19.50 ISBN 0-306-40660-8. CIP
A comprehensive collection of papers and discussions from a national symposium on the care of dying and handicapped children. The authors, who represent parents as well as professionals from such fields as medicine, nursing, social work, psychology, and religion, write vividly and cogently about the coping efforts of children and families faced with serious crises, death, and/or handicaps; the impact on professionals working with them; and the rich range of creative helping approaches and strategies. This compelling, useful volume should be particularly meaningful for anyone involved in programs to help not only children but also parents and other family members trying to cope with a child's death, dying, or handicap. Excellent bibliography and index.

Coping with
Crisis and Handicap

Coping with
Crisis and Handicap

Edited by
Aubrey Milunsky, MB.B.Ch., M.R.C.P., D.C.H.

Harvard Medical School, Eunice Kennedy Shriver Center, and
Massachusetts General Hospital
Boston, Massachusetts

Plenum Press • New York and London

Library of Congress Cataloging in Publication Data

Main entry under title:

Coping with crisis and handicap.

Proceedings of a national symposium.
Includes bibliographies and index.
1. Handicapped children—Family relationships— Congresses. 2. Handicapped children—
Care and treatment—Congresses. 3. Terminally ill children—Family relationships—
Congresses. 4. Terminally ill children—Care and Treatment—Congresses. I. Milunsky,
Aubrey. [DNLM: 1. Counseling—Congresses. 2. Death—In infancy and childhood—
Congresses. 3. Handicapped—Congresses. 4. Terminal care—In infancy and childhood—
Congresses. WS 200 C783 1979]

HV888.C67	362.8'2	81-2565
ISBN 0-306-40660-8		AACR2

Proceedings of the National Symposium on Coping with Crisis and Handicap,
co-sponsored by the Eunice Kennedy Shriver Center, the Boston University
School of Nursing, the Boston College School of Nursing, and the
Massachusetts Nurses Association, and held September 10—11, 1979, at the
Copley Plaza Hotel, Boston, Massachusetts

© 1981 Aubrey Milunsky

Plenum Press is a Division of Plenum Publishing Corporation
233 Spring Street, New York, N.Y. 10013

Printed in the United States of America

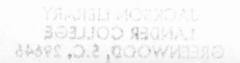

Contributors

GEORGE J. ANNAS, J.D., M.P.H. • Associate Professor of Law and Medicine, Boston University School of Medicine, and Chief, Health Law Section, Boston University School of Public Health, Boston, Massachusetts 02118

HENRY A. BEYER, J.D. • Associate Director, Center for Law and Health Sciences, Boston University School of Law, Boston, Massachusetts 02115

ALAN J. BRIGHTMAN, Ph.D. • Executive Director, Educational Projects, Inc., Cambridge, Massachusetts 02139

SUE S. CAHNERS, A.C.S.W. • Director, Social Service Department, Shriners Burns Institute, Boston, Massachusetts 02114

JOHN A. CARR, B.D., S.T.M. • Chaplain, The Yale–New Haven Medical Center, Yale Psychiatric Institute, New Haven, Connecticut 06511, and Gaylord Hospital, Wallingford, Connecticut 06492.

NED H. CASSEM, Ph.L., M.A., M.D., B.D. • Associate Professor of Psychiatry, Harvard Medical School, and Chief, Psychiatric Consultation–Liaison Service, Massachusetts General Hospital, Boston, Massachusetts 02114

MARY S. CHALLELA, R.N., D.N.Sc. • Director of Nursing and Training, Eunice Kennedy Shriver Center, and Assistant Clinical Professor, Boston University School of Nursing, Waltham, Massachusetts 02154

WALTER P. CHRISTIAN, Ph.D. • Director, May Institute for Autistic Children, Inc., Chatham, Massachusetts 02633

ALLEN C. CROCKER, M.D. • Director, Developmental Evaluation Clinic, Department of Medicine, The Children's Hospital Medical

Center, and Associate Professor of Pediatrics, Harvard Medical School, Boston, Massachusetts 02115

GUNNAR DYBWAD, J.D. • Professor Emeritus of Human Development, Florence Heller Graduate School of Social Welfare, Brandeis University, Waltham, Massachusetts 02154

GENEVIEVE V. FOLEY, R.N., M.S.N. • Head of Pediatric Nursing, Pediatric Service, Memorial Sloan-Kettering Cancer Center, New York, New York 10021

EARL A. GROLLMAN, D.D. • Rabbi, Temple Beth El, Belmont, Massachusetts 02178

HAROLD S. KUSHNER, D.H.L. • Rabbi, Temple Israel of Natick, Natick, Massachusetts 01760

IDA M. MARTINSON, R.N., Ph.D. • Professor of Nursing and Director of Research, School of Nursing, University of Minnesota, Minneapolis, Minnesota 55455

AUBREY MILUNSKY, MB.B.Ch., M.R.C.P., D.C.H. • Director, Genetics Division, Eunice Kennedy Shriver Center; Assistant Professor of Pediatrics, Harvard Medical School; and Medical Geneticist, Massachusetts General Hospital, Boston, Massachusetts 02115

MARGARET A. O'CONNOR, B.A., M.Ed. (candidate) • Hospital Teacher, Shriners Burns Institute, Boston, Massachusetts 02114

HARRIET SARNOFF SCHIFF • Birmingham, Michigan 48010

LUDWIK S. SZYMANSKI, M.D. • Director of Psychiatry, Developmental Evaluation Clinic, and Associate, Department of Psychiatry, The Children's Hospital Medical Center, Boston, Massachusetts 02115

ROBERT M. VEATCH, M.S., B.D., M.A., Ph.D. • Professor of Medical Ethics, The Kennedy Institute of Ethics, Georgetown University, Washington, D.C. 20057

J. WILLIAM WORDEN, Ph.D. • Assistant Professor of Psychology, Harvard Medical School, and Research Director, Omega Project, Massachusetts General Hospital, Boston, Massachusetts 02114

Preface

For over 20 years I have accepted the challenge and had the privilege of caring for sick children, agonizing with their parents during periods of serious illness, which were sometimes fatal. Because of my particular interest in and concern about birth defects and genetic disease, many of these children had severe disabling handicaps, which were often genetic and included mental retardation. Hence care of these children and their families was often complicated by the presence of serious or profound genetic defects. The initial realization of the nature of the disorder invariably led to emotional difficulties and inevitably later spawned chronic distress. For some children inexorable deterioration led to untimely deaths, while the parents agonized over their handicapped, chronically ill, or defective—but nevertheless loved—children.

The personal pain I experienced in having to communicate diagnoses about irrevocable or fatal disease or handicap soon after birth was surpassed later only by the agony of caring for child and family through to the end of a fatal disorder. While these cumulative sad experiences may have conferred no special abilities in caring for sick and handicapped children or their distraught parents, I soon recognized the critical importance of care by an interdisciplinary team. Most would concur about the importance of a team approach to the care of the dying or handicapped child and agree that we all have much to learn. Conversations with only a few bereaved parents will confirm how inadequate our support systems really are. The majority of parents with dying or seriously handicapped children will also, in my experience, be able to be correctly critical of at least some aspects of the management of their children.

As a consequence of these realizations, we organized a National Symposium on Coping with Crisis and Handicap. This volume constitutes a record of those proceedings. Symposium discussions evolved as a consequence of the multidisciplinary involvement of those concerned about the care of dying and handicapped children. We were fortunate in being able to benefit from the participation of a unique set of individuals, all of whom brought not only acknowledged expertise but also special and often personal insights into these particularly difficult problems.

The first day was devoted to a consideration of coping with death and dying in childhood. Discussion about the management of the dying child in the hospital initiated the proceedings. Ms. Genevieve Foley, who has had extensive experience in the care of dying children, particularly those dying from various malignancies, shared her special insights, gained from long experience at the Massachusetts General Hospital in Boston. Professor Ida Martinson has done some outstanding research studies concerning the management of the child dying at home; she communicated many of the lessons and guiding principles of care observed in her studies. Mrs. Harriet Sarnoff Schiff, who lost her 10-year-old son to congenital heart disease, articulated with singular clarity the parental perspectives that were so instructive in her book on bereavement. The eternal question "Where was God?" was eloquently addressed by Rabbi Harold Kushner, who had also lost a child from a prolonged genetic disorder.

Drawing upon extensive experience in the care of children with chronic fatal genetic disease, I next shared my insights and perspectives on this subject, with special attention to cystic fibrosis and genetic disorders characterized by mental retardation and/or a fatal outcome. Rabbi Earl Grollman discussed the child's perception of death and, for this volume, kindly allowed us to reprint his outstanding paper "Explaining Death to Children." In addition, I am grateful for his paper on "The Clergyman's Role in Grief Counseling." Dr. William Worden outlined his impressive experience, gained at the Massachusetts General Hospital, in the managment of individuals who have attempted suicide and their families. Dr. Edwin Cassem, who was unable to attend the meeting, was most gracious in allowing us to reprint a major paper he has written on "Treating the Person Confronting Death." Through his unique involvement with the critically ill and dying at the Massachusetts General Hospital as Chief of the Psychiatric Consultation–Liaison Service, our understanding of the dimensions required for sensitive and superlative care has been broadened.

The first day concluded with a careful examination of the medical ethics involved in the care of the child with terminal illness. This subject was addressed by Professor Robert Veatch, who has made extensive contributions to the literature on death and dying. Professor George Annas rounded out an emotionally enervating day by considering various legal rights of the dying patient. He is especially well known for his writings on the rights of hospital patients.

The second day focused on the problems concerned with coping with congenital or acquired handicap in childhood. We were especially fortunate to learn from the personal experience and theological insights of Reverend John Carr. His accomplishments and perspectives about living with physical handicap were especially enlightening. Dr. Mary Challela, who has had outstanding experience with the nursing management of mentally retarded children, outlined her detailed approach to helping parents cope with profoundly retarded children. The frequently neglected subject of the siblings of children with handicaps, especially mental retardation, was thoughtfully addressed by Professor Allen Crocker. We were all able to benefit from the lessons of his long experience in the Developmental Evaluation Clinic at the Children's Hospital Medical Center in Boston.

The enormous and stressful problems associated with the management of autistic children were discussed by Dr. Walter Christian. His expertise, gained partly from his experience as Director of an institute for autistic children, added a valuable dimension to the symposium. Dr. Ludwik Szymanski is a child psychiatrist whose special knowledge of retarded children reflects his involvement in the Developmental Evaluation Clinic at the Children's Hospital Medical Center. His subject was the frequently forgotten one of the sexual vulnerability of retarded individuals.

Societal failure to teach children early in life about the acceptance of normal and abnormal human variation is evidenced daily by children's cruelty to their disabled or defective peers. Dr. Alan Brightman, who is recognized as an outstanding educator, skillfully guided us through the approaches necessary to more sensitively educate our children about their handicapped peers. The awesome injuries suffered as a consequence of burns in childhood have required the help of individuals with unusual mettle. At the Shriners Burns Hospital in Boston, Ms. Margaret O'Connor has not only taught many burned children during their long process of recovery but also helped teachers and their pupils to cope with the return of the profoundly burn-disfigured child to the classroom. Her work has also necessitated educating public health agencies about their responsibilities vis à vis the rights of these burned children to return to school. The coping strategies of these children and their families, as observed and analyzed by a most experienced social worker, were discussed by Mrs. Sue Cahners, also of the Shriners Burns Hospital.

Mr. Henry Beyer, in the penultimate address, considered various legal aspects concerned with the care and rights of the handicapped. He has done legal work for years on behalf of the handicapped. The final presentation was made by Professor Gunnar Dybwad, who is internationally recognized for his many decades of work on behalf of retarded citizens. His special societal perspectives, with particular reference to government responsibilities for the mentally handicapped, are a distillate of his distinguished experience and prolific writing.

This book should prove valuable for all those concerned about coping with death, dying, and handicap in childhood, including members of the many professional disciplines that focus their attention or care on such children. The list is endless, and includes nurses, physicians, social workers, psychologists, teachers, theologians, ethicists, lawyers, health educators and administrators, public health department personnel, and staff members in preventive and social medicine departments. This book would be especially meaningful as well to all parents and families faced with the necessity of coping with death, dying, or handicap in childhood.

The gamut of emotions experienced by every parent with a dying or seriously handicapped child includes guilt, despair, chronic grief, and even rage. Unfortunately, coping with such overwhelming emotional burdens is not only very personal but often very lonely. None of us are taught or prepared beforehand to cope with the loss of a child or with having a seriously handicapped one. Moreover, the support systems that we have developed are still invariably inadequate. This book is therefore dedicated to the effort of helping parents cope by focusing on their needs and on those of their children and by increasing the sensitivity, knowledge, and insight of their caretakers.

AUBREY MILUNSKY

Acknowledgments

This national symposium was made possible by the fruitful and excellent collaboration between the Eunice Kennedy Shriver Center, the Boston University School of Nursing, the Boston College School of Nursing, the Northeastern University College of Nursing, and the Massachusetts Nurses Association. I am indebted to all these groups, whose combined efforts yielded such a successful symposium. Special appreciation is due to the program committee, co-chaired by Mrs. Babette R. Milunsky, and whose members included Mrs. Barbara Gilman, Ms. Judy Heck, Ms. Marcia C. Hehir (Associate Director, Department of Programs and Divisions of Practice, Massachusetts Nurses Association), Mrs. Suzette E. Kushner, Ms. Elizabeth H. McAnulty (Nurse Coordinator, The Massachusetts Sudden Infant Death Syndrome Center, Boston City Hospital), Ms. Sheila A. Packard (Education Coordinator, Visiting Nurse Association of Boston), and Ms. Judith A. Shindul (Clinical Nursing Supervisor of the Psychiatric Intensive Care Unit, McLean Hospital, Belmont, Massachusetts).

AUBREY MILUNSKY

Contents

1

GENEVIEVE V. FOLEY is the Head of Pediatric Nursing at Memorial Sloan-Kettering Cancer Center in New York. She received a bachelor's degree from the Boston College School of Nursing and a master's degree in maternal–child health care from the University of Pennsylvania, where she was inducted into Sigma Theta Tau, the National Honor Society of Nursing. From 1970 to 1980 she was the pediatric nurse clinician at the Massachusetts General Hospital in Boston, where she specialized in pediatric cancer nursing. She was a member of the MGH team who planned and implemented the therapeutic regimen for the late Chad Green.

Ms. Foley is the author of several articles and co-editor of *Nursing Care of the Child with Cancer*. She is a charter member and past national president of the Association of Pediatric Oncology Nurses. In 1980 she was named one of Boston's Ten Outstanding Young Leaders by the Boston Jaycees. She is the first nurse so honored by the Boston Jaycees.

The Child Dying in the Hospital

GENEVIEVE V. FOLEY

In *To Live Until We Say Good-Bye,* Elisabeth Kübler-Ross tells the story of Linda and her daughter Jamie, age 5. Jamie is dying of a brain tumor. Linda, a single parent, decides to have Jamie die at home. Kübler-Ross relates how Linda telephoned her and "proudly announced 'We did it!' 'We did it' referred, naturally, to her own courage and her own pride that she had decided to take Jamie home."[1] The narrative continues with Kübler-Ross and Linda describing the many positive aspects of the decision. Linda identifies the unique opportunity: "to do all that was left to be done for her [Jamie]—to make her comfortable, to provide her with familiar things, and most important to surround her with the love of her family and friends."[2]

Although this episode in the book is concerned with a child dying at home, it says much about the child who dies in the hospital. The idealist may like to think that every dying child could be at home if he/she wished. The pragmatist may know that thought is sentimental and unrealistic. Cancer and other chronic illnesses of childhood give the child and family time. Time to plan. Time to prepare. Time to investigate alternatives. For many, though, time is a luxury. Accidents continue to be the leading cause of death for all age groups except infants. Neonatal difficulties and sudden infant death are problems that claim the lives of many children under one year of age. Some children die from suicide and abuse, others from chronic illness. The emergency room, the operating room, the neonatal or children's intensive care unit, the inpatient

GENEVIEVE V. FOLEY ● Pediatric Service, Memorial Sloan-Kettering Cancer Center, New York, New York 10021.

units—any of these may be the place in the hospital where the dying child and his/her family will be found.

For some children, then, there will be no alternative. This statement may fill some with dread. The phrase "dying in the hospital" has come to mean, at worst, dying physically or emotionally alone, in a cold-to-alien environment over which the individual has little control. At best, "dying in the hospital" means "it was all we could do and everyone was nice."

That last statement tells you, "I do not think we have found our best yet." An editorial in *Cancer Nursing* gives us a clue where to start. The editorial describes a community meeting at which the benefits of hospice care were being related. At the conclusion of the discussion a man in the audience asked, "Why do you have to wait till you are dying to get this kind of care?"[3] The man posed the question most in harmony with the lessons concerning life that the dying teach us. I therefore urge you to return to your practice setting and look at how family-centered your pediatric nursing service is. How welcome are families? How much are families included in planning care? Do only the dying have sibling visits? I suggest that the first method to improve care for the child dying in the hospital is to improve care for any hospitalized child and his/her family.

Where to begin the improvements is a difficult question. Allow me to share with you how the Pediatric Nursing Service at Massachusetts General Hospital approached this question. Seven years ago the Pediatric Nursing Practice Committee wrote a statement of philosophy and a set of objectives. These documents pledged the service to family-centered care based on principles of child growth and development and to accountable nursing care through the provision of a primary nurse for each patient. The years have passed and we are still struggling to achieve our goals, but the commitment has been made and the goals set.

Last year the Pediatric Nursing Practice Committee wrote "Standards of Care for the Dying Child." The committee debated the content of the standards for months, and our implementation is imperfect. The goals are clear, however. The staff in all units know the department's agreement on the importance of continuity of care, on the need for sibling involvement, and on the advisability of follow-up care after the child dies. The standards also recognize that the staff's grieving must be worked through if they are to be able to continue to invest themselves emotionally in their patients.

As we professionals work to improve our service to others we will need to do more than make written commitments. I believe we must listen to the families and the children who are experiencing chronic illness. This is a group that is very knowledgeable about the hospital system and our interventions. If we have the humility to learn from them I believe they can help us change our methods of care delivery. I project that changes will first occur in the inpatient pediatric units, the intensive care areas next, and finally the emergency room.

What are the elements of health care needing change? Let us recall Linda's words: "to do all that was left to be done for her—to make her comfortable, to provide her with familiar things, and most important to surround her with the love of her family and friends." Four areas are identified—participation and control, comfort, personalization of the environment, and provision of a loving atmosphere that invites family and friends.

Before I begin to look specifically at each of these points I should mention that my own interest is in chronic illness. My practice primarily involves children with malignancies. I am based in the inpatient pediatric units as a member of the Pediatric Nursing Service. For the past six and a half years I have also been involved with pediatric ambulatory chemotherapy patients.[4] I follow families from the day of diagnosis through the good and bad times of their treatment and finally to their triumph over cancer or its triumph over them. I have kept in touch with many of the families of the children who have died. Sharing with these families over a long period of time through the illness–wellness continuum has led me to certain biases. I shall share these with you.

Let us turn to the first area where we need to improve our care. I believe that we are less inclined to make decisions for families if we really know them. The chronically ill child again points the way. From the minute the family learns that a child has a life-threatening illness the family must be helped to become active partners in care. It is the responsibility of nurse and physician to educate the child and family about the disease and treatment plan. A family accustomed to sharing in decisionmaking will have less difficulty in making decisions about how to handle the child's death.

A key concept for all of us is that there is no one right way to die, nor one right place. The acceptance of individual differences is essential. It is not our responsibility to sit in judgment on the family who decides to stay home (notice I did not say who is sent home) while their child dies. It is not our responsibility to condemn the family whose culture demands weeping and wailing. It is our responsibility to prepare the family for the range of responses which we have seen in the months and years after a child's death. It is our responsibility to act as the child's advocate and bring the child's worries and fears to light. It is not our job to decide in isolation how those worries and fears should be met.

It is not necessary that we decide what is the appropriate way for parents to act when confronted with their child's body. It is our task to provide privacy and to facilitate their wishes.

An especially important area where the staff can help parents stay in control concerns the needs of siblings. No matter where a child dies, or at what age, parents need help in telling their other children. Some nurses, particularly in emergency rooms and intensive care areas, have been hesitant to discuss the needs of siblings with recently bereaved parents. I have more often experienced a reduction in parental anxiety after a discussion concerning siblings. The point is that the family may accept or decline the offer. We need to make it.

A second important area involves the decision as to where the child is to die. The hospital nurse must have the knowledge and skill to offer a family the full spectrum of alternatives. Last week a mother and I talked about where her two-and-a-half-year-old was to die. Her questions ranged from the impact of a home death on the child's brother to the laws of her state regarding pronouncement of death. She did indeed decide on a home death on the basis of information provided by someone who knew both her and her sons.

The provision of an environment in which the child feels comforted but which also reflects the child as an individual is a major challenge. Children feel the most comfort in an environment that welcomes their parents. Those of us who care for the chronically ill learn daily that almost all parents can be present for treatments and procedures if adequately prepared and supported. I remember Lisa, a three-year-old dying of central nervous system leukemia. Her father always held her for her spinal taps. His closeness to her at that time added to her comfort enormously. Tommy's mother was different. She did not even want to see the bone marrow needles. For her a chair was placed so that her back was to those performing the test and she was side by side with her son, looking at him, holding his hand, offering encouragement.

Comfort is also concerned with pain relief and physical well-being. Many parents and children who decide that the hospital is the best place to die do so out of concern for pain relief. In reality this is an infrequent problem with proper management of outpatient pain medications. Occasionally an extreme pain control measure is required and hospitalization is necessary. Reliance on medication alone is a mistake, however. Comfort is enhanced by expert physical care, by positioning, by good oral hygiene, and so on. Many comfort activities can be taught parents so that they are the ones bringing physical comfort to the child. At minimum parents can be helped to assist if they wish.

A child's environment is personalized by people and things. The inclusion of significant others in addition to parents is crucial. Karl was a teenager whose daily bath was given by his favorite brother—even when Karl lapsed into unconsciousness. For Mark a special teacher was needed in his last hours, for Mary a grandmother and favorite uncle.

The child's special toys, pictures of family members, classmates, girl- or boyfriends, the child's own clothes—all of these contribute to personalizing the child's space. This is a very important task for child and parent and should be accomplished no matter what the space limitation. Recently in the intensive care unit I saw a comatose child whose family had taped a religious prayer card, a Red Sox schedule, a family picture, and a Star Wars trading card to the head of his bed. Thus the family communicated some of the child's individuality. No one should die known solely as a motor vehicle accident.

The last issue is caring and involves our making the hospital a more human place not only for patients and families, but for all of us as well. How can dying in the hospital be humanized? On the surface this is a deceptively easy question. The answer involves no money, no equipment, no additional staffing. Death is more human when you and I stop being strangers, when you and I allow ourselves to tear down the boxes labeled "parent," "child," "professional" and treat each other as people united by a common purpose.

Humanization of care begins with identifying yourself and by asking the child and parents, "What would you like me to call you?" Humanization continues with a thoughtful assignment of staff as a minimum and the provision of a primary nurse as tbe optimal. It is our saying no to caring for a different patient every day. You and I know that we do not share ourselves with people who continually run in and out of our lives. Yet we expect parents and children at a vulnerable time in their lives to relate to and take support from the parade of people who interact with them daily. Consistency is essential for the establishment of trust. Consistency of persons should be the goal, consistency of approach the unalterable norm. Utilization of nursing care plans, nursing histories, and patient progress notes are all readily available ways to personalize care.

Again, if all the way along we have tried to know the child and family as people we will not have to worry about an emotionally isolated death. We will have fewer concerns about his/her being surrounded by strangers. We will know the family members and friends in the room.

Dying more humanely in the hospital is possible. The suggestions offered here are not new. They are rooted in common sense, growth and development theory, and the principles of interpersonal relations. Why then are we still fighting the fight? I think it is because we have not yet solved the riddle of supporting ourselves. I sincerely believe that the loss I feel and the staff feels when one of the children dies is never equal to the grief of a parent. Yet time after time we are faced with reinvestment and loss. Harriett Sarnoff Schiff has said of the husband and wife that "it is impossible to give comfort when you feel an equal grief."[5] I believe her statement is also true when cumulative grief becomes overwhelming. The popularity of "burn-out" lectures attests to this very serious problem. I believe one way to help is primary nursing, for while the emotional investment is enormous, so too are the rewards. Opportunities for the staff to express their grief as individually as the patients are also essential.

Healing ourselves, formalizing our commitment, listening to parents and children, working to make the hospital as homelike as possible—these are the challenges as we share with dying children and their families their most valuable commodities—time and love.

REFERENCES

1. Kübler-Ross, E.: *To Live Until We Say Good-Bye*. Englewood Cliffs, New Jersey: Prentice-Hall, 1978, p. 64.
2. Ibid., p. 74.
3. Paulen, A.: "Why Wait?" *Cancer Nursing* 2:267, 1979.
4. Foley, G.V., and McCarthy, A.M.: The child with leukemia: In a special hematology clinic. *Am. J. Nursing* 76:115, 1976.
5. Schiff, H.S.: *The Bereaved Parent*. New York: Crown Publishers, 1977, p. 6.

DISCUSSION

MODERATOR: AUBREY MILUNSKY

C. AGUILAR (Birth Defects Evaluation Center and Genetic Counseling Center, San Antonio, Texas): A trend that we are seeing in our Neonatal Intensive Care Unit (NICU) is taking the child who has been on the respirator out of its isolette or cubicle and in the last hours allowing the parents to hold the child. Do you feel this personalization of care is important and can we expand it?

G.V. FOLEY: I hope that the participation of the parents started before then. I think in some ways it's almost more difficult to determine how to help a family get through the tubes and the A-lines and everything else that's involved during the acute phase. Parents tend to be terrified of all that equipment.

I think we are more often seeing dying children being placed in their parents' arms to die. I think we are trying to do that in both our Neonatal Unit and our older children's unit as well. I think that it is something to be encouraged.

N. CALLAN (Springfield, Massachusetts): My question concerns the other children in the ward at the time of a death. How can we help them and their families?

G.V. FOLEY: This is a difficult issue for us, because as I explained, our system enables inpatient nurses to go to the Outpatient Department to take care of the ambulatory chemotherapy patients. I have learned how closely families keep in touch with each other. For us it is not only the inpatients but also the outpatients who have known a particular child who has died. The primary nurse or I tell a family who has been close to a child who has died. We share information with them in case of a relapse, for example, and try and prepare them. We do in fact tell them when a child they know has died. The children in the inpatient units who are in the same room or who have known the child are also told. The nurse usually tells them.

B. FOSTER (Neonatal Intensive Care Unit, Birmingham, Alabama): I was wondering whether you have any suggestions on how to personalize care for a mother who is still in the hospital and has a baby who's dying? The obstetrician will not release her from the hospital which is 200 miles from the referring hospital.

G.V. FOLEY: One thing you can do is take a picture of the baby. Parents have often said that they had nothing by which to remember the child who died. What we are trying to do is have our primary nurses telephone the mother every day and share with her and the family information about the child. Ask if the father is able to come at all. In many instances, he is not. You could suggest the mail. One can mail a toy or take a picture of the child in the isolette or on the warming table with the toy right by him. Many people feel comforted by having religious medals placed right near their child. Family pictures are helpful.

E. Scott (Maimonides Medical Center Neonatal Intensive Care Unit, Brooklyn, New York): We have the parents take a picture with the baby if we think the baby is not going to make it. We had a recent case where the mother now has about six or eight pictures of herself, her husband, and the baby before he died.

G.V. Foley: It's very important. The other thing we have all learned in treating cancer patients is that nobody really knows who is going to live and who's going to die. Therefore, I would suggest getting at least one picture ahead of time just in case. I am very happy to hear that that practice is becoming more widespread.

2

IDA M. MARTINSON is Professor of Nursing and Director of Research at the School of Nursing at the University of Minnesota. She received a diploma in nursing from St. Luke's Hospital in Duluth, Minnesota in 1957 and a B.S. in Nursing Education from the University of Minnesota in 1960 with a Master's degree in Nursing Education from the same institution in 1962. In 1972 she received a Ph.D. in Physiology from the University of Illinois at the Medical Center, Chicago, Illinois. She has been affiliated with the University of Minnesota School of Nursing since 1972.

Dr. Martinson began her research with dying children in 1972 and recently completed a federally funded grant from the National Cancer Institute on "Home Care for the Child with Cancer." She has edited *Home Care for the Dying Child: Professional and Family Perspective,* which received an American Nurses' Association Book of the Year Award in 1977. She and her staff have also been instrumental in producing a documentary film, *Time to Come Home.*

Dr. Martinson has served as a member of the Cancer Control Grant Review Board of the National Cancer Institute and on the Commission for Nursing Research of the American Nurses' Association. In 1980, Dr. Martinson was appointed to the Advisory Council of the National Institute of Aging.

Care of the Dying Child at Home

IDA M. MARTINSON

The purpose of our study, "Care of the Dying Child at Home," is to examine the feasibility and desirability of a home care alternative to hospitalization for children dying of cancer. Home care is defined as "the delivery of services, coordinated by nurses with physicians and other health care professionals as consultants, to enable parents to give comfort care required by a child at the end stage of life." The feasibility and the desirability issues are broad, recognizing the state of knowledge at the time we began, in that we were even unsure whether parents could provide the necessary care or if they desired such an alternative.

Some background would be helpful. A pilot study was done from 1972 to 1975 in which home care was offered to eight families. In five the children did die at home. Based on this nonfunded pilot study a federal grant proposal was submitted to the National Cancer Institute of the Department of Health, Education and Welfare and the project was funded in 1976. During the first three years, there were two research phases. For the first two years the grant staff provided and directed the nursing care of children with cancer at the end stage of life. During this time collaborative arrangements were being developed with public health nursing and three hospital/clinic-based institutions. The grant staff essentially organized and provided the actual delivery of care as well as collected data on this care. During the third year, the coordination of the care, both directly and indirectly, was essentially turned over to three already existing health care organizations, and to the public health

IDA M. MARTINSON • School of Nursing, University of Minnesota, Minneapolis, Minnesota 55455.

nurses utilized by these institutions. The grant staff then devoted its full attention to the question of the desirability of home care and to the observation of what was happening in the three institutions to help answer questions regarding feasibility of institutionalization of this home care alternative.

The criteria for referral to the study included the following: (1) the patient was 17 years of age or younger, (2) the patient had some form of cancer and was expected to die fairly soon as a consequence, and (3) no inpatient hospitalization-requiring procedures were planned.

The services available for the family were as follows:

1. The nurse would be on call 24 hours a day, 7 days a week.
2. The nurse would be available to help family members, who were the primary care givers, and to deal with problems that might arise.
3. The nurse was available to make home visits whenever and wherever the family desired.
4. The option of the child returning to the hospital was always open.
5. The child's physician could be called at any time.

During the first two years, 64 children were referred to the project; 58 of those children died. Sources of referrals for these 58 children were as follows: More than 50% were from the University of Minnesota; St. Louis Park Medical Center in suburban Minneapolis provided the next largest number; and 15 children were referred from 8 other hospitals. A total of 23 physicians were involved: 14 from the University of Minnesota, 2 pediatric oncologists at St. Louis Park Medical Center, and 7 other physicians representing 8 other hospitals.

The places of death for the 58 children were as follows: 46 (79%) at home, 12 (21%) in the hospital, with one of these children dying in a hospital in Mexico, and one child dying in an ambulance while returning to the hospital.

At the beginning of the study we had a number of questions:

1. Would the age of the child make a difference?
2. Would the diagnosis determine whether home care could be provided?
3. Would the length of time from diagnosis to death be a factor?
4. Would the length of home care involvement have an effect?
5. Would families be able to give specialized care that a dying child needed?
6. What would determine whether a child returned to the hospital?

In our analysis we looked at the age of the children with cancer who participated in our study. The range of ages of the children who died at home was 1 month to 17 years, with the largest number being in the age range of 15 to 17, where 13 children died at home. The ages of children with cancer who died in the hospital ranged from 3 to 17 years, and the data indicate that there is no one age range that stands out, based on this study. I would suggest that the age of the child is not a significant factor in determining the feasibility of home care.

The type of cancer diagnosis for the children who died at home included 19 children with leukemia, 4 with lymphomas, 10 with central nervous system cancers, 4 with neuroblastomas, 6 with bone tumors, and 3 with other types of cancer. Based on this data it appears that type of cancer is also not a determining factor.

The period of time from diagnosis to death for the children with cancer ranged from less than 3 months to over 9 years. Again, it would appear from preliminary analysis of our data that the time from diagnosis to death was not a determining factor as to whether the parents could care for their child at home.

The length of home care involvement through death varied. Fifteen families were involved with home care for less than one week; 4 families, 1–2 weeks; 7 families, 3–4 weeks; 16 families, 1–3 months; and 4 families, over 3 months. The direct professional nurse involvement for the 46 children who died at home was an average of 13.8 home visits with a range from 1 to 110. The hours per family had a mean of 31.5 with a range of 1 hour to 305.6 hours. This was supplemented by telephone calls, which ranged from one family who had no phone calls to another family who had 101 phone calls, with a mean number of calls per family of 22.7. Hours on the telephone for the duration of home care average 4.1 per family with a range from 1 to 23.5 hours. The total number of professional nurse home visits for the 46 families who had a child die at home was 634 visits.

The hospital-oriented equipment that was required by these children included the following: Five of the children required oxygen; the parents were taught how to handle the oxygen administration, and these five families did exceedingly well. Four of the children required IVs; in all cases the families did not start the IVs, but they did change the bottles and supervised the use of the IV. Two children needed indwelling urinary catheters which the nurse inserted, but the parents gave the continuing care. Three children required a Texas urinary catheter. One child had a gavage feeding tube. Four of the children required suctioning. One child had a gastrostomy tube. An interesting item is that 25 of the children required none of this special equipment.

The reason for final hospitalization for the 12 who died in the hospital was varied. Four we classified as supportive care; in other words, there was no one medical reason why these children were hospitalized. In these four cases, the child was at home longer than in the hospital with supportive care. Two children were admitted for pain; one of these children had been at home for 25 days without pain and was admitted 1 day prior to death. Two of the children were admitted because of respiratory distress. In one of these instances, the family called the nurse, who did not immediately respond to the situation, and the parents felt it was necessary to admit the child to the hospital. One reason where we felt it was best to recommend hospitalization was continuing seizures. One child in our study had been at home for 36 days and had experienced seizures, but when he started experiencing seizures on a continuing basis, he was admitted to the hospital where he continued having seizures and died 4 days later.

Other questions we considered regarding the feasibility of this home care alternative were factors regarding the family, which might make dying at home an impossibility: Would the geographic location make a difference? The rural or urban areas? Would the social status of the household be a factor? Would the number and/or age of siblings be a factor? Would marital status have an effect? Were special living arrangements required?

Families participated in our project who resided in areas throughout both urban and rural areas of Minnesota, North Dakota, and Wisconsin. Using the Hollingshead Two-Factor Index of Social Position, we classified the families of the children who died at home from the highest, category 1, to the lowest, category 5. Of the families, 48% were inthe two lowest categories while 22% were in the two highest categories.

The ages of siblings in these families ranged from 17 siblings between the ages of 1 and 5, with the largest number of siblings between the ages of 6 and 10. There were 107 siblings in the families of the 46 children who died at home. As we look at the number of siblings involved in the families, we had 5 families where the dying child was the only child, and we also had 5 families where there were 9 siblings in the family. From this we conclude that it appears that the number or age of siblings is not a critical factor in home care.

Marital status is also of interest. There were 54 families where there were two parents in the home and 4 families where there was one parent in the home. Three of these families were mother-only, and one was a father-only family. Three of the children with one parent at home died at home, including the one with the father only.

Place of death in the home for 31 of the 46 children was in the living/ family room, essentially the center of family activity.

Questions were raised regarding the qualifications or requirements for the home care nurse. What type of specialization was needed? Were age, experience, and educational level factors? Also, were physician home visits necessary? What other health professionals were required to make home visits?

Of the 58 nurses who worked with the families, 24 were hospital-based, 22 were involved in public health nursing agencies, 2 were nurses on the grant staff, 5 were unemployed, and 5 were others, such as school nurses. We looked at the number of families cared for by these nurses and found that 13 families were assisted by a hospital nurse, either from the referring institutions or a local hospital. Eighteen of the families were assisted by a public health nurse, 6 of the families by the home care staff nurses, 7 of the families by unemployed nurses, and 4 families by others.

A few of the families had two co-primary nurses; 3 families had two hospital nurses; 1 had two public health nurses; and 6 had a combination of a hospital nurse and a public health nurse. We noted with the combination hospital nurse and public health nurse team that less consultation with the project staff was required. The hospital nurse was able to handle the emergency-type questions, and the public health nurse was able to handle situations requiring knowledge of local resources. An interesting observation that has evolved from this is the need for more of a nurse/nurse network arrangement.

The age and experience of the home care primary nurses ranged from 23 to 63 years of age, with from 1 to 44 years since they received their nursing degrees (RNs). Seven of the nurses had Master's degrees; 29 were baccalaureate nurses, 4 were nonregistered nurses, and the balance had hospital diplomas. The 4 nonregistered nurses included 3 licensed practical nurses and 1 student nurse.

The number of physician home visits through the time of death and immediately after the death of the child for the 58 families were as follows: 44 of the families did not have a physician visit at home, 9 of the families had 1 physician visit, 1 family had two physician visits, 2 families had four physician visits, and 1 family had 17 home visits including 12 visits by a psychiatrist.

Home visits by other health care professionals for the 58 families included: a laboratory technician who made 1 visit to 3 families and 2 visits to 1 family, an X-ray technician who made 1 visit to 1 family, an occupational/recreational therapist who made 1 visit to 1 family, a chiropractor who made 7 visits to 1 family, a Home Health Aide who made 1 visit to 1 family and 43 to another, and a homemaker who visited 1 family 16 times.

Although we had no social worker make a home visit during the time of home care, we have data that indicate social work involvement before referral to the home care project as well as with the family following the death of the child.

With regard to the cost effectiveness of home care, we have looked at cost figures as considered by insurance companies. For 46 children, the duration of final care days was a mean of 38.9 days with a cost estimate of $1218, a median of 20.5 days with a cost estimate of $705. This cost estimate is based on the cost of nursing service at the rate of $10 a day to be on call 24 hours a day and for telephone consultation, $45 per home visit, and $10 for a clinic visit. In discussions with insurance companies, we were urged to use a comparison group. The first group we utilized was a group of 22 children who had died at the University of Minnesota Hospital prior to 1976 before our project was funded. The 22 children who died of cancer at the University of Minnesota Hospital had a mean duration of final care of 29.4 days with a cost estimate of $5880, based on the cost of nursing service and room and board at the rate of $200 per day. The median was 21.5 days, with a cost estimate of $4300.

We have recently attempted to update these cost figures. We have estimated a daily cost of $51.79. This includes nothing for room and board, as the family provided this; $40.04 per day for nursing care, based on $35 for the first hour of a visit and $10 for each additional half hour; $2.57 for room furnishing; $3.49 for equipment; $2.99 for supplies; and $2.54 for medications. Laboratory tests accounted for $0.14 per day.

Contrasting this cost to a group of children who died in the hospital while receiving comfort care only, the cost per day for these children was $279.91. This included $158.09 for nursing care, room, and board; $27.69 for supplies and equipment; $12.94 for medications; and $81.19 for laboratory tests.

We have attempted to discover whether the parents were satisfied with the home care services provided. One of the ways we looked at this was to have the parents rate their choice of care if they had to choose over again. Of the mothers and fathers, 97% said they would definitely choose home care, one might choose home care, and one mother said she would definitely choose hospital. Out of the 46 families whose child died at home, there is one mother who said that although she cared for her child at home, she would definitely choose hospital if she had to do it again. Of the mothers and fathers whose child died in the hospital after having home care services, 6 said they would definitely choose home care, 1 might; 4 were not sure, 1 might choose hospital, and 4 parents, representing 2 families, would choose hospital again.

Ratings by parents of satisfaction with home care services provided followed some of the same pattern: 97% were very satisfied with the nursing services provided and 3% were somewhat satisfied. Of the mothers and fathers of the children who died in the hospital, 11 (79%) were satisfied and three (21%) were not satisfied. The three parents who were not satisfied represent 2 families who would definitely choose hospital care if they had to choose again. It is of interest to note that the 2 nurses who worked with these 2 families state that they would not be willing to provide home care services in the future. Examining these instances more closely, we note that there were several areas with these families in which severe communication problems existed between the parents, nurses, coordinators, and physicians.

The institutions who assumed the care delivery aspects during the third year of the grant are the University of Minnesota Hospital Home Health Services Department, Minneapolis Children's Health Center, and St. Louis Park Medical Center, along with the public health nursing agencies throughout the state. The institutionalization of this model of health care delivery for the dying child has now been expanded to include children dying from causes other than cancer at both the University of Minnesota and Minneapolis Children's Health Center.

ACKNOWLEDGMENTS. I would like to acknowledge the cooperation and help received from the current staff: Gordon D. Armstrong, Ph.D., and D. Gay Moldow, R.N., M.S.W., research associates; Emily J. Kulenkamp, B.A., research specialist; Nancy Olson and Sue Hwuse, research assistants; and John H. Kersey, M.D., and Mark D. Nesbit, Jr., medical consultants.

This study was funded by the Department of Health, Education and Welfare, National Cancer Institute, Grant Number CA 19490.

Discussion

MODERATOR: AUBREY MILUNSKY

RABBI H. KUSHNER: What have you found is most helpful after the child has died? Concerning the procedure with the body for instance, do you prefer to have the child home dead for awhile, or do you call Emergency immediately?

I. MARTINSON: We have a manual for nurses on this very important point about what to do when the child dies at home. First, let me speak from a family's point of view. In general, we have found families who do not want their child's body to be rushed away. It works out best that the funeral home is contacted *before* the death of the child and explanations given about expectations for the child dying at home. The families may wait and have the child's body at home 1–3 hours before they are called. Maybe we should realize as professionals that we should not rush at this time. I think we tend to rush this too much at the hospitals where we need to spend more time. We had one case where after a one-month-old baby died, the mother rocked the child's body for a long while. This bothered me initially, but then I realized that this is what the family wished to do. So there should be flexibility. But you also have to be careful because there are rules, regulations, and laws. If, for example, the child's body needs to cross your county or state lines, be sure that this is cleared ahead of time. It is so important that we have health professionals involved so that difficulties can be anticipated and prevented. In a large city, you have to clear things with the Medical Examiner or Coroner's Office because it is so unusual to have a child die at home in our society that they begin to think of child abuse. We have now in our larger cities preregistered children. Remember that the Medical Examiner is only trying to do his job. When there is an expected death there should be no difficulty and no investigation.

P. MARINO (Staff Nurse, Massachusetts General Intensive Care Unit for Children, Boston, Massachusetts): I would like to ask you about your preparation for the nurses involved in this program and the ongoing support systems for them.

I. MARTINSON: There have been phases. When it began I was doing it, and I don't know how I prepared myself. The first nurse that I worked with was 250 miles away and my instructions to her were given via a 45-minute phone call. Over a period of time, we developed this manual, which has been a help to the institutionalized phase. Probably even more important than this, however, is the fact that the staff and the graduates were on 24-hour, 7 days a week call to the nurse out there, and I was on call to my staff. In the institutionalized phase the nurse coordinators in these three institutions have also been willing to have that type of commitment to the nurse out in the rural areas. We have to realize that a dying child, at least in our geographical area, will probably always have a new nurse working with the family. You do then need a supportive system, and I think that being on call is really important to them. We have also produced a manual for parents which many of our nurses find helpful.

E. BUCKMAN (Northeastern School of Nurses and private psychiatric practice): How do your nurses deal with "burn-out" because of the commitment they have to their families?

I. MARTINSON: I think that's an interesting area. One of the things that I am beginning to find with home care for the dying child is that "burn-out" might not be a factor because they are just caring for that one child. Where "burn-out" will probably occur with this project will be with the coordinators. From a personal point of view, I have been in this since 1972 and have avoided "burn-out." The reasons are probably complex. My husband's background is history of religions, and we have discussed the impact of this work on me and whether I had hardened. He told me this story of Quanion, a goddess, who had really suffered in this life and had a chance to leave, but instead she chose to remain. The term in Chinese is translated "a separate compassion," and I think that's probably the reality of it. I still hurt— you still must have that compassion to be in this work, and yet you also realize that it is a separateness. I wonder if that won't help to keep us in focus. Our professional background should help us somewhere. There are ways we can cope with "burn-out." It is a new field and certainly one worthy of a great deal of investigation and will challenge us all. More study is needed on how to prevent "burn-out" in parents.

A. MILUNSKY (Medical Geneticist, Harvard Medical School): Many parents have great difficulty in telling their child that he/she will die soon or in the near future. Many actually want to tell because of the feelings of truth, honesty, closeness, and being at hand, but have difficulty doing that. What is your view in helping parents who want to communicate that information? Is there a way that you can help them do that and, secondly, do you have a special way in dealing with the many physicians who find that if the parent tells the child, the care that the physicians has to render is made so much more difficult.

I. MARTINSON: I would think the care needed by the physician should be made easier by the child knowing and the parents knowing more. At least that would be my experience in Minnesota. Almost without exception our children did know that they were dying. Now they didn't always tell this to the parents, but in the interviews with either the siblings or the grandparents or the clergy or the physician, we have usually found out that the dying child had been talking to someone. I work mainly to get the parents ready so that they think they can tell. I think our responsibility as health professionals is to work with the parents because they are the ones who are going to be there when the child is going to want to talk. It takes all we have to be able to get the parents ready and able to do that. And that is one of the greatest challenges that I face in this work.

A. MILUNSKY: The basic reason why the pediatrician in particular is on the burner when the child knows explicitly from the parents that he/she is going to die is that the physician's helplessness is brought into the strongest relief. The child thinks, "Well, what on earth are you going to do now? Here I am. I am going to die. I know it, and you know it. And here you are making mumbo jumbo. I'm not going to take your medicines."

I. MARTINSON: I think that's where we need to change not only our nurses' attitudes but also physicians' attitudes. I think we've got one of our greatest health professional challenges ahead of us to see that comfort care is what we need to be able to do. There should not be a sense of helplessness at all at this stage. I would rather say that it should be a challenge. I mean, to really give comfort care that a child and a family needs takes the best of us. A helplessness comes because we don't yet know what to do. I think my study has just begun to open up an area, and we need all of you to work into it. And I think in a few years time, instead of anybody feeling helplessness, it would rather be, "My word, there is so much to do, what do we do first." It is our basic attitude at that point that must be changed— the quality of care.

3

HARRIET SARNOFF SCHIFF was a Detroit newspaper reporter whose 10-year-old son died from the complications of congenital heart disease. In reconstructing her life after a period of almost unbearable grief she realized that "it is a tragedy that must not be compounded by allowing everything around you to die also." The experiences of Mrs. Schiff and her husband led her to return to writing in an effort to help others in similar tragic circumstances. The result was the now well-known book *The Bereaved Parent* (Crown). She has lectured widely on the subject of death and dying.

The Bereaved Parent

Harriet Sarnoff Schiff

There is a story about a prince fleeing revolutionaries who wanted to kill him. Terrified, he sought refuge in a peasant's cottage. Although the peasant did not know the frightened man was a prince, he sheltered him by hiding him under the bed. No sooner was the prince safely hidden when his pursuers battered down the cottage door and began to search for him. They looked everywhere. When they came to the bed, they decided it was just too cumbersome to move. Instead, they poked their swords through it. At last they left. The prince, pale but alive, crawled out from under the bed and turned to the peasant. "I think you should understand you have just saved the life of your prince. Name three favors and they will be yours." The peasant, a simple man, thought for a while and said, "My cottage is in great disrepair, and I have not had the money to fix it. Can this be done?"

"Fool!" cried the prince. "Of all things in the world why did you ask so small a favor? I will honor your wish."

"My next request is to have my neighbor's stall moved further from mine in the market place."

"Idiot! Of course, I will do as you wish. But you could have had riches. Take care you do not anger me with your third request."

No longer able to contain his curiosity, the peasant said, "As my third request, I ask only that you tell me how you felt as the swords were being pushed through the bed."

The infuriated prince shouted, "How dare you offend majesty by asking a prince of his emotions! For this you will be beheaded in the morning." So saying, he called his guards and they led the weeping man away to the local jail.

All through the night the man wept for his folly. When the sun rose

Harriet Sarnoff Schiff • Birmingham, Michigan 48010.

his jailers came to him and led him into a courtyard where an executioner in a black hood stood awaiting the terrified man.

Forced to kneel on the block, he heard a soldier call, "One, two," but before he could say "three," another soldier came galloping into the courtyard calling, "Stop! Halt! The prince commands it!"

With these words the executioner, whose blade had been resting on the peasant's neck, withdrew his sword. The shaking man rose and faced the soldier who had saved his life.

"His highness gives his pardon and orders me to give you this note."

The peasant, relieved to the point of tears, read these terse words: "As your final favor you wanted to know how I felt under that bed when the revolutionaries came. I have granted your request because now you know."

This story illustrates why my book, *The Bereaved Parent*, is in its seventh printing and is also doing well in paperback. Like unto like. The power of a shared experience. And that, of course, is where many in the helping professions feel the problem. Many have not shared this sort of tragedy, and yet are considered qualified to help people in crisis. How can they best feed their skills into the sort of help needed? Mainly by really understanding the problems of bereaved parents.

When Robby, our ten-year-old son, died in 1968 following open heart surgery, my husband Sandy and I felt like we were walking uncharted paths. Certainly we knew that others had experienced the death of a child. But we felt, despite being surrounded by friends and family, isolated and alone. There was no one who could come forward and say, "My child is dead. I know what you are feeling because I have felt the same things."

After I began to heal, I felt a great need to do something challenging and creative with my life, so I decided to pursue an early love, writing. Ultimately I became a reporter for the *Detroit News*.

Incidentally, in interviewing the initial hundred families for my book, I found a striking similarity in our patterns. Many bereaved parents, as I myself had done, sought creative outlets such as painting, sculpting, even paint-by-number. There is a great need to prove you still have some power, even if it is only the power to bring something forth that is small.

As a reporter I found over and over again people who felt as Sandy and I had felt. If you picture the role of a reporter, it is generally that of somebody intruding upon tragedy. I am certain many people have had that experience. During my reporter days I covered many deaths where there were surviving parents. Drownings, accidents, motorcycle mishaps, killings. I would go to the home, introduce myself and immediately offer my condolences, always adding something about also having a dead child. Instead of trying to shove me back out the door, people would beg me to sit down. They would ask, "But how did you survive?" In fact they questioned me more intently than I did them. I would leave their homes reluctantly because there was a deadline to meet. I would have to elbow my way out just as others had had to elbow their way in! I would leave feeling a sense of having things to impart that I had lived through, and always there was a sense of something undone.

What of course was happening was that I was able to understand because I had been there. This sense of a timeless, shared experience stayed with me, and when I left the *News* I began work on what ultimately became *The Bereaved Parent*.

I never had a doubt that my book would be bought. What came as somewhat of a surprise to my publisher, who felt he was providing slightly more than a public service in agreeing to publish it, was the demand. He ended up with a book, demand for which kept growing. The reason is a basic case of statistics. Each year 600,000 people die who are under the age of 54. Many of them leave surviving parents. Also, I didn't lie. I talked about my feelings, warts and all, as well as the feelings of others.

My time spent interviewing was sometimes exhausting, sometimes painful, but always it was interesting. I was never turned away. All I had to say was that I too had a dead child, and it was the *Detroit News* all over again. I felt sorry when the interview was over.

I found that people held diverse impressions about funerals, and that customs are as varied as can be imagined. It is like the story of the Buddhist servant who wanted the afternoon off to attend the funeral of a relative. His occidental employer jokingly asked if the custom of placing a bowl of rice at the grave would be followed.

"Certainly," said the servant.

"When will your aunt eat the rice?" the employer asked derisively.

"Oh, at the same time your cousin who died last week smells the flowers you sent," was the unabashed reply.

When Robby died, we simply had enough to do holding our wits together just to follow our religious traditions. Other families have planned funerals down to the last hymn. As I say, to each his own.

I feel strongly that funeral directors and clergy must offer guidance in the case of an "embarrassing death." By that I mean a death where a child has been engaged in a criminal activity, a child who commits suicide, or a daughter who is raped and killed. It is important to dis-

courage a private funeral in order to help keep the channels of communication with friends and family open so that the grieving will not
seem to end when the funeral is over. If the funeral service is private
the people who make up our support system will assume the grieving
is also private, and then we have disaster compounded with disaster
because grieving must never be private. It is a community affair and a
community responsibility.

An additional difficulty that is perhaps unique to this country is our
transiency. Where in England people hand homes down from generation
to generation and families have known one another since the beginning
of time, this is not the general American pattern. I interviewed families
who, along with the horror of having a dead child, had the added burden
of trying to figure out where to have it buried! After all, where is home
when one travels from city to city annually as IBM and GM people are
likely to do? How does one develop relationships that are knowing and
instinctive on the basis of such short residences?

My editor wrote me a note when he received my completed man-
uscript. He questioned the wisdom of leaving in my nonappreciation of
heart fund donations at the time of Robby's death. It made him uncomfortable. I guess people feel that sort of donation implies, "so that he
did not die in vain." When a ten-year-old child dies, he dies in vain. At
the time Robby died, I did not give a damn about anyone else living or
dying. All I knew was that my son, my precious child, was dead. Later,
of course, I did not want anyone to walk where we had walked. But
that was later. The large volume of mail I have received has reinforced
the truth that many of us are angry and it takes a lot to overcome the
rage! Since I was willing to speak the truth, many people have written
to say they felt the same but were so ashamed!

The death of a child is truly the ultimate tragedy because it is unnatural. It defies the sequence of events we have been taught should
occur. We are raised to believe we will one day bury our parents. No
one has ever cautioned that we run the risk of burying our children.
After all, they are immortality!

Second, no one ever tells us about the negative dynamics of parenting. We are told to love our children. No one ever tells us there will
be times they are annoying or infuriating. Perhaps delivery-floor nurses
could help people by acknowledging this truth.

We are told children are a reward for goodness yet nobody is angel-
wing perfect. We are told our job as parents is to protect and nurture
our children. We all sin and make mistakes. When a child dies, all too
often parents see the death as a punishment for their own shortcomings.

These are the things most parents feel, along with a sense of power-lessness and a sense that events are beyond their control.

The cause of death is another factor. In the case of illness, such as in our home, we felt we at least put up a battle.

We lost it and that was a tragedy, but at least we had the time to fight, the time to say and do many things. An early bicycle; early trips. *I love you's.* Contrast this with the father who has a fight with his son. The son slams out the door. Two hours later there is a police officer waiting on the steps to tell the father that his son was killed.

Quite a contrast. Who is better off? Perhaps the parents of an ill child. But how does one stack up the number of years I grieved and feared my son's death? Like the country music song tells us, "there ain't no easy road."

"Well, at least you have each other." I imagine every bereaved parent has been told that! My husband and I found that despite attempts to console us with these words they had a negative effect. We discovered that not only did we not have each other to lean upon, but, to compound the problem, we thought we were supposed to, too. There were no studies back in 1968 telling us that within three to four months, 90% of all bereaved couples are in marital crisis. Thirteen years ago all we knew was that we were supposed to be closer than ever and yet we weren't. We were told we could grieve together and yet we couldn't. We were told many things and truly there were only platitudes offered by people who had given themselves the hopeless task of trying to comfort us.

The problems are complex and many. Often husbands and wives cannot discuss the dead child. One woman wrote me who had lived with this problem since 1944 when her son was killed on a bomber mission over France. She said the only time she could grieve was when she was alone. Her husband refused to accept the son's death. He died recently, and she wrote to say that now at last she was going to France to visit her son's grave. Before she went she explained, "This I must finish before I die." Now the woman, who is 80, is back home. "My affairs are now in order," she wrote again. "I have seen my son's grave."

Sexual dysfunction is another major problem. Men claim they are impotent. Women refuse to have relations because they feel that had it not been for sex, their child would not have been conceived, and they wouldn't now be in such pain. Cheating on spouses is not uncommon.

There is a stage many bereaved parents experience when they ques-tion whether it is really better "to have loved and lost than never to have loved at all." After many years, I decided my love outweighed my pain, but it took years to make that decision.

Serious problems result when husbands and wives blame one an-other. One husband I spoke with bears great anger toward his wife although many years have passed since their child was killed by a car. He was an intense father. She was a relaxed mother. She would say, leave him alone. He would say, watch him. One day the child was playing out of doors, watched over by the mother who suddenly re-membered she had not turned off her stove. She was indoors for just a few moments, but when she came outside the youngster was lying in the street. He had been hit by a car and was dead. The couple did not divorce. He, unfortunately, felt vindicated rather than compassionate. He was that kind of man. They live a bleak, meaningless existence. The mother lives in a shell. The father ignores the wife.

Drinking and drugs become problems in many homes.

Often, parents undergo personality changes. One man dropped all his successful business situations, including his car. He now rides a motorcycle and wears jeans daily. Perhaps he wanted to return to an earlier time in his life when things were relatively carefree and easy and pain was minimal. His wife divorced him. She said she simply could not bear living with a man-child.

Some women also change radically. Many go from immaculateness to slovenliness. The husband forces himself to go off to work. He leaves a woman with her hair hanging, bathrobe dirty, house a mess. He comes home to the same thing. This is a common pattern.

One man insisted his wife get dressed. In fact, he literally ordered her upstairs. She cried and begged, but he was adamant. They were going out to dinner.

"I cried all during that first meal," the wife told me. "But when my husband came home the next night and insisted again, it was a little easier." Gradually she began making demands upon herself, and their home situation improved.

Many times bereaved parents become convinced that their marriage is star-crossed and doomed to failure. They question why they should continue together. It is similar to the story of the 80-year-old woman who divorced her husband after more than 50 years of marriage. When the judge asked her why, after half a century, she had decided upon such a drastic measure, she replied, "Enough is enough!"

The pity, of course, is that it is not always the case that things will not get better. Very often, by meeting with fellow bereaved parents in groups where problems are aired in an open, noncritical atmosphere, people can hold onto their relationships. They need counselors who are sympathetic and who will guide them toward meeting other bereaved parents.

Realistically, though, all of us know that not every marriage will be saved. Whatever the reasons—powerlessness, loss of respect, and so on—parents will divorce. In this area also, meeting with fellow parents is critical. The need to bounce off one's frustrations and pain is enormous, especially for a single parent. Also, if a bereaved parent chooses to remarry, it is imperative that the new mate be made aware that certain times like holidays will be painful. The parent, on the other hand, must not expect more than anyone is capable of giving. A new spouse cannot feel your degree of torment. Many second marriages have broken up because of this unfulfilled and unrealistic expectation. A bereaved parent has a right to expect comforting and solicitousness, and that is all. Be glad there is someone to share your life. Some of the grieving will have to be alone. Sitting in on a group situation can help the new mate to really understand the dynamics of this sort of bereavement.

Aside from the very serious marital problems that develop when a child dies, other problems and their potential for great anguish can be found in surviving siblings. In all of my interviewing I never came across a surviving sister or brother who felt their parents were helpful during the grieving period until I appeared on the Phil Donahue show. There I found one daughter, who did not feel abandoned or unsupported by her family. Her brother had been in an accident. Instead of shunting her off to some overly cheerful relative as her brother lay dying in the hospital, her parents kept her with them through the long night of waiting in the hospital corridor. She paced with her parents and was with them when the doctor came out with his negative prognosis. When he died she was there. She also went to make funeral arrangements. Everything became a shared family experience. She did not feel shut off from her own or her parents' grief.

This is so different from most bereaved family patterns where parents frequently

- Allow a breakdown in communication to occur between the children and themselves.
- All too often fail to talk individually with each child in order to let the children express their true feelings, even if those feelings are negative and include anger at the dead sibling.
- Fail to remind children of the good acts—and there are always good acts.
- Do not reinforce the fact that quarreling is not unique. It is instead a normal function of being a sibling.
- Do not tread carefully in discussing God. Too many youngsters

are completely turned away when parents, in answer to the inevitable, oft repeated "why" say, "Because God wanted her," and
walk away.
- Do not realize that it is a great act of kindness to allow an honest
  memory of the dead sibling, warts and all.

What happened in our home, although sad, is fairly typical of the
families I found. Robby was born with a heart condition. He spent a
childhood full of hospital visits and surgeries. He endured all this fairly
well, but he hit bottom when at the age of eight he was told he could
not play little league ball. Remember, he had attended school and had
functioned fairly normally. There had been no great distinction between
him and his friends, but suddenly everyone was playing ball but him.

Our son Dale was ten at this time and Robby was eight. When Dale
heard how unhappy his brother was, he went to his own coach, explained the circumstances, and asked that Robby be made bat boy on
Dale's team. The coach agreed. The look on Robby's face, the gratitude
for the uniform, the purple shirt that said "Termites," and the cap,
should have lasted Dale a lifetime, but it did not. He did not remember
this. He did not remember playing GI Joe with Robby on cyanotic days.
He did not remember pedaling him on his own bicycle when Rob was
too tired to pedal himself. All these things were wiped from his memory.
Dale only remembered the resentment caused when his parents spent
so much time with Robby in the hospitals. Only remembered that, like
two normal brothers, they had quarreled, and Dale had punched Robby.

None of these things even surfaced until I interviewed him for my
book. Dale was 19 when he told me he felt responsible for Robby's
needing surgery. Now he knew that his brother had had a heart condition. This guilt, though, was something he carried for years.

At the age of 16 Dale nearly died. He was bleeding internally from
an ulcer. His hemoglobin had dropped to 3.8. It was then that we slowly
began to pick through the years when, in our grief, we ignored his. It
was then we learned for the first time that our rabbi, who had told Dale
his brother was dead (the surgery had been done out of town), had
cautioned him to be strong for his parents! We were so immersed in our
own grief that we did not recognize that he was not mourning. His grief
was simply overlooked, as is the grief of most surviving siblings.

I wonder what bubbles beneath the surface of many such people.
How many children know that wishing a sibling dead does not make
a sibling dead? The problem is that the children need parents to help
make things all better when the parents themselves need consoling and
when they feel young and vulnerable, too.

It can also be dangerous to take one's dead child, who was sometimes good and sometimes bad, and turn him into a saint. How, after all, can anyone live up to the image of a canonized sister or brother? How many youngsters turn instead to the negative sides of their makeup in order not to compete?

Parents should also understand the necessity of an honest discussion. They should open the topic if it has not been really thrashed about even once—and that is not uncommon. It is possible to control your tears if your children are frightened by them, but there is a great need in general to lessen the awe people feel about tears.

Bereaved fathers are often ignored by society and suffer from many of the problems that siblings encounter. People in general did not view my husband as a co-parent. He did not receive equal support. He received instead a pat on the shoulder and a few mumbled words like, "Take care of Harriet." I was the bereaved parent and the injured party, not both of us equally. It is a very cruel and unfair burden for a co-parent to be responsible for supporting everyone emotionally, and yet this is the norm and it is one that must change because it can lessen the marital crises that so often occur at this time.

Everyone has heard the old expression, "a friend in need is a friend indeed." Sometimes that may be true. Unfair as it may seem, however, it is imperative for bereaved parents to take initiatives, to call people, have them over, and not to sit back counting noses and complain about all those who supposedly abandoned you in your hour of need. We must speak about our dead child openly in order to allow those near us to do the same. After all, they too are grieving. We must ask about friends' living children in order to diminish their embarrassment about still having living children. This embarrassment is all too common.

We must understand we still have the right to enjoy lives. We often stay deep in our mourning because of the "umbilical cord of grief." When we cry we feel close to the dead child, almost like we can touch him, but when we laugh it is a shock to our systems, and we feel we are abandoning him.

I have one major goal these days. I want to make it socially acceptable to say, "I have a dead child." No one flinches, turns away, or fails to pick up on a conversation when someone says, "I am a widow, a widower, or my mother died." People simply say, "I'm sorry," and go on. That is not what happens when a child dies.

I remember all too well at the age of 33 when I would reply "two" to the question of, "how many kids do you have?" When I said "two," I would cross my fingers behind my back and say silently, "Robby, I'm not denying you." I used to say "three," but people would invariably

ask what school they went to. Then the moment of revelation and then
the pulling back. This would happen with casual acquaintances, and all
they needed to know was how to say, "I'm sorry," just as they would
in any other death.

That is why the subject of the bereaved parent is so vital, the need
to discuss it so great. Because if it is open and not hidden in the closet,
someday someone faced with a loss such as ours will have steady foot-
steps and know the path has been paved and at its end is survival.

DISCUSSION

MODERATOR: AUBREY MILUNSKY

B. SLOCUM (Providence, Rhode Island): I'm a bereaved parent, too; everything you have said just makes so much sense. There is so much inside that I would like to say. But first, thank you for the book. I am a co-leader of a bereaved parents group in Providence, and we keep passing it around. One basic aspect of the problem is the whole communication process—the fact that people are not comfortable talking about death at all, not only to children. Death is just a taboo word, and it's very encouraging to see it changing in the hospital and that professionals are willing to meet with parents and with groups to talk about it. And all of you who are professionals here, keep doing it, because we need all that help. Another point you made and which bears emphasis is that grieving does not end at one month.

H.S. SCHIFF: Some people have a concept that it ends with the funeral. That's what I thought all those years that I grieved and cried in my heart. For two years I never slept more than a few hours because I kept picturing Robby dead and picturing him in his casket. It was exactly what I thought it would be but I never thought a day beyond the funeral. Somehow it would all stop then. That's what all the books had shown me, and that's what I had believed. It was a horrible shock to learn that there's a long period of grief afterwards.

R. BARATZ (Staten Island, New York): I had a stillborn 15 years ago. I think you just put into perspective things I have thought about and have not been able to speak about because there had not been a bereaved parents group at that time. I have not totally realized until now exactly how I felt.

H.S. SCHIFF: Thank you. Parents of stillborns have severe grieving problems very often unacknowledged. I think what you generally hear is, "Well, you didn't know the child and you'll have another." "How do you feel about having another child?" I'm often asked. And I always say, if you are going to have another child to replace a dead one, then please don't. It's not fair to be a replacement person. If you want that child, it's because you want another baby. But I recognize that a stillborn can cause a terrible pain.

SPEAKER: As a nurse and as a parent of a child with cancer, I have been involved in a group called Candle-Lighters. I know we meet, as I'm sure many of you do, with resistance on the part of parents in reaching out to a group like Bereaved Parents. I was wondering if you would share the ways you would use to approach parents to help counteract some of the resistance to discussing problems they have faced in their bereavement or of ways of preparing personally for the child's death?

H.S. SCHIFF: I don't know why there would be resistance. I get mail asking for suggestions where to go. It's important to make contacts and take initiatives and to state that there is such a group. It's important for hospitals to share this information. The parents have to accept the fact that they have a real problem. And that's where the difficulty is. It's not you they are denying, it's the problem more than anything else. Once they realize the difficulty, they're willing to enter counseling. Parents of children with heart disease tend sometimes to be resistant to joining support groups because that could include some very sick children and the parents don't want to get that close to it. Education and discussion about beneficial aspects are useful as is a place to pour out frustrations. Person to person contact initially enables the formation of a group.

B. SHERMAN (Needham, Massachusetts): I'm a bereaved parent and a leader of a Compassionate Friends chapter in Needham. There are about 25 Compassionate Friends' members here today. We are willing to talk to any of you during the intermissions. We are willing and able to share our experiences with you.

4

RABBI HAROLD S. KUSHNER has been a Conservative Rabbi for 20 years, the last 14 of them at Temple Israel, Natick, Massachusetts. He is a graduate of Columbia University and of the Jewish Theological Seminary, and holds a doctorate in Bible. He is the author of *When Children Ask About God* and two volumes of sermons, and has edited several contemporary prayerbooks. Much of his theological thinking and writing was prompted by the serious illness, and ultimate death at age 14, of his son Aaron. The material presented here is drawn from Rabbi Kushner's book *When Bad Things Happen to Good People* (Schocken, 1981).

Where Was God?

The Legitimacy of Religious Anger

HAROLD S. KUSHNER

Some of us can remember a time when movies were less sexually explicit than they are today. At such a time, the manager of a certain movie theater once noticed that there was a man who came every day to see the same movie. Day after day, he would show up as soon as the box office opened and go in for the first showing. One day, the manager said to him, "I notice you're here every day to see the movie again. You must really enjoy it." The man answered, "Oh, the movie's all right, I guess. The part I come for is one love scene toward the end. The hero and heroine are alone in a field, and just as they get to the intimate part, a train comes by and you can't see what they do. That's why I keep coming back. I figure one of these days, that train is going to be late."

I feel a little like that man sometimes. Every year at the beginning of the Jewish New Year, the assigned Scriptural reading in the synagogue is the twenty-second chapter of Genesis, the story of God's commanding Abraham to offer his only son Isaac as a sacrifice. God tells Abraham to take his beloved son and offer him on an altar at the top of a mountain. Abraham meekly obeys, and sets out to do it early the next morning. And like the man who went to the movies every day, waiting for the train to be late, every year I read that story and I say to myself, "Maybe this time Abraham is going to say, 'No!' Maybe one of these times Abraham is going to stand up to God and say to Him, 'I'm sorry. If You want a world where young children die, *You* make it happen. Don't ask me to be Your accomplice.'"

I am concerned with the issue of the validity of religious anger, the

---

HAROLD S. KUSHNER • Temple Israel of Natick, Natick, Massachusetts 01760.

right of people who have lost a child to say, "This is unfair. This is not how the world is supposed to be." I am concerned as a clergyman who has had to counsel parents in their bereavement, and I am concerned even more as a parent who has lost his first-born son to a rare ailment. Both personally and professionally, I believe in the right of a bereaved parent to be angry.

Those of us who work with the critically ill, with the dying, and with their families know how much anger parents feel when their child dies. What is a parent supposed to do with all that anger? Sometimes they displace it onto doctors and hospital personnel. Those who work in hospital settings are familiar with this phenomenon. I know people whose children died several years ago, and you can still hear them tighten up when they talk about it. You can hear the anger in their voices as, years later, they insist that a more careful diagnosis or a speedier exit from the golf course would have saved their child's life. Incidentally, all too often, the doctors play into their hands. A lot of doctors have trouble handling death. They define their job as making sick people healthy, and if a sick child does not get well, they think they have failed. At some level, perhaps subconsciously, they agree with the parents that it was their fault. Had they been the superdoctors they sometimes encourage people to believe they are, they would have pulled the patient through. Sometimes the result of this attitude is that the doctor cannot face the parents, or is manifestly uncomfortable facing them after the child has died, and that is not fair to either side.

If the parents do not vent their anger at the doctor, they may vent it at God or at the priest, minister, or rabbi who comes to comfort them in God's name. When I was a young, inexperienced rabbi, that used to bother me. I would walk into a house of mourning reluctantly, and I would feel this hostility, this bitterness, directed against me as if I personally had been responsible for their tragedy. The first few times I became very defensive. If the mourners bristled at me, I bristled right back. I argued with them; I criticized them for being so hostile. As you can imagine, I was a big help. Incidentally, and I think this is important, over the years this has stopped happening. I think it has stopped, not because I became a better pastor, but because more of the families I called on knew that I had a dying son, and for the last two years since I lost my son I think that made a difference.

When parents lose a child and get angry at the doctor or at the minister, that is not good, because it complicates communication with the individual who is there to help them. It makes it harder for them to be helped. But there is another pattern which is even more frequent and much worse, and that is for parents to turn their anger inward upon themselves, to blame themselves and saddle themselves with feelings of guilt.

As a close student of the grieving process, I can tell you some things that you may already know. First, whenever there is a death, the survivors feel guilty. They feel guilty for being survivors, and they feel guilty because they think that had they done things differently, the person would not have died. I remember, many years ago, conducting two funerals in one week in January for two elderly women in my congregation. I visited the family of the first one afterwards, and her son said to me, "If only I had sent my mother to Florida, gotten her out of this ice and freezing cold, she would be alive today." I visited the second family, and her son said, "If only I hadn't sent my mother to Florida—the long plane trip, the change of climate—that's what killed her." Fact number one: Whenever there is a death, the survivors feel guilty.

Fact number two: One of the important things that religion has to do at a time of death is to help the survivors work out that guilt. We do it with mourning rituals in which people make themselves a little bit uncomfortable. We do it by structuring opportunities to give charity or offer special prayers to perpetuate the name of the deceased. And we do it by having friends and neighbors come by to talk and mostly to listen.

Fact number three: Although it is the job of religion to help the survivor work out his guilt, all too often religion and well-meaning religious people conspire to make him feel even more guilty. How do they do this? They do it by finding it necessary to defend God, and to insist that it was right and necessary for this person, for this child to die. By implication, and sometimes even explicitly, we say to the bereaved, "You have to show yourself as a man of faith by taking this with good grace. You're in enough trouble with God already; don't make it worse by complaining." That is why I keep wishing that, one of these years, when I read the story of the sacrifice of Isaac, Abraham would talk back to God. I would feel that my faith was a lot more like his if he would do that.

When tragedy strikes, when a child dies, when parents are told that their child will have to spend his life crippled or retarded, they are bound to ask, "How could God let this happen?" How we answer that question will have much to do with how these people feel about themselves and their tragedy, as well as their ability to be comforted by religion rather than be estranged from it. If we are too zealous in defending God, if we insist too strenuously that God had good and sufficient reasons for letting a child die, we may make the parents so resentful of God that they won't let religion comfort them. When we do that, we picture God as the One who inflicts suffering rather than the Friend of those who suffer. I know bereaved parents who have not been able to set foot in a church or synagogue since the day their child died because their resentment of God runs too deep.

If we try to tell them that it was in some way good for their child to suffer, as in the funeral oration that said "Let us rejoice that young Michael was taken to Heaven with his soul yet unstained by sin" (which, I assume, means before he was old enough to be interested in girls), when we tell the parents of the retarded, "God must have loved you to have selected you for this special task," what we are asking people to do is to repress and deny their emotions until they are no longer sure what they feel. There is an epidemic in America today of people who are estranged from their real emotions because others have told them how they are *supposed* to feel. That is why there are psychiatrists who earn $40 an hour saying to people, "You're not in touch with your feelings. You're going to have to get in touch with your feelings."

Let me emphasize that what I am going to say now is a psychological, not a political, comment: It bothers me that a recent President of the United States smiled when he was angry. I think that is another case of a person censoring his feelings, telling himself that he is not supposed to feel what he does in fact feel, like the parent who says, "I'm not angry with you; I'm just very disappointed," when in fact she is angry and has every right to be.

When we, in the name of religion, deprive the hurt and the bereaved of their right to be angry, we double the burden of their guilt. To the guilt of surviving, we add the guilt of doubting and questioning God, at a time when the task of the religious person ought to be to lift the burden of guilt from the shoulders of the afflicted.

Let me tell you a true story out of my experiences as a rabbi. It is my favorite religious atrocity story. Some years ago, I officiated at the funeral of a young woman who had died of leukemia, leaving a 16-year-old son. I returned to the house afterwards, and I knew that this was going to be a tough situation, because 16 is a hard age for a boy to lose his mother. But as I walked in to the house, I got an idea of how hard my job was going to be. I heard a sister-in-law of the deceased woman saying to the boy, "Don't feel bad, Barry; God took your mother because He needed her more than you do."

Now, I count at least three serious mistakes in that one sentence, three things that should not be said to a person who has just suffered a loss. First: *Don't feel bad.* Why should a boy not feel bad on the day of his mother's funeral? It is just another instance of people being asked to hide from their honest, legitimate, and perfectly understandable feelings.

Second: *God took your mother.* I do not believe that. I will say more about that later, but I do not believe that God goes around distributing His daily quota of malignancies, deciding who should be sick and who should be healthy. I think we do God a very dubious favor when we give Him "credit" for all the terrible things that happen to people. The only thing we accomplish is to make the people who need God the most feel more distant and resentful of Him.

And last: *God took your mother because He needed her more.* What the boy's aunt was trying to say was that his mother's death was not just an arbitrary accident. There was logic and reason to it. It served a good purpose. One week ago, I got a letter from a woman who listens to me on the radio. She sent me a poem trying to explain my son's death by pointing out that Heaven would be an awfully dull place if only old people went there, so God took my son to make Heaven more enjoyable. You will excuse me for not finding that comforting. If I took that line of thought seriously, I would hate God for taking my son away for His own purposes.

In addition to that, the aunt was saying to the boy, though I am sure she did not mean to, "It's your fault. Had you needed your mother more, she would still be alive." What is that going to do to the guilt that the boy must already feel? Do you remember what it was like to be 16 years old and wish your parents would disappear and let you be independent? To be 16 and have to eat the food they served, and wear the clothes they bought for you, and ask them to do you a favor and drive you where you had to go? If we come along and say that "your mother died because you no longer needed her", we once again double the burden of guilt for an already unhappy young man. We try so hard to make God look good at a time like that that we end up making His innocent victims look like villains.

Why do children suffer and die? Why do bad things happen to good people in God's world? I am not sure I know, but I do not believe that God wants these things to happen. Sometimes bad things happen because human beings are cruel to each other. Sometimes they happen because the laws of nature do not make exceptions for nice people. If I eat too much, smoke too much, or cross the street without looking where I am going, I will suffer the consequences, no matter how moral or useful to the world I may be. But I do not think God causes it. I think that religious people sometimes get so desperate to make sense of the world, to believe that Somebody is in control of it, that we attribute all kinds of cruel and immoral things to God. And then what happens? We either hate God for doing it, or we hate ourselves for deserving and then resenting it.

But suppose we stopped holding God responsible for the unfair things. Would He really be a diminished God, a smaller God? If, in the face of the world's unfairness, which we cannot deny, we have to choose between an all-powerful God who is not kind or just, or a kind and just God who is not all-powerful, is there any question in which direction our Judeo-Christian tradition directs us? We may *fear* an all-powerful God who can strike down good and bad people alike and who demands unquestioning loyalty from all His subjects, even those He strikes down. But we cannot love or revere a God like that. We can love and worship a God who reaches down to the broken in heart and binds up their wounds, a God who associates Himself with the victim of an accident or illness, who is the source of our anger at unfairness and not the object of it.

If God does not cause tragedies and cannot prevent them, what good is He at all? Let me answer this very carefully. I have a theological argument with the insurance companies. I object to their calling earthquakes and hurricanes "acts of God." Not the devastation but the ability of a human being to bounce back and rebuild his life after the devastation—that is the act of God. The leukemia, the unexplained crib death—I don't call those acts of God. Rather the ability of parents to survive the worst thing that could possibly happen to anyone and still not give up on life—that, for me, is God at work. Where do we get the strength, where do we get the faith, where do we get the sense of a purpose to living? That, and not the illness or accident, is what God does for us.

People whose children are hurt, people whose children get sick and die, are going to be angry, and no one has a right to tell them not to be. If they get angry at God, it will be hard for them to find strength in religion, in prayer, in the religious community. If they get angry at what happened to them, but do not blame God for it, it may be easier for them to find sources of help in religion. That is why the biggest mistake we can make would be to defend God for what He allegedly did. I think we have to validate their anger, agree with them that what happened to them was rotten and unfair, but also tell them that God did not cause it. Instead, He is ready to help them survive it.

When a high school student who is taking a course with me asks, "How can I believe in God?" that is a theological question, and I answer that student in the traditional theological terms. But when the parent of a dying child asks me the same question, "How can I believe in God?" that is no longer a theological inquiry. That is a cry of pain and outrage, and it requires a very different answer. Remember the story of Job from the Bible: When Job began to suffer, his friends came to comfort him and make him feel better. But they really did not know how to comfort him. They heard him ask what sounded like a theological question— "How could God do this to me?"—so they tried to explain God to him. They explained that God controls everything in the world, and that he and his children must have committed some sin they were not aware of to cause God to punish them. But that was not what Job wanted or needed to hear. He did not need theology. He needed sympathy and human companionship. He needed someone to hold his hand and tell him that he was a good person who deserved better, not someone to point a finger at him and tell him he must have done something pretty bad to deserve this.

The same holds true for our conversation here this morning. The title of my paper was "Where was God?" I could have spent this half-hour talking about God, giving you all sorts of learned theological explanations. Believe me, I know all the words: The argument about plentitude, the argument about ontological necessity, and so forth. I chose instead to deal with the needs of the bereaved and battered family because I thought that was really more relevant.

I summarize with three points, three mistakes we have to watch out for when we deal with victims of tragedy. I speak from experience, because I have made all these mistakes and I have had them made on me.

The first mistake is to keep away from the suffering person because we do not know what to say, or because we are afraid that our friend will be hurt or jealous to see us with our intact families, to hear us talk about our children making the football team or getting into college, when his child never will. Our intentions may be the best, and it may in fact be painful for him to meet with us, but the result of avoiding him is immeasurably worse. It makes him feel shunned and lonely, and only reinforces his suspicion that all this happened to him because he is a bad person.

Second, we make a mistake when we are so intent on defending God's good name that we defend even the most horrible of tragedies as being His will. We would be better off identifying God not as the author of suffering but as the Source of our opposition to suffering and our ability to survive it and go on living.

And the last mistake we too often make is to pass judgment on people's feelings, to tell them that they should not be angry or bitter, to say that we have no right to doubt or question God just because calamities happen to us rather than to other people. That is not fair.

Even someone who has gone through a similar loss himself has no right to impose his reaction pattern on others.

Ultimately, I have no doubt that God will survive our anger at Him and even our boycotting of Him in anger if necessary. But what will happen to us if our anger estranges us from the Source of our hope and strength? The final truth, it seems to me, is the one offered by the psalmist many centuries ago: *I lift mine eyes unto the hills. From whence does my help come? My help comes from the Lord, Maker of heaven and earth.*

DISCUSSION

MODERATOR: IDA MARTINSON

A. MILUNSKY: We are all fortunate today to be experiencing a remarkably enriching time together. For the parent faced with the question "Where is God?" how do you respond?

RABBI H. KUSHNER: When I talk about God I have often been accused of what one writer calls "religious gerrymandering." That is, I draw the lines of my definition of God to include all the good qualities and exclude all the bad ones, and I plead guilty to that. As you heard, I do not hold God responsible for accidents, for tragedies, for malignancies, except in so far as the world has built into it laws of nature which make no exceptions for nice people. God does not cause malignancies, God does not cause Down syndrome, God does not cause automobile accidents or drowning accidents. Biology causes some of them, human cruelty and thoughtlessness cause some of them, and I find God not in the tragedy but in the capacity of the person to survive. I'm prepared to say, and this is my theological break with religious orthodoxy, that many things happen in the world which God does not cause. You know, it's always a little dangerous to ask a rabbi a theological question, especially when lunch is on the fire, because I could really give you more of an answer then anyone is prepared to hear. There is a school of theological thought known as predicate theology. Very briefly that means when you find a statement about God, read it backwards. Focus not on the subject, but on the predicate. To say God is love, God is truth, God is justice is not a psychological profile of the Being that lives in Heaven by the name of God. What is He like, what are His habits, how does He spend His time. Focus on the predicate, read those sentences backwards. Love is godly, truth is godly, honesty is godly. When you practice predicate theology on this kind of conversation, you end up identifying certain kinds of events and behavior as godly, as manifestations of God in human life. Other kinds of behavior—dishonesty, cowardice, illness—are not godly. They are things that happen in life, moments in which God is not present. What we can do when tragedy strikes, if we are wise enough not to blame God because we are not so compulsive about having to find a reason for everything, is invite God to come into our lives and strengthen us at that point precisely because we don't hold him responsible for the malignancy or for the accident.

A. MILUNSKY: In families of deep religious fervor of all persuasions is there a better coping mechanism or is it the opposite?

RABBI H. KUSHNER: I've not found a pattern. It varies very much because it's very hard to identify what we mean by deeply religious fervor. Some people are very much involved with religious life for mature healthy reasons. Some people are very much involved with religious life out of certain personality weaknesses and immature needs of their own. Some people stay away from the church and synagogue for valid reasons because they find the rabbi boring, some people stay away for invalid and immature reasons. So you can't really generalize. I have found, for example, some people of deep personal faith are able to accept tragedy. They take it in stride. They see it as one of the things that is dealt out to them in life, and it does not shake their faith. They respond, they cope in very healthy ways drawn from their religiousness. Other people of equivalent external piety are very troubled by this. They do a lot of denying. They are afraid even to raise the question. I remember in Chaim Potok's second novel, *The Promise*, there was a character who was a man of zealous orthodox piety and whose wife and children were killed in the Holocaust. He will not open the question of why God permits this to happen. And I suspect the reason he will not open the question is because he is afraid of the answer he will find. I think a lot of people who have suffered tragedies will not raise the issue of how could God let it happen because they can't stand the possibility that they will be led step by inexorable step to the conclusion that there is no God or God does not take care of them, or the world is not as secure as they believe. I have seen a lot of people choose to blame themselves when they were totally innocent because the alternative would be to say the world is random and chaotic, and they can not handle that. By the same token I have found people who do not involve themselves with organized religion at all who show immense depths of serenity and of calmness and faith, and who call upon these depths and resources which I would call profoundly religious. And I have found others who could not turn to religion under normal times and simply feel they can not turn to it in the moment of need.

H. S. SCHIFF: Rabbi, I would like to ask you about something that struck me very badly at the time Robby died, and every year since, when we do these particular prayers, it strikes me badly again. I don't know if it's translated into the Christian liturgy or not. That is, that God never gives you a burden greater than one you can bear. And every time I think of that I say, "Dear God, had I been weak Robby would be alive. Is that what that means?" And I'm very troubled by this, and it's a basic premise that does come up.

RABBI H. KUSHNER: I think the Christian equivalent, if I'm right, is that God tempers the wind to the shorn lamb. I've got to tell you, when you remarked about how your own rabbi in Detroit mishandled things at the time of Robby's death, for a moment I felt very defensive and apologetic for rabbis. All I can say is that I hope one does not hold all rabbis guilty of that, as I don't hold all laymen guilty of the mistakes that they make. I happen to disagree totally with the point of view that you've expressed and that you quoted. I don't think that God gives us only the burdens that we can handle for two reasons. First, I don't think God gives us burdens. I think they fall on us. Second, my experience, and I suspect yours, has been to meet all sorts of people who were not up to handling the burdens that life laid on them. I have seen people crack under the strain. I have seen marriages go asunder. I have seen all sorts of dysfunctional behavior—people not able to handle what was dealt them. People become very bitter. People become terribly depressed. I have seen lives terminated. The ultimate tragedy is not that one child dies of cancer but that the five other members of his family die at the same time. I don't see how anybody with minimal clinical pastoral experience could claim that God never gives us more than we could bear.What does happen sometimes is that, when we are struck by tragedy, we discover resources within ourselves we never would have believed we had. The widow or the bereaved parent who says to me on the day of funeral, "How can I go on living?" finds the capacity to go on living. The man who comes home after Hurricane David and sees his house has been washed away and says, "I simply don't have the strength to start over again" starts over again. Where do they get that strength? Had someone said to me before our son was conceived and born, you are going to have a child with progeria who will live 14 years, who will be stared at and laughed at, who will have all manner of pain practically every day of those 14 years, can you handle that? I would have sworn on a Bible that I could not handle that. I'm glad I didn't have that choice. I'm glad none of us have the choice, because I think we would have chosen too easy a path for ourselves. Somebody once quoted the aphorism, "It's taking the course of least resistance that makes river and human beings crooked." Bad things happen to us, some of us break under the strain, most of us discover resources that we would have been just as happy not to have to find, but we are heartened to discover what we have within us.

5

RABBI EARL A. GROLLMAN, a pioneer in family crisis intervention, is internationally known for his work concerning divorce, death, dying, and bereavement. He has appeared on numerous national radio and television shows, and his publications have won him acclaim from both religious and professional counselors.

His book *Talking About Death: A Dialogue between Parent and Child* received the Trends Citation by UNESCO at the International Children's and Youth Book Exhibition in Munich, Germany. He has also written *Explaining Death to Children, Concerning Death: A Practical Guide for the Living, Talking About Divorce: A Dialogue between Parent and Child, Living—When a Loved One Has Died, Living Through Your Divorce,* and *Caring For Your Aged Parents.*

Rabbi Grollman is a contributor to many religious and professional publications and has spoken at many universities and seminaries, including Yale Divinity School, the Medical College of Ohio, Hebrew Union College in Los Angeles, and Ohio State University.

For more than 25 years he has been rabbi at Temple Beth El in Belmont, Massachusetts, and is past president of the Massachusetts Board of Rabbis. A native of Baltimore, Maryland, he attended the University of Maryland, received his B.A. from the University of Cincinnati, and earned his Doctor of Divinity degree from Hebrew Union College. He is a member of the Professional Advisory Board of the Foundation of Thanatology at Columbia–Presbyterian Medical Center in New York and an editorial advisor to *Omega,* a journal devoted to thanatology.

The Clergyman's Role in Grief Counseling

EARL A. GROLLMAN

There is no time when the minister is more sorely needed than during the crisis of grief. He is usually the first one to be called. He officiates at the funeral. He visits with the family after the service. In Erich Lindemann's words, the clergyman may be instrumental in emancipating the survivors from their bondage to the deceased, assisting them in their readjustment to the environment in which the loved one is missing, and aiding them in the formation of new relationships.

The pastor is most effective when he acts as a pastor and not an amateur psychiatrist. He should not forsake his own traditional resources and spiritual functions. His represents a fellowship with the past, present, and future tied together by rites, theology, and a religious ethic. He has his own unique framework of viewing and handling guilt, forgiveness, conflict, suffering, and hostility. The practice of psychotherapy as the only real ministry to their congregation has led the Suffragan Bishop of Washington, Paul Moore, Jr., to write: "Too many priests forget their priestliness when they learn some of the basic skills of counseling—or perhaps they have not been trained properly in the use of priestly techniques and therefore are not confident in their exercise. These clergy become 'clinical therapists' who happen to have the prefix 'Reverend' in front of their names."

Despite the fact that he is not a medical therapist, the clergyman may be of unique assistance. He represents a concerned religious community. His truest function is revealed in terms of years and decades, as he watches children grow, marries them, and teaches their children

EARL A. GROLLMAN • Temple Beth El, Belmont, Massachusetts 02178.

in turn, and as he stands beside loved ones around the death bed of a patriarch whom he has come to admire and respect. Beyond a crisis of faith, people turn to him with marital questions, parent–child problems, and a variety of inter- or intrapersonal needs. He is minister, pastor, and counselor to individuals and families in joy and adversity.

In death, the pastor does not ascertain the emotional state of the bereaved through cross-questioning but rather through empathy and understanding. The spirit of caring and compassion must communicate itself to the bereaved, although not always on the first visit. Very often the most effective counseling comes after the funeral service. Before the ceremony itself, the bereaved may experience credulous disbelief and need to spend their time arranging the burial.

The funeral itself offers an opportunity to comfort the mourners. It is the rite of separation. The "bad dream" is real. The presence of the corpse actualizes the experience. The process of denial is gradually transformed into the acceptance of reality.

When the funeral is over, the religious leader has the task of reconciling the bereaved to their loss. In the process, he should pay particular attention to possible grief reactions which could lead to personal disintegration and mental illness. When the grief work is not done, the bereaved person may suffer morbid grief characterized by delayed and distorted reactions. He may show great fortitude at the funeral but later develop symptoms of somatic disease or agitated depression. This may include the actual denial of the death, the development of schizophrenic tendencies or psychosomatic diseases such as hypochondriasis, ulcerative colitis, rheumatoid arthritis, and asthma. Obsessive–compulsive behavior may manifest itself where the bereaved appeases his guilt through extreme cleanliness, or he may be unwilling to terminate the funeral service, e.g., "Tell me the eulogy again." He may exhibit self-punitive behavior detrimental to his social and economic existence.

It must be noted that the line of demarcation between "normal psychological aspects of bereavement" and "distorted mourning reactions" is thin indeed, just as it is between "normality" amd "neurosis." Each symptom must be viewed not as a single and decisive entity but as part of the total picture.

The family may still feel guilty and very much to blame even after a terminal illness. The guilt may take the form of self-recrimination, depression, and hostility. One tendency is to look for a scapegoat, e.g., the minister, physician, or funeral director. Inwardly this person may accuse himself but instead turns the anger outward in an attempt to cope with his own guilt.

In this area, the clergyman can be extremely effective. The helplessness that assails the grief-stricken often leads them to envisage the role of the religious leader in a symbolic aspect as the representative of God. As such, the clergyman can assuage intense feelings of guilt by offering a meaningful concept of forgiveness as well as help them to transform the errors of the past into a loving memorial by living more nobly in the future.

The importance of ritual is dramatically portrayed in the French film *Forbidden Games*. A girl's parents die in an air raid. She receives comfort by constantly playing a game of "funeral" and providing every dead creature with an elaborate internment of flowers and ornate casket. "Playing" at burying things helps her to relive, digest, and ultimately master the shock of her parents' death. The child has succeeded in bringing relief to herself.

Ceremonials play an important part in helping people of all ages face death and then face away from it. Religious rituals are community rituals. They are performed only by those who share a religious sameness and by no one outside it. The traditions create a sense of solidarity and belonging, the feeling that one is a member of a group with all the comfort, gratification, pride, and even pain that such a sense brings. For the ceremony is the same for all. It is definite and prescribed. Here the bereaved can be made to understand in clear-cut, unmistakable terms what is desired of him. Perhaps by carrying out the religious ceremonies, he will feel that he has regained the love he has lost, and that he comes to peace with his own conscience. Even rituals which might seem irksome and pointless to others may be heartily welcomed. They may represent the sought-for punishment, the neutralizer, the deprivation that could balance off the imagined indulgence at the bottom of the guilt.

MOVING INTO ACTION

Certainly, the one-to-one relationship between individual and pastor is necessary and beneficial. However, this is not the only approach. Grief work in groups is also therapeutic in assisting the bereaved to overcome their separateness and abandon the prison of aloneness. Just as there is a Golden Age Club for the aged, an Alcoholics Anonymous for the alcoholic, why not a place for those who have suffered the grief and dislocation of death? They especially need to share with like-minded people their feelings, troubles, and hopes.

A pilot program is being pioneered in Boston as a result of the efforts of Dr. Gerald Caplan, Director of Harvard Community Mental Health Program. He has enlisted an ecumenical bereavement team where clergymen of all faiths meet to discuss their own varied experiences with grief and death.

A widow-to-widow program has been established. Women whose husbands died within the past year meet periodically to participate in discussions of their personal loss. To assess the accomplishment, one must first remember how they appeared at the first meetings. Their faces were drawn. Most were nervous and shy. Some just sat and stared and were totally uncommunicative. As one of the leaders, I observed an amazing metamorphosis. They became animated. They discovered that one touch of sorrow had made them all kin. They were able to participate with others who were also suffering the emotional trauma of bereavement. They belonged to the largest fraternity in the world—the company of those who had known suffering and death. This great universal sense of sorrow helps to unite all human hearts and dissolve all other feelings into those of common sympathy and understanding. Problems may be singular to each person but there are others present who have faced similar troubles and together they seek to work them out. Each gives of her own understanding, compassion, and supportive concern.

I have met with widows of varying faiths and even participated with women of particular denominations such as at Catholic Retreat Houses. I am convinced that each church and synagogue could well form an organization to assist widows and widowers in reconstructing their lives and their homes. Together with interested members of his congregation, the clergyman could be of inestimable value in personal as well as group counseling during their difficult period of loss and aloneness. The bereaved should be claimed for the useful citizens that they are and their talents utilized as part of the larger fellowship of their respective faith and house of worship.

The religious leader's primary objective is to aid in the emergence of a new self which has assimilated the grief experience and grown because of and through it. With the aid and encouragement of a pastor who is kind, sympathetic, and understanding, the bereaved may be helped to accept death through a more profound and meaningful religious approach to life. By the clergyman's evaluation of the experience, by catharsis, confession, remembrance, and release, the members are guided to new purposes. Their introspection may bring new value judgments to the meaning of life and love. The synagogue or church would no longer be an impersonal entity since its members have extended their hands in warmth and affection. But most important is the comfort they will gain from a new and abiding concept of God, as stated in this prayer from the *Union Prayerbook:* "Even though I cry in the bereavement of my heart when my beloved is taken from this earth, may it be as a child cries who knows his father is near and who clings unafraid to a trusted hand. In this spirit, O Thou Who are the Master of My Destiny, do I commit that which is so precious into Thy keeping."

Thus, the clergyman has an important part to play in grief counseling. As he helps others, he is himself helped in this exchange. The I and Thou relationship is, according to Martin Buber, "a religious experience" beyond the most knowledgeable theology and psychology. In the parable of the Chassidic Rabbi, Moshe Leib of Sasov:

> How to love man is something I learned from a peasant. I was at an inn where peasants were drinking. For a long time all were silent until one person, moved by the wine, asked a man sitting beside him, 'Tell me, do you love me or don't you?' The other replied, 'I love you very much.' The intoxicated peasant spoke again, 'You say that you love me but you do not know what I need: if you really loved me, you would know.' The other had not a word to say to this and the peasant who put the question fell silent again. But I understand the peasant, for to know the needs of men and to help them bear the burden of their sorrow, that is the true love of man.

A CHILD'S PERCEPTION OF DEATH:
A DECALOGUE FOR THE CONCERNED CAREGIVER

I. Do take the word "death" off the taboo list. Allow it to become a concept that can be discussed *openly* in the home and the school. The question is not whether children should receive death education, but whether the education they are receiving is helpful and reliable.

II. Do understand that mourning and sadness are appropriate for people of all ages, *and children are people.* Grief now walks by their side. Numbness, denial, anger, panic, and physical illness are variations of their experience of pain.

III. Do allow them to release their emotions. Let them call their feelings by the rightful names: "I am *angry.* I am *sad.* I am *hurt.*" If they wish, they can scream it out or put their thoughts into words—in the form of poetry or a story. Sorrow, like a river, must be given vent, lest it erode its bank.

IV. Do contact your children's school and inform them of the loss in the family; otherwise, teachers might not understand any change in your youngster's grades or sudden sullenness or regressive behavior.

V. Do go for help if you feel unable to deal with your children during this crisis. There are times when even the best informed and well-intentioned adult is simply inadequate.

VI. Do NOT tell a child that he or she is now the man or woman of the house or a replacement for a dead sibling. Never say, "You remind me so much of . . . " Do not tell him that he is a substitute adult or surrogate relative, or a friend—lover—companion—confidante. It is difficult enough for youngsters to lose a loved one. Do not deprive them of their childhood.

VII. Do NOT give stories and fairy tales as an explanation for the mystery of death. Never cover up with a fiction or a confusing interpretation that you will someday have to repudiate.

VIII. Do NOT teach your children that you have the final answers. Keep the door ajar to their doubts, questions, and differences of opinion. Adults demonstrate their maturity when they say: "Are you surprised that I don't know everything about death? Don't be. That's why we must talk together. Let's help one another." Respect their individuality, for in the long run they must find their own answers to the problems of life and death.

IX. Do NOT be afraid to express to your children your own emotions of grief. If you repress your feelings, they are more likely to hold their own emotions at bay. Children receive permission to mourn from adults.

X. Do NOT forget that, even though the sorrow of loss continues past the funeral, when you have sorted out your own feelings you will be better able to understand your troubled children who come to you laden with questions and beset with fears. The real challenge is not just how to explain death to your children but how to make peace with it yourself.

BIBLIOGRAPHY

1. Allport, Gordon W.: *The Individual and His Religion.* New York: Macmillan, 1960.
2. English, O. S., and Pearson, G. H.: *Emotional Patterns of Living.* New York: Norton, 1963.
3. Feifel, H. (Ed.): *The Meaning of Death.* New York: McGraw-Hill, 1959.
4. Freud, S.: *Mourning and Melancholia. Collected Papers,* Volume IV. London: Hogarth Press, 1925.
5. Fulton, R. (Ed.): *Death and Identity.* New York: Wiley, 1961.
6. Gorer, G.: *Death, Grief, and Mourning.* Garden City, New York: Doubleday, 1965.
7. Jackson, E. N.: *For the Living.* New York: Channel Press, 1965.
8. Johnson, P. E.: *Psychology of Pastoral Care.* New York and Nashville: Abingdon Press, 1953.
9. Liebman, J. L.: *Peace of Mind.* New York: Simon and Schuster, 1946.
10. Lindemann, E. Symptomatology and management of acute grief. *Am. J. Psychiatry* **101**:141, 1944.

11. Linn, L., and Schwartz, L. W.: *Psychiatry and Religious Experience.* New York: Random House, 1958.
12. Osborne, E.: *When You Lose a Loved One.* New York: Public Affairs Committee, 1958.
13. Ostow, M., and Scharfstein, B. A.: *The Need to Believe.* New York: International Universities Press, 1954.
14. *Union Prayerbook. I.* Cincinnati: 1948.

6

AUBREY MILUNSKY was born and educated in Johannesburg, South Africa. He completed his medical specialist training in London and became a member of the Royal College of Physicians. He is board certified in both internal medicine and pediatrics. At present he is an Assistant Professor of Pediatrics at the Harvard Medical School and Director of the Genetics Division at the Eunice Kennedy Shriver Center. He is also a Medical Geneticist at the Massachusetts General Hospital, where he was the Associate Director of the Cystic Fibrosis Clinic. He is a member of the Society for Pediatric Research and the American Society for Human Genetics. Dr. Milunsky serves on the editorial boards of the *American Journal of Medical Genetics*, the *American Journal of Law and Medicine*, and *Prenatal Diagnosis*.

For 20 years he has had an in-depth interest in genetic diseases and in particular the care of children with cystic fibrosis or mental retardation. His laboratories for the prenatal diagnosis of genetic diseases are among the most experienced in the world.

He is the author, editor, or co-editor of eight previous books. His writing has focused exclusively on the early detection or prevention of genetic diseases. His books include *The Prenatal Diagnosis of Hereditary Disorders* (Charles C Thomas, 1973), *The Prevention of Genetic Disease and Mental Retardation* (W. B. Saunders, 1975), *Genetics and the Law* (with G. J. Annas) (Plenum Press, 1976), *Know Your Genes* (Houghton Mifflin, 1977; Avon, 1979), *Genetic Disorders and the Fetus: Diagnosis, Prevention, and Treatment* (Plenum Press, 1979), and *Genetics and the Law II* (with G. J. Annas) (Plenum Press, 1980). *Know Your Genes*, written for the lay public, is being or has been translated into seven languages.

Care in Chronic Fatal Genetic Disease

AUBREY MILUNSKY

Robert Louis Stevenson, who had pulmonary tuberculosis at the time, wrote that life is not a matter of holding good cards but of playing a poor hand well. Perhaps that was a typical perspective of someone with chronic illness, albeit nongenetic. The treatment of the fatally ill child calls for special perspectives and insights into normal child development. A clear perception of expected and achieved milestones of development is critical to care, whether it be provided by the nurse, physician, or any other health personnel. It is important to recognize, for example, the stage at which a child becomes aware of death and how he/she thinks of death.

The management of a four-year-old preoccupied with fantasies of death is totally different from the management of the same child with the same illness at the age of twelve. The care of the child with chronic fatal genetic disease is one of the most demanding tasks that exists in pediatrics. After 20 years I find the task even harder, as experience has engendered in me the ability to recognize the many nuances that require care and finesse in caring for children who each leave their own unique and deep imprint.

Proper care of these children requires careful orchestration. The arrangements with nurses and health care teams both inside and outside the hospital require careful coordination. Care of children with fatal genetic disease is all encompassing, all demanding, and totally ener-

AUBREY MILUNSKY • Genetics Division, Eunice Kennedy Shriver Center, Harvard Medical School, and Massachusetts General Hospital, Boston, Massachusetts 02115.

vating to both parents, nurses, and doctors. The provision of such care is especially demanding because the problem is threefold. First, the parents and then the child must recognize and acknowledge the irreversibility of the disorder. Second, the inherent issue of genetic culpability must be treated. This is no better exemplified than in established sex-linked disease, when it is indeed the mother who transmitted the deleterious gene to the child, who, for example, may have muscular dystrophy, hemophilia, or one of the at least 200 other diseases in that category. Guilt is equally burdensome to both the mother and father when the child inherits a dominant disease from either of them, especially where the affected parent has a mild form of disease but the child contracts a devastating form of the same disorder—a variability so characteristic of dominantly transmitted disease. Third, the inexorable effects of chronic serious tension resulting from living for a long period on the brink of disaster must be dealt with. These three features combined make the care of children with fatal genetic disease a unique and enormous challenge to all those who try.

In this chapter, I will focus on those individuals *primarily* involved in the care of such affected children—the parents, the affected child, the siblings, the physician, and the nursing and paramedical staff. An effort will also be made to elucidate those principles that I believe guide the provision of optimum care for the child with chronic fatal genetic disease. For this purpose I will mainly use cystic fibrosis as the prototype example. Some may not be entirely familiar with this genetic condition. It is an autosomal recessive disease with chronic lung infection associated with intestinal malabsorption due to pancreatic insufficiency that leads to prolonged suffering over many years until an invariably fatal outcome (about 80% of its victims die by the age of 20).

I am also cognizant that some parents of my own patients may be among my readers. It is an exacting exercise having the parents of one's patients bear witness to a discourse on what I think optimum care is all about. Since they are in many ways my teachers, I believe that they will be able to assess how much we have all learned, taught, and grown together.

THE PARENTS

The parents of the affected child bring their own life circumstances to bear on the tragedy that confronts them. To begin with, their ages may be important. Just recently I had a 17-year-old patient who had a baby with an anencephaly. The baby died within 48 hours of birth. I saw this young woman with her mother in counseling. She was totally bewildered, having had a child with, as she put it, "almost no head." Despite careful counseling she seemed oblivious of the recent past and without plans for the future. Emphasis that prenatal tests would be available in all future pregnancies seemed to fall on deaf ears. Anxiety blocked her ability to assimilate the counseling information provided. The importance of a return visit was emphasized.

The educational status of the parents may be important. There are some who feel that really sophisticated university-educated people can cope with all tragedies, whereas those without such education may be unable to effectively cope with any such serious problems. I do believe that education allows the parents to intellectualize and recognize that they are intellectualizing but makes little difference to the coping process. However, educated parents more often learn to be more discreet about communicating their inner emotions to the affected child. They would be less likely, for example, to make utterances in the child's presence, such as, "If we had been more careful we wouldn't have had you."

"Parenting" ability is highly variable. None of us have been formally taught how to parent. Many parents have very diverse views about their roles and responsibilities. Certain fathers, for example, would think that simply bringing home the bread is all they have to do, leaving all the other burdens to the wife. The state of the marriage is a key factor. The unstable marriage will obviously compound the problems of care and is more likely to come apart in the face of chronic fatal illness of a child.

The financial stability of the home is important, especially when threatened by long and costly illness. The birth order of the affected child may be important. It is clearly easier to take care of a first and only child with a defect than one who is the tenth child. The health knowledge of the parents will influence the way they view their ill child and affect their care. Their own emotional stability is obviously critical and would bear heavily on their ability to cope with unexpected tragedy. Some have extreme neuroses and on occasion may become latently or overtly psychotic. About one person in 100 in this society has latent or overt psychosis. The basic personality makeup of both parents is equally important. The depressive personality will clearly succeed less well with the care of an affected child than one who has an optimistic or more normal bent. One parent may be able to cope much better than the other. One may be maladaptive.

All of these and other factors in the background may potentially complicate matters as the tragedy unfolds or is suddenly thrust upon the parents. Anger for various reasons is often encountered in these families. In some instances paradoxical parental anger is directed at the child. Parents may be angry about the constant illness and care required for the child. They often are inwardly angry at themselves for having these feelings, which result from their assumed burden of guilt and enormous personal sacrifice. Their anger not infrequently extends to include particular views of the physician. Parents may be angry that the diagnosis was incorrect initially or that referral to a specialist was not offered or recommended.

Some parents have strong feelings about how they were informed of their child's fatal genetic disorder. Their often justifiable complaints focus on the failure of the doctor to share his/her concern about a probable diagnosis, and many weeks or sometimes many months elapse before an actual diagnosis is made. Some mothers have communicated their anger about obstetricians who informed them while they were still in the delivery room of the hospital that the child was seriously defective and then promptly disappeared from the ward.

Certainly the response from such grave news is well documented. The initial shock and disbelief may lead to panic and later to denial. Such denial may last for a short period or for many years. Acceptance and understanding sometimes only comes slowly.

In recent years I have increasingly seen parents wracked by guilt or anger after having declined prenatal genetic studies or after not having had these tests offered. Problems arise after the birth of a defective child when prenatal diagnosis would have enabled avoidance of such a birth.

Frequently one parent (most commonly the mother) is sharply focused on an imagined wrong-doing in pregnancy. She may have taken a tranquilizer or other drugs, been exposed unnecessarily to someone with an infection, had X-rays, gone on a long trip, or even taken LSD, marijuana, or cocaine years before pregnancy.

A common experience is to see parents "shopping around" at different hospitals, hoping to find a diagnosis where none had been made or seeking a treatment or cure where none existed. These activities should be recognized as the desperate efforts of loving parents to do as much as they possibly can and are, on occasion, the consequence of inadequate medical care.

It is a common observation that parents with chronically ill children tend to overprotect them. This overprotectiveness and the infantilizing of the child by parents in these situations is understandable and difficult to avoid or change.

Steady deterioration in the child's condition is sometimes associated with parental emotional "distancing." Parents subconsciously begin to prepare themselves for the loss of the child and by virtue of this mechanism attempt to diminish their own emotional pain. Evidence of this phenomenon may be a new pregnancy, mother taking on a full-time job when it is obvious she is already overcommitted, remarriage, adoption of another child, father initiating a major new time-consuming project, separation, or divorce. The child might begin coming to the clinic alone or with someone other than the parents. This emotional divestment by the parents must be understood since it is to some extent a normal defense mechanism. I believe that chastising the parents is not an appropriate response in these situations.

This distancing or detachment essentially reflects mourning before death—the "living dead" syndrome. Sometimes parental grieving is almost as strong as if the child had died years before the fact. Strangely, the parents may be assisted in this normal process by their child's response in subconsciously (perhaps) recognizing both their own prognoses and their feelings. Children may respond by recognizing this grief and, in a way protective of their parents, withdrawing to become less demanding.

Strong marital unions are likely to become stronger and closer when faced with adversity. In contrast, disharmonious marriages are likely to fall apart as a consequence of the stresses that flow from having a child with a major birth defect or genetic disorder. There are data to suggest that there is a higher frequency of divorce, separation, alcoholism, abandonment, and other social ills in families who have an affected child. When divorce and marital disharmony are so frequent in society at large, it would take an enormous study to demonstrate that these conditions occur with significantly higher frequency in families with affected children than in those without.

THE AFFECTED CHILD

The basic character and personality of the child are important not only in coping with disease but also in influencing family, friends, and medical assistants. The querulous, miserable, demanding, angry, impossible to please child clearly represents management problems in contrast to the pleasant, fun-loving, lovable, and accepting child. The child's relationships with the parents may be close, distant, dependent, or different with each parent. Reactions by the child to the diagnosis may be crucial and are often age-dependent.

Coping with serious symptoms such as persistent breathlessness, constant and paroxsysmal severe coughing spells, persistent vomiting, and frequent episodes of separation anxiety on recurrent hospital admissions all take their toll on both the child and the surrounding loving family. Conflict with parents is neither surprising nor infrequent. The child may deny the illness and refuse to take the many necessary medicines, reject the need for chest therapy, refuse to put on warm clothing, go swimming when it is still too cold, and so on.

Then there is the uncompromising peer pressure coupled with rejection by peers which is so difficult to cope with, advise about, or manage. Not infrequently there are cruel reactions. I have teenage patients with cystic fibrosis who, for example, have been told directly by their "dates" that they were too sick or puny. Younger children are sometimes told by their even younger siblings or neighbors that "they were going to die anyway."

The self-consciousness of the chronically ill child is equally hard to handle. Being aware of their small size, emaciated appearance, and persistent cough is distressing enough, but is more so when everyone in the classroom looks at them. Chronic illness, which may delay the child's sexual development, causes even further separation from peers. Their developing isolation and loneliness often compounds an already hopeless situation.

The child's resentment may be painfully apparent. "Why me?" is the frequent question to both parents and doctor. The anger at the parents for transmitting the genetic disorder to the child breeds a resentment which is not easily stilled. Some of these children have already experienced the death of a previously and similarly affected sibling.

Finally, there comes the acceptance and resignation associated with depression. Notwithstanding all possible reassurance, the emaciated, breathless child who believes (or knows) that death may be just a question of time, may be almost impossible to provide with enduring encouragement. The management of the depressed child in this context represents among the strongest challenges to the pediatrician.

The Siblings

Both the younger and older siblings grow up in an atmosphere of chronic tension and crisis. Often they feel relatively neglected because of the total preoccupation of the parents with the affected child. They always notice the frequent exhortations to the sick child to "put on a sweater," "don't go outside it's too cold," or "have another helping."

Not infrequently, behavioral problems or deviancy arise in the siblings, possibly a consequence of such relative neglect. School problems, drugs, nocturnal enuresis, and running away from home are some of the responses we have seen.

Some older siblings might become extremely sensitive about bringing their friends to the house for fear of being personally ostracized or stigmatized in the context of dating or marriage.

Parents sensitive to the many confounding and compounding problems usually marshal the entire family as a team. Special care is taken not to neglect the healthy siblings. By example and by lesson they are taught compassion, sensitivity, and understanding. It is these families, for some of whom it has been my privilege to care, who, despite overwhelming emotional and physical enervation, have enriched their family unit. Unfortunately, it is not the rule that families are enhanced and enriched in these circumstances.

THE PHYSICIAN

Many factors will influence the effectiveness of the physician in caring for the child with chronic fatal genetic disease. The age, experience, personality, training, and natural ability of the physician are all obviously important. Many of us know the warm, competent physician, while so many more of us know that cold, distant, and awkward doctor. Has that physician come to grips with coping with the possibility of his/her own death, the death of a loved one, or the death of a patient? Can physicians be taught to communicate in a warm, sensitive, and understanding fashion, or is it an inborn ability?

The religious dictates and ideas of the physician relating to euthanasia and the need for resuscitation may be additional critical factors.

The helplessness of the physician in doing any more than palliating the child may on occasion prove personally difficult. We have all experienced the sad weariness and feeling of hopelessness when faced with such an insoluble problem. Despite every effort to encourage, to palliate, and to comfort, there is the nagging sensation of distress, failure, and loss of a child with whom there has been close contact, often for many years.

In this context I should mention how difficult it is for physicians to attend the funerals of their young patients. Despite the crushing sadness, I have made it a rule to always attend. It is, I believe, integral to optimal and continued care of the entire family for the pediatrician to attend the funeral.

The Nursing and Paramedical Staff

Everything that I have described as pertinent to physicians is even more applicable to the nursing and ancillary medical staff. It is the nursing staff who take on the brunt of responsibility in caring for children with incurable genetic disorders. The stress experienced is not imagined but very real. Indeed, considerable effort should be expended in training nurses to cope with these special challenges. Nurses, I believe, will be the first to agree that physicians in particular also require training, not only to cope with their dying patients (including children), but to recognize that such care is optimally provided by a team. In the pediatric ward such a team is constituted by the nursing staff, the physical and occupational therapists, the teacher, the social workers, and chaplains, and it is directed by a (one hopes) sensitive and understanding pediatrician.

The effects of chronic emotional tension and work in a crisis atmosphere require anticipation. In this way the so called "burn-out" of nurses can be prevented, using nursing rotations to other services or by planned patient allocations, or other such measures. No matter how competent the pediatrician, success or failure of inpatient care of the dying child will inevitably and directly reflect the excellence or otherwise of the nursing staff.

Guiding Principles for the Care of the Child with Chronic Fatal Genetic Disease

There must be many principles which could be elucidated to guide those involved in the care of the child with chronic fatal genetic disease. The following constitute a set of briefly elaborated guiding principles I use, which while not totally encompassing, probably include the most important.

The Comfort of Competence

Seriously ill patients will invariably confirm their feelings of comfort and reassurance when they recognize the competence and ability of their physician. Parents and child must feel assured that they are receiving the best available medical care. This does not imply that their physician must be either omniscient or an acknowledged superspecialist. The patient (or parent) is often reassured by the competency of the physician's judgment, especially when the physician suggests the help of a consulting specialist. Indeed, it is the competent physician who anticipates the patient or parent's concern and offers or suggests a second opinion without waiting to be prodded by the patient or family.

HUMANITY AND COMPASSION

No elaboration is required on this principle. The need for a warm, sensitive, understanding physician must be obvious to all. On occasion this role is best fulfilled by a family practitioner working hand-in-hand with a superspecialist.

TRUTH AND HOPE

Where possible, direct truth in the care of all patients is to be recommended. However, in the care of the dying child very little is to be gained by removing all hope by revealing the bare truth. There are some cases in which the maturity of the child, the circumstances of the family, and the presence of other factors may properly incline the physician to discuss the actual prognosis with the child. There is no doubt that some children and their families benefit from this direct approach. It has been my experience however that the removal of hope by truth-telling has rarely been of obvious benefit to the dying child, although certainly these families have been relieved of the burden of having to tell themselves or from having to masquerade a prognosis that is not consonant with the child's condition.

LISTENING AND COMMUNICATING

The best communicator is the careful listener. The child is often unable or unwilling to accurately articulate even overwhelming fears or anxieties. Children do, however, signal their innermost feelings in many nondirect ways. Hence the perceptive listener may detect nuances or direct clues to the child's innermost problems. In this way the team can minister most effectively to the unspoken needs of the affected child. Time, caring, and prolonged empathetic contact is an essential ingredient to helping by listening. It is the nursing team that makes the most important contribution to care in this respect.

RECOGNITION OF ANXIETY BLOCK

The assimilation of new information about serious or fatal illness of oneself or a loved one is usually incomplete in the face of overwhelming anxiety. The medical care team, especially the physician, needs to recognize this common phenomenon and routinely indicate to the patient how difficult it is to remember all the facts being communicated when great anxiety is present. It would be natural to require repetition of most of this information in a subsequent visit, and there should be encouragement that specific questions and concerns be committed to paper for discussion the next visit. Parents of children with serious or fatal genetic disease invariably find that their upset has prevented them from remembering much of the details communicated at the time the diagnosis was first made.

CARE AND ANTICIPATION

Totally enveloped in the often overwhelming problems of caring for the dying child, it is easy to overlook the attention that is required to anticipate more serious problems ahead or completely differentiate complicating aspects. For example, the decision to allow the child to die at home requires the added consideration of the possible effects on the other siblings. Indeed, optimal long-term care of a fatally ill child requires that the physician constantly focus the parents' attention on the needs of their children. Those parents who have lost a child will often bear witness to their own emotional and physical enervation and how, through retrospective analysis, they recognized that despite their every best effort, the other children were relatively neglected. With anticipatory care parents can be helped to avoid what otherwise becomes an almost inevitable additional family problem.

AWARENESS AND SENSITIVITY

The highly charged atmosphere surrounding the care of the dying child requires special awareness and sensitivity from the entire medical caretaking team. The various parental and family behaviors that the team will witness is another important subject. There are endless examples: the nonvisiting mother or father or sometimes both parents; the angry and vindictive father holding all physicians culpable for not having diagnosed his child's disease earlier or for being unable to cure the child; the remarkable family who rally to the dying child's support and are brought closer and made more compassionate individuals through a devastating experience; the parents who only halfheartedly administer to the needs and medications of the affected child; and the persistently denying family who maintain an outward facade of a belief that the child will get better despite every ominous sign to the contrary.

Great care should be exercised not to adopt a critical attitude toward these families who undergo often indescribable personal pain. It is required of the medical caretaker not to chastise but to provide noncritical support without demanding any specific behavioral response.

CONTINUITY OF CARE

The care of children with chronic fatal genetic disease is, by definition, continuous. For most such children care is provided in specialty centers or clinics. An unfortunate facet of such care in specialty clinics is that the child is likely to be treated by constantly changing junior physicians and a consequence of such care is that the child and parents do not establish a solid rapport with a particular physician. Often the senior physician in charge of such a specialty clinic will see parents briefly on each visit. My belief is that this arrangement does not constitute optimal care.

RESOURCEFULNESS AND RESPONSIVENESS

The child with chronic illness often requires and benefits from the help of various agencies. The physician should be aware of these resources and not only respond to the spoken needs but be sufficiently resourceful to anticipate needs. Knowing about the available local facilities may enable the physician to assist the parents in turning anguish into action. Channeling the anxiety of both patients and parents is helpful to both them and the specific disease Associations with whom they begin to work for their mutual benefit.

The burden and the challenge of caring for children with chronic fatal genetic disease as well as their extended families is a demanding task. The highest standards of care can only be achieved through a team approach. Care can be dispensed in many ways but there is little place for dogma other than that which demands sensitivity, warmth, and deep understanding from the entire team. In our combined efforts to prolong life, to palliate, and to comfort, the feelings and dignity of the child must remain uppermost in our minds.

BIBLIOGRAPHY

1. Bluebond-Langner, M.: *The Private Worlds of Dying Children*. Princeton, New Jersey: Princeton University Press, 1978.
2. Boyle, I., diSant'Agnese, P., Sack, S., Millican, F., and Kulczycki, L.: Emotional adjustment of adolescents and young adults with cystic fibrosis. *J. Pediatr.* **88**:318, 1976.

3. Easson, W.: *The Dying Child*. Springfield, Illinois: Charles C Thomas, 1977.
4. Fassler, J.: *Helping Children Cope—Mastering Stress Through Books and Stories*. New York: The Free Press, 1978.
5. Friedman, S., Chodoff, P., Mason, J., and Hamburg, D.: Behavioral observations on parents anticipating the death of a child. *Pediatrics* **32**:610, 1963.
6. Fulton, R., Markusen, E., and Owen, G. (Eds.): *Death and Dying: Challenge and Change*. Reading, Massachusetts: Addison-Wesley, 1979.
7. Gath, A.: *Down's Syndrome and the Family*. London: Academic Press, 1978.
8. Green, M.: Care of the child with a long-term, life-threatening illness: Some principles of management. *Pediatrics* **39**:441, 1967.
9. Green, M., and Solnit, A.: Reactions to the threatened loss of a child: A vulnerable child syndrome. *Pediatrics* **34**:58, 1964.
10. Grollman, E.: *Talking about Death: A Dialogue between Parent and Child*. Boston: Beacon Press, 1970.
11. Grollman, E. (Ed.): *Explaining Death to Children*. Boston: Beacon Press, 1967.
12. Grossman, M.: The psychological approach to the medical management of patients with cystic fibrosis. *Clin. Pediatr.* **14**:830, 1975.
13. Kerner, J., Harvey, B., and Lewiston, N.: The impact of grief: A retrospective study of family function following loss of a child with cystic fibrosis. *J. Chron. Dis.* **32**:221, 1979.
14. Lawler, R., Nakielny, W., and Wright, N.: Psychological implications of cystic fibrosis. *Can. Med. Assoc. J.* **94**:1043, 1966.
15. McCollum, A.: *Coping with Prolonged Health Impairment in Your Child*. Boston: Little, Brown, 1975.
16. Milunsky, A.: *Know Your Genes*. Boston: Houghton Mifflin, 1977.
17. Milunsky, A.: *The Prevention of Genetic Disease and Mental Retardation*. Philadelphia: W. B. Saunders, 1975.
18. Moos, R. (Ed.): *Coping with Physical Illness*, New York: Plenum Press, 1970.
19. Sahler, O. (Ed.): *The Child and Death*, St. Louis: C.V. Mosby, 1978.
20. Turk, J.: Impact of cystic fibrosis on family functioning. *Pediatrics* **34**:67, 1964.

DISCUSSION

MODERATOR: IDA MARTINSON

D. TODRES (Director, Pediatric Intensive Care Unit, Massachusetts General Hospital, and Assistant Professor of Medicine, Harvard Medical School): I think that one of the essential points that arises in this is the whole concept of not talking to the parents but talking with them. I'm always interested in hearing when one of the residents is going to talk to the parents. I know what that means. I know that the parents are going to have to sit and *listen* to what the resident has to say. I think we've learned a great deal today. In communicating with families the essence is in learning how to listen and talk to parents. We have to learn from the parents. We can't learn from ourselves. I think we've learned that beautifully from talks by Harriet Schiff and Rabbi Harold Kushner. I would add three T's as a mnemonic. One is the *time* it takes to talk to a family. I think that abbreviated discussions occur all too frequently. You certainly must develop *trust*. I think that's where we have to work hardest. We have a great deal to do as far as the health care professions go and to provide the families of these critically ill and dying children with a sense of trust. The third T would be *touching*, which was very nicely brought out by our last speaker.

J. HARTMAN (Nurse Coordinator, Genetics Clinic, Manchester, New Hampshire): I'm so glad Dr. Milunsky brought up the fact that it's so important for the family to recognize the other siblings when there is a child in the home that is ill. I have had a personal experience with this, and I think that part of my problem was also being in the health care profession and overemphasizing the problem that the child had. It has affected one of our other children. I hope you will all take heed of this emphasis.

S. JANCOSEC (student at Harvard Divinity School): Rabbi Grollman, my father died when I was six and I came from a family where I didn't even know he had cancer. He was away for a long time and I just never saw him again. My mother took away all of his pictures, and it was a very complete death. When you were talking I realized that when I think of him now I can never not think of him and not be six-years-old. That used to scare me but somehow there is something rich about that experience. But sometimes I wonder if I will respond to the death of other people as a six-year-old, and I wonder whether as a clergy person you have suggestions about making distinctions.

RABBI E. GROLLMAN: I wish I could just say something very profound and abstruse and say this is where you would be. No one knows where he or she would be under any kind of situation. One of the real problems that I have in terms of speaking in different places is that I often do the opposite of what I hope to do. I induce guilt. I will say many things and somebody will say "I did it wrong." It might have been right for you, but it wouldn't necessarily be right for me. A few weeks ago I was on a television program and we were discussing my book on divorce and somebody called up and said, "My son is being divorced and he wants to come and live with me. Do you think he should?" I became the psychologist and I said, "Do not allow the return to the womb. I mean if he is going to be divorced, let him stand up by himself and for himself." That night the telephone rang. "Rabbi Grollman?" "Yes." "I understand my mother called you." I said, "What?" He said, "You told her that I should not come to live with her." I said, "Now I recall." "I have no money, tell me where to go." So I think that the fact that you are here means you will learn from your own experience. Where you were is now past history.

We don't know what is right for you, but I think that *you* will know because I can hear the compassion and the empathy in your own voice and you are there to listen and to touch. Be careful of theologizing. You don't argue theology in the house of bereavement. And be careful of the word. Sometimes physicians find it easier to use all of their medical jargon. What they are really saying is, "I don't want you to be privy to this information." Sometimes we clergy people are just as guilty. Especially when we talk to children. We talk about the immortality of the soul. I went six years to the seminary and I still don't know what it means. But it's easier to use big words than it is to come to a person and hold his hand and to look into his or her eyes. Just be the kind of person you are.

M. HARRIS (Social Worker, Burlington, Vermont): I was wondering whether, Rabbi Grollman, there is ever a situation where one might feel that there is justified guilt. I'm thinking of perhaps a car accident where alcohol was involved and where a child dies. The relative who was driving is then saying, "It's all my fault." And there is some element of feeling, "Yes, there was some fault there." How do you help a person deal with that?

RABBI E. GROLLMAN: There is no question in my mind. I wasn't trying to talk about guilt as something only pathological. I'm trying to say that guilt happens because often we do fail. Often a person is drinking and he is in a car accident and his wife dies. These are things that do happen. What I find is that very often the people who are most guilty (and this is something that I have never been able to understand, maybe somebody can explain it) are the most giving. These are the ones (I guess the technical words are the "punitive superego"). The funeral that I had today was for somebody who for 16 years was being taken care of by her sister. When I went to make the call yesterday the surviving sister said, "You know, sometimes I yelled at my sister." These are the problems that I have. And the answer is yes, we do things. But the word for *sin* in Hebrew is *'chet'*. It denotes an archery term. It means we fall short of the mark. We don't always do the right things and sometimes we do cause hurt for others. I think, first of all, we have to help the individual as a social worker, as a nurse, as a physician, as a clergy person, to understand is it real guilt or not? Some of the studies taken at Harvard show that people who have cancer of the throat do not say, "It's because I'm a heavy smoker," but rather they will remember a sexual indiscretion that happened three years before. I think it's important to say that this has nothing to do with the cause. I think what I do is what most people have done. A large segment of people are working through their own feelings by joining groups. I'm not helpful to people. I don't have the answers. I am most helpful when I can say, "I know a group in which you might be interested." We had a widow and a widower's group. I remember once meeting somebody and I thought I was saying all the right words. The woman wasn't being insolent when she said, "Is your wife living?" and I said, "Yes." And she said, "Then what the hell do you know?" I have never forgotten that. Now when somebody has died, I will make a visitation. I will not only make one visitation, but several. They expect it initially. But if you make the second visit it shows your friendship. When you go to the funeral it shows that you are more than just a person who sits with his white coat in the hospital. But I will also send somebody else, if that person wants to go and will be received by the mourner. The widow will often say, "You know what is the most difficult time for me?" And the lady will say who has gone as the visitor, "When you wake up in the morning." And she said, "How do you know?" And she said, "Because that's how I felt when Ray, my husband, died." And then one touch of sorrow makes the whole world kin. The Society of Compassionate Friends— they know what it's about. They don't need any kind of theologizing. They are able to share experience. I am most helpful not when I come with answers but when I know the right resources in my community. Let me conclude with something which Dr. Milunsky has given as one of his final points—you are not alone. It's not easy taking care of people who have gone through crisis. It hurts me, and it hurts you. But you are not alone. There are other people who can also support and who also care.

There is a story about an old orthodox rabbi who said to his disciple, "Do you love me?" and the young student said, "Yes, I love you very much, Rabbi." And then the rabbi said, "Do you know what gives me pain?" and the young student said, "How do I know what gives you pain?" and the rabbi responded, "If you do not know what gives me pain, you do not really love me." So, Martha Harris, after all the i's have been dotted, and the t's have been crossed, you're not alone. By your presence, by your empathy, by your caring, you're showing you love God by loving people who feel pain.

7

RABBI EARL A. GROLLMAN, a pioneer in family crisis intervention, is internationally known for his work concerning divorce, death, dying, and bereavement. He has appeared on numerous national radio and television shows, and his publications have won him acclaim from both religious and professional counselors.

His book *Talking About Death: A Dialogue between Parent and Child* received the Trends Citation by UNESCO at the International Children's and Youth Book Exhibition in Munich, Germany. He has also written *Explaining Death to Children, Concerning Death: A Practical Guide for the Living, Talking About Divorce: A Dialogue between Parent and Child, Living—When a Loved One Has Died, Living Through Your Divorce,* and *Caring For Your Aged Parents.*

Rabbi Grollman is a contributor to many religious and professional publications and has spoken at many universities and seminaries, including Yale Divinity School, the Medical College of Ohio, Hebrew Union College in Los Angeles, and Ohio State University.

For more than 25 years he has been rabbi at Temple Beth El in Belmont, Massachusetts, and is past president of the Massachusetts Board of Rabbis. A native of Baltimore, Maryland, he attended the University of Maryland, received his B.A. from the University of Cincinnati, and earned his Doctor of Divinity degree from Hebrew Union College. He is a member of the Professional Advisory Board of the Foundation of Thanatology at Columbia–Presbyterian Medical Center in New York and an editorial advisor to *Omega,* a journal devoted to thanatology.

7

Explaining Death to Children

EARL A. GROLLMAN

"Before I can teach children about death, someone has to straighten me out!"
So said a sophisticated religious school teacher when asked how she explains death to her students. Her comment was understandable. Death touches the ebb and flow of the deepest feelings and relationships. Especially in Western culture, man has tended to seek refuge in euphemistic language. Movies and plays often treat death as a dramatic illusion. One of the reasons why many persons reject the aged is that they remind them of death. Discussion about death is relegated to a tabooed area formerly reserved for sex and dread diseases like cancer.

Fear of death is not only a cultural phenomenon but a part of being human. The knownness of life and the unknownness of death, the termination of the natural joys of living, the conclusion of the relatively controllable activity of life all create a pervasive dread of death touching every man.

Since no mortal has ever pierced the veil of mystery surrounding death, the teacher and particularly the parent are frightened at the thought of finality. Perhaps for this reason many adults try in every way to keep from children the idea of death. Parents wish to protect their offspring. Often because of their own anxieties, they rationalize by saying: "The youngsters are really too little to understand." Yet, the parents who try to hide their grief for the sake of their youngsters are rarely able to do so effectively. The gap between what the adults say and do and the underlying feeling that the children sense is likely to cause more confusion and distress than if the parents tell the child the truth.

There are no simple, foolproof answers to death, the most difficult of all questions. Not only children but adults differ more widely in their

7

Reprinted with permission from *Explaining Death to Children*, edited by Earl A. Grollman, Beacon Press, Boston, Massachusetts, 1967.

EARL A. GROLLMAN • Temple Beth El, Belmont, Massachusetts 02178.

reactions to death than to any other human phenomenon. But in trying to help the children see death as an inevitable human experience and in sharing grief with them, the parents may be able to diminish in the process their own bewilderment and distress. Instead of feeling inadequate because they do not know what happens after death, adults should welcome their children's questions as occasions to explore the problem with them.

While insight is a gift, parents must first place themselves in a position to receive it. They must prepare themselves for it. They must be quiet and learn to listen to their children. They must sit down and watch the youngsters while they work and play. They must observe them in action and hear the tone and timbre of their voices. Let the youngsters tell the adults how they feel about death, what they think, what they know, where they want to go. Parents should respond by trying to let the youngsters know that they understand what the children are trying to say. Adults should try to answer the question in the spirit in which it is asked.

CHILDREN CAN UNDERSTAND THE MEANING OF DEATH

Language has a different meaning for adults and children. Philosophical meanings are far too abstract for the very young. But this does not mean that they do not reach out for an understanding of death. The small child tries out the word "death" and rolls it around his tongue. He experiments with the appearance of death much as he experiments with his daddy's hat and briefcase to get the feel of "going to work." All children are concerned about death and frightened by it. Some have even said they do not wish to grow up, "because if you grow up, you get old and die."

Dr. Jerome Bruner, Director of the Harvard Center for Cognitive Studies, begins with the hypothesis that any subject can be explained effectively in an intellectually honest form to any child at any stage of development. Understandably, the knowledge may be imparted with symbolic imagery or intellectual reasoning.

It cannot be precisely determined what concepts of death can be understood at a given age. Children differ in behavior and development. Some are responsible and stable; others are more immature and younger in relation to their years. Girls are generally more mature than boys. The ability to cope with the material will depend upon the maturity of the individual and his ability to cope with his problems.

In addition, environments vary greatly. The child's individual thinking is influenced not only by his biological equipment but by his adjustment to the world as understood through his family and social hierarchies. A young Jewish child in Nazi Germany would have a clearer insight into the reality of death than the small American lad whose parents believe that he should be shielded from any knowledge which might arouse painful emotions. The mourning task is initially dependent on this ability to have a concept of death and would differ from individual to individual as well as society to society. Age is but one of the factors affecting an understanding of death, to be contrasted to the unique religious, political, and social attitudes of the person and his group's dynamic cultural life.

"Some two-year-olds will have a concept of death, while five-year-olds will not," Dr. Robert A. Furman reports in a continuing study that is now in its twentieth year. "It is fundamental, however, to make a sharp distinction between a child's not mourning and his incapability of mourning." Dr. Furman dismisses the notion that a child may be incapable of grief. He also quotes a colleague who stresses that "mourning as a reaction to the real loss of a loved one must be carried through to completion."

In another recent study in Budapest, Dr. Maria Nagy studied 378 children between the ages of three and ten to determine their feelings concerning death. She learned that between these ages children tend to pass through three different phases. The youngster from three to five may deny death as a regular and final process. To him death is like sleep: you are dead, then you are alive again. Or like taking a journey: you are gone, then you come back again. This child may experience many times each day some real aspects of what he considers "death," such as when his father goes to work and his mother to the grocery. When the late President Kennedy's son John-John returned on a visit to the White House following the demise of his father and saw his father's secretary, he looked up at her and said: "When is my daddy coming back?"

Since the mother and father usually return from an absence, the child comes to believe that the people he loves will always come back again. Consequently, the youngster, when told about the death of a member of the family, may seem to be callous and possibly express an immediate sorrow and then seem to forget about it.

Between five and nine, children appear to be able to accept the idea that a person has died but may not accept it as something that must happen to everyone and particularly to themselves. Around the age of nine, the child recognizes death as an inevitable experience that will occur to him. Of course, these are all rough approximations with many variations, but they may prove to be of some value when children raise questions. Nagy's investigation also demonstrated three main questions in the child's mind: "What is death?" "What makes people die?" "What happens to people when they die; where do they go?"

What Should Parents Say?

It is easier to suggest what *not* to say. Do not tell a child what he is incapable of understanding. Do not be evasive, but modify explanations according to the child's understanding. What is said is important, but *how* it is said has even greater bearing on whether the child will develop anxiety and fears or accept, within his capacity, the fact of death. The understanding of the very young child is, of course, limited. He doesn't ask for or need details. His questions should be answered in a matter-of-fact way, briefly, without too much emotion. Too complicated a reply often confuses the child.

Always remember to avoid philosophical interpretation. Even adults find it difficult to comprehend abstruse concepts such as "ultimate reality" and "the absolute." Children, too, easily mistake the meaning of words and phrases, or take literally what is only an idiom. The result is that fantasies are inadvertently formed. One boy, hearing that God was high and bright, assumed that the weathercock on the barn must be God, for it was the highest and brightest object he knew. Difficult terms may slip by the child as he takes only the familiar words to weave them into a meaning of his own. Dr. Jean Piaget, the great developmental psychologist, points out that even when an adult supplies an answer, it is frequently only partially heard and inaccurately comprehended.

Even if the child does not fully understand the explanation given him, death will be less of a mystery and therefore less frightening to him. And he will, in time, ask further questions that permit further rational explanations. If his questions are answered frankly, he may even drop the whole matter for the time, satisfied that nothing is being hidden from him, relieving him of doubts and disturbing fantasies.

Often the children ask questions to test the parents. In answering, the parent needs to understand the train of thought leading to the question. Otherwise the reply he gives may be misleading. What is the child really asking? What does he mean when he asks: "What is going to happen to you, Mommy, when you die?" Is he in truth seeking a theological answer or is he searching for security in his anxiety? Perhaps he wants to know: "Who will take care of me?" Or, "Will Daddy die, too, and leave me all alone?" The best answer may be mostly nonverbal.

The parent might hold the youngster close and say: "All of us hope to go on living together for ages and ages—for a longer time than you can even think of."

Parents who have their own religious convictions will naturally share them with the child. They may explain in an understandable way their belief in a life after death. Those adults who do not believe in a personal hereafter might say: "We really don't know. But this we do believe—the kind of lives we lead will continue to be reflected in the lives of future generations. People the world over have different approaches to this question which has no final answers. Each one learns day by day that there is more to be discovered about this important subject." This type of reply prompts the child to go on thinking and probing. And even if there are no definitive answers, he comes to understand that unanswered problems are also part of life.

In explaining death to children, the parent may proceed from two areas of concern. There is the *interpretative* area, where the religious concepts are explained. "There is a God; God is a loving spirit. God's love transcends human experience; God can be trusted." In addition, there is the *factual* area, where the adult draws upon scientific sources. In this sphere, as far as the physical body is concerned, the parents should make it clear that with death, life stops, the deceased cannot return, and the body is buried. A less explicit explanation is apt to result in more confusion and misinterpretation.

SHOULD PARENTS INDICATE THEIR RELIGIOUS CONVICTIONS, IF ANY?

Of course, the approach will depend upon what the parents believe. Since religion deals with the meaning of life and death, those with deep faith in God may find real comfort in their beliefs when one of their loved ones dies. Those who believe in personal immortality often find the problem of death less painful since they believe they may one day rejoin their loved ones. Explanation of death will vary according to whether it is seen as a beginning of a new experience or an end to existence. Be honest and be open. Children adjust most quickly when they are in on things and have the help of those they love.

Essentially, the adult's aid to the youngsters will depend very largely upon his own resources, his own attitude, and the social culture and religious traditions affecting him. One can answer the children's questions with sincerity and conviction only if one responds with feelings that are real.

How Should the Subject of Death Be Introduced?

Since the subject matter is so sensitive, the first discussion should not concern the death of the child's mother or father. Nor should the explanation revolve around dogma, belief, or theology. Death and its meaning should be approached gently, indirectly, and unsentimentally. It might involve a discussion of flowers and how long they last. Or whether bees and wasps that sting humans should be destroyed. Or the springboard may be an immediate experience such as the death of a pet or a biblical story where death is mentioned. One should proceed slowly, step by step, according to the understanding of the child and the personal beliefs of the parents. Fear will be lessened when the discussion is focused not on the morbid details of death but on the beauty of life.

What Not to Say About Death

Would It Not Be Wise to Answer a Child with Stories and Fairy Tales about Death?

Childish descriptions of death as eternal sleep are subject to much change and revision with advancing age. Adults must tailor fundamental knowledge and belief to the interests and capacities of the children. But the inability of the adult to tell the young the whole truth about death, even if it were available, does not give him the prerogative to tell a lie. The parents should never cover up with fiction that they will someday repudiate. To state, "Now Grandmother is up in the sky with a beautiful pair of shining wings so that she can fly away," is not faith. It is imaginative fancy. And it is not helpful to the child. It simply gets in his way when he is having trouble enough distinguishing the real from the pretend.

Why Do Parents Feel It Necessary to Give the Child Explanations of Death That They Themselves Do Not Accept?

A parent may feel his own beliefs are too stark for a child. He therefore attempts to affirm a conventional conviction about death which he does not hold. A mother, for instance, may comfort her young son on the loss of his father by saying that he will live eternally, while she herself is mourning a husband irretrievably dead. Or a father may spin a tale of the heavenly happiness of the children's mother, while hopeless finality fills his own heart. In such cases the confused child, not able to bring awareness of the double-talk to the surface, develops a double dread of death. He panics not only at the loss but also at the agonized reaction of the bereaved parent.

Some pragmatic parents, while not exactly believing in the survival of personality, operate on the theory that they have nothing to lose and perhaps something to gain in answering, *as if*, for example, life after death were a fact. Sometimes because the adults' own emotional reactions to death are so painful, they make statements to their children that are misconceptions of reality. They add to the child's own fantasies representing a realm halfway between the fully conscious and the unconscious thought of dreams.

What Are Some Unhealthy Explanations of Death?

Mother has gone on a long journey. Again an untruth is stated that the child must later unlearn for himself. The parents are catering to misconception and fantasy. Freud remarked that, to the child, death means little more than departure or disappearance, and that it is represented in dreams by going on a journey.

To say, "Your mother has gone away for a very long journey for a very long time," is geared to provide some comfort for the child and to ease the strain of his mother's disappearance. But the child interprets this explanation to mean that his mother has abandoned and deserted him without telling him good-bye. Far from being comforted and holding the memory of the deceased dear, the child may react with anxiety and resentment. This is not surprising, for the adult's pattern of mourning is similar. How often has a widow castigated a deceased husband: "How could he do this to me? How could he leave me alone?" A euphemism of death is "when my husband *left* me."

The child may develop the delusion that someday the mother will return. Or unconsciously he may assume, "Mommy didn't really care enough about me so she stayed away." And if the mother only went away on a journey, why is everyone crying?

God took Daddy away because He wants and loves the good in heaven. The mother, seeking to ease the burden for her child, explained the death of the father: "We can't be selfish about it. If God wants Daddy because he is so good and so wonderful, we must be brave and strong and accept God's will and request." Despite the best of intentions to bring comfort, a questionable theology is introduced. Adults may understand that "the

Lord giveth and the Lord taketh away" in the sense that many believe that the God who makes birth and life possible also makes death necessary. Despite hopes and fears, they cannot alter the unalterable: "Daddy is dead!" But to assert that God took the loved one because "he was good" is to question the posture of both religious and scientific reality. Why be good if God may reward us for our piety by death? If one lives to an older age, does this mean that he was not good? The righteous one may surely die young but may he not live to a ripe old age? To equate longevity with goodness is hardly a solution to a vexing problem. The child may develop fear, resentment, and hatred against a God who capriciously robbed him of his father because the man was loved by God. The youngster may become even more frightened when he thinks: "But God loves me too; maybe I'll be the next one He will take away!"

Daddy is now in heaven. "Mother, if Daddy is supposed to go to heaven, how is it that they buried him in the ground? He's heading in the wrong direction. How can God take Daddy to heaven if Daddy is in the ground?" And the child demanded in an active voice that the mother disinter the remains of the father. The parent in this case did not herself believe in heaven but took a deceptive and dishonest path. Many parents realize that the introduction of the traditional idea of heaven creates far more problems than it solves.

Grandma died because she was sick. The psychologist Dr. Chloe has asserted that small children equate death with illness and going to the hospital. People do become sick and die, but most everyone becomes ill many times and yet survives to live a long life. Will the child himself die if he has a cold, the mumps, the measles? How does the youngster make a distinction between a serious illness and one not quite so grave? The comparison of sickness to death only prolongs and intensifies the fear of death.

To die is to sleep. It is only natural to draw the parallel. Homer in the *Iliad* alludes to sleep (Hypnos) and death (Thanatos) as twin brothers, and many of our religious prayers entwine the ideas of sleep and death. Traditional Jews, for example, on arising from sleep in the morning thank God for having restored them to life again.

But one must be careful to explain the difference between *sleep* and *death:* otherwise, one runs the risk of causing a pathological dread of bedtime. There are children who toss about in fear of going to "eternal sleep," never to wake up again. Some youngsters actually struggle with all their might to remain awake, fearful that they might go off to their deceased grandfather's type of sleep.

Understandably, it is easier for the parent to respond with fictions and half-truths that also make him appear to know all the answers. But the secure adult has no need to profess infinite knowledge. It is far healthier for a child to share the joint quest for additional wisdom than for his immediate curiosity to be appeased by fantasy in the guise of fact. The child may be shocked to discover there is something his father or mother does not know, but he has to discover this as he grows. Deception is worse than the reality factor that the parents are not all-knowing. Why not admit the lack of understanding of this mysterious area of life?

How Does the Child Experience Grief?

Do Children Experience Grief?

Of course, the child experiences a sense of loss and with it, sorrow. His grief is a complicated mechanism. He feels remorseful that a loved one is dead. On the other hand, he feels sorry for himself because he was picked out for personal pain. He is faced with many problems about which he is helplessly confused. He may believe that the departed has run out on him. His fears often give rise to anger with hostile feelings toward those who are closest to him. The knowledge that from now on there is nothing the child can do to "make up" to the deceased can be a very heavy burden. Yet, the parents who understand that a variety of reactions may possibly occur are well on their way to helping the child toward a more positive and more mature approach in dealing with the loss.

According to Dr. John Bowlby of Tavistock Clinic, London, each child experiences three phases in the natural grieving process. The first is protest, when the child cannot quite believe the person is dead and he attempts, sometimes angrily, to regain him. The next is pain, despair, and disorganization, when the youngster begins to accept the fact that the loved one is really gone. Finally there is hope, when the youngster begins to organize his life without the lost person.

How Does the Child Face the Loss of a Pet?

When a pet dies, the child is brought face to face with some of the implications of death—its complete finality and the grief and loss it inflicts. The youngster may experience some guilt because he feels he had not cared well enough for the pet and was in some way responsible for its fate. He may conduct in great secrecy an elaborate burial. To the adult, the ceremony may seem to be a thoughtless mockery of a very sacred religious ritual, but to children these burial rites are far from prank or ridicule. The youngsters engage in them with as much zest and enthusiasm as a mock wedding ceremony or secret initiation. This kind of "game" has profound meaning. Here is their real opportunity to work things out for themselves and play out their feelings and fears. Sometimes the "play" is accompanied by real sadness and tears which afford an opportunity to help them put the experience somewhat behind them. Parents may well suggest such a ceremony to a child whose pet has died.

There is the example of a small boy whose pet dog was killed by an automobile. His first reaction was one of shock and dismay. This mood was followed by outrage against his parents, who he felt were guilty of the death because they did not take proper care of the pet. The boy behaved like the adult who rages against God for neglecting His charges. Yet, the anger against the parents was but a substitute for his own guilt, for the youngster had on occasion expressed the wish to be rid of "that awful pest." The child then insisted that as part of the burial service one of his favorite toys be buried with the dog. The toy served as a kind of peace offering to the offended pet. Now, the lad was freed of his own anxiety and could continue to function effectively in his everyday activities. Thus the ritual combined the dynamics of guilt, assuagement, and reparation, to the mourning behavior of adults.

Usually the child accepts the loss of a pet, constituting a step in acceptance of one of the many unpleasant realities of everyday living. More often than not, a new pet replaces the dead one, and life goes on much as it did before.

THE CHILD'S LOSS OF A BROTHER OR SISTER

Although deprived of a sibling who had played an important role as caretaker or playmate, the child still has the security of the parents' presence. Yet it would be an error to assume that the death of a brother or sister is relatively unimportant, even though few conscious and readily discernible reactions may be detected. For the death of a child invariably affects the parents. Whether they turn more closely and protectively to the surviving child, or are so disturbed by their grief that they are unable to maintain a healthy parent relationship to him, the child will experience some modification of his life situation.

An older child's reaction to such a death may be the frightening realization that this could happen to him! Would it occur tomorrow, or next week, or next year? If the cause of death of an older sibling is not made clear, the younger child may take on babyish behavior to prevent himself from growing to that age when he, too, might die.

The youngster may try to replace the deceased person. He sees his parents grieving and he wants to make everything all right again. He may suddenly try to act like the lost brother in ways not suited to his own capacities and well-being. He may be burdened by the feeling that he must take the other child's place.

If he and his brother or sister were close, death may bring a long-lasting feeling of loss. His parents' grief, the many reminders around the house, and the abrupt cessation of a relationship that had been an important part of his life, all combine to make readjustment slow and painful. The situation may be further complicated by strong feelings of guilt because of past anger or jealousy toward the dead person, or because of failure to make the brother or sister happier while he or she was still alive.

WHAT ARE SOME REACTIONS TO THE DEATH OF A PARENT?

One of the greatest crises in the life of a child is the death of a parent. Never again will the world be as secure a place as it was before. The familiar design of family life is completely disrupted. The child suffers not only the loss of a parent, but is deprived of the attention he needs at a time when he craves that extra reassurance that he is loved and will be cared for. Here, too, the child's reactions are complicated by guilt feelings. Sometimes guilt evolves from earlier hostility toward the dead person and also from the feeling that the survivor bears some responsibility for his death.

Anna Freud points out that a child's first love for his mother becomes the pattern for all later loves. "The ability to love, like all other human faculties, has to be learned and practiced." If this relationship is interrupted, through death or absence, the child may do one of four things: remain attached to a fantasy of the dead person; invest his love in things (or work); be frightened to love anyone but himself; or, hopefully, accept his loss and find another real person to love.

If the boy loses a mother, he may regress to an earlier stage of development. His speech becomes more babyish. He begins to suck his thumb. He whines a great deal and demands the attention of adults. He says in effect: "Dear Mother, see, I am only a very little baby. Please love me and stay with me." Later on in life, because he was injured by his mother, the prototype of all women, he may believe that all women have a tendency to hurt men. To avoid being wounded by them, he loves them and leaves them before the women can do what his mother did to him: hurt and abandon him. However, it cannot be overemphasized that these dynamics need *not* occur. For example, in a home where there is no mother, there are almost always mother substitutes— a housekeeper, an aunt, or an older sister. Even in time of death there can be an exposure to intimate relationship with some significant person.

The small boy whose father dies will feel the loss of a male person to imitate, a masculine foil with whom he can learn to temper his feelings of aggression and love. The mother, however, may contribute to the boy's difficulty. Deprived of a husband, she may try to make up for her own deprivation by trying to obtain gratification from her son. The boy feels he now possesses his mother and she will continue to gratify him completely. Therefore he need not look for pleasure elsewhere. From observations, one can cite many examples and consequences of doting mothers and spoiled sons. (The reverse may be observed in girls who lose their father or mother.)

In general, if a child allows himself to find someone else whom he can come to love and trust, it is a good indication that he has worked out his grief. If he has not, he may spend the rest of his life searching, consciously or unconsciously, for an exact replica of his childhood relationship with the lost person and be disappointed over and over again that someone else cannot fulfill his original needs. For example, we all know people who are very successful at their jobs but who are unable to maintain a devoted and continuous love relationship.

IN TIMES OF DEATH, SHOULD PARENTS DISCOURAGE THE CHILD FROM CRYING?

Children's capacity for grief is often not recognized. The thought of death does bring fear. Paul Tillich, the theologian, who has been a strong influence in American psychiatry, based his theory of anxiety on the belief that man is finite, subject to nonbeing.

Yet, is this the child's first experience with death? He has heard the word used before, and he may have been exposed to death. He has seen dead animals and dead insects. He knows flowers wilt and die. And he is afraid. Death strikes his own family or pet, and he cries as he expresses painful emotion. When the child gives vent to tears and sorrowful words, he feels somewhat relieved. By the display of emotions he makes the dead person or pet seem more worthy.

Too often, well-meaning people say: "Be brave! Don't cry! Don't take it so hard!" But why not? Tears are the first and most natural tribute that can be paid to the one who is gone. The child misses the deceased. He wishes the loved one were still with him.

The son and daughter whose father dies should express their grief. It is natural. They loved him. They miss him. To say, "Be brave!" and especially to the son, "Be a man!" sounds as if one were minimizing their loss and places an impossible burden upon the boy. Be realistic enough to say, "Yes, it's tough!" Make them feel free to express themselves. Otherwise the adult deprives them of the natural emotion of grief.

Don't be afraid of causing tears. It is like a safety valve. So often parents and friends deliberately attempt to veer the conversation away from the deceased. They are apprehensive of the tears that might start to flow. They do not understand that expressing grief through tears is natural and normal.

Tears are the tender tribute of yearning affection for those who have died but can never be forgotten. The worst thing possible is for the child to repress them. The child who stoically keeps his grief bottled up inside may later find a release in a more serious explosion to his inner makeup.

Crying is the sound of anguish at losing a part of oneself in the death of another whom one loves. Everywhere and always, grief is the human expression of the need for love and the love of life.

Just as the parents should not deny the child the opportunity to cry, they should not urge him to display unfelt sorrow. He is likely to feel confused and hypocritical when told he ought to express a regret he does not honestly feel. There are many outlets for grief and the child must utilize those openings that most naturally meet his needs.

WHAT ARE OTHER POSSIBLE REACTIONS TO DEATH?

Death is an outstanding example of a traumatic event which threatens the safety of all the surviving members of the family. It could bring in its train these well-marked symptoms:

Denial. "I don't believe it. It didn't happen. It is just a dream. Daddy will come back. He will! He will!"

The child may frequently look as if he were unaffected because he is trying to defend himself against the terrible loss by pretending that it has not really happened. The adult may even feel that the youngster's apparent unconcern is heartless. Or the parent may be relieved and feel, "Isn't it lucky! I am sure he misses his father, but he does not seem to be really bothered by it." Usually, this signifies that the child has found the loss too great to accept and goes on pretending secretly that the person is still alive. This is why it is so necessary to help the child accept reality by not conjuring up fairy tales and compounding the problem with: "He went away on a long journey."

Bodily Distress. "I have a tightness in my throat." "I can't breathe." "I have no appetite at all." "I have no strength." "I am exhausted." "I can't do my homework." "I can't sleep." "I had a nightmare."

The anxiety has expressed itself in physical and emotional symptoms and is often brought on by visits from friends or the mention of the deceased loved one.

Hostile Reactions to the Deceased. "How could Daddy do this to me?" "Didn't he care enough for me to stay alive?" "Why did he leave me?"

The child feels deserted and abandoned. "Bad Mommy—she's gone away!" There may yet be another aspect. Think of a child's anger after he has been left by his mother for a day or two. Although he may not show much reaction just after her return, later he may turn on her angrily saying: "Where were you, Mommy? Where were you?" Similarly, the child uses this protest to recover the lost person and ensure that she never deserts him again. Although no amount of anger will make the loved one return, the youngster may still use it simply because the protest worked successfully in the past.

Guilt. "He got sick because I was naughty. I killed him!" Guilt is often coupled with the expectations of punishment.

Hostile Reactions to Others. "It is the doctor's fault. He didn't treat him right." "Maybe he was murdered." "It is God's fault. How could He do this to me?" "The minister doesn't know anything—he keeps saying God is good."

The anger is turned outward usually in the attempt to cope with guilt. The youngster may even be angry at sympathetic friends simply because they are not the deceased. He doesn't want any substitutes— even as the very young child does not want anyone but his mommy.

Replacement. "Uncle Ben, do you love me, really love me?"

The child makes a fast play for the affection of others as a substitute for the parent who had died.

Assumption of Mannerisms of the Deceased. "Do I look like Daddy?"

He attempts to take on the characteristic traits of the father by walking and talking like him. He tries to carry out the wishes of the deceased. Or the boy tries to become the father as the head of the family and the mate of the mother.

Idealization. "How dare you say anything against Daddy! He was perfect."

In the attempt to fight off his own unhappy thoughts, the child becomes obsessed with the father's good qualities. The falsification is out of keeping with the father's real life and character.

Anxiety. "I feel like Daddy when he died. I have a pain in my chest."

The child becomes preoccupied with the physical symptoms that terminated the life of the father. He transfers the symptoms to himself by a process of identification.

Panic. "Who will take care of me now?" "Suppose something happens to Mommy?" "Daddy used to bring home money for food and toys. Who will get these things for us?"

This state of confusion and shock needs the parent's supportive love: "My health is fine. I will take care of you. There is enough money for food and toys."

These are some of the reactions of children as well as adults. Some may never appear. Some come at the time of crises. Others may be delayed, since so often the child represses his emotions and attempts to appear calm in the face of tragedy. At one moment he may express his sense of helplessness by acting indifferent. A moment later the feeling of loss will take the form of boisterous play. The parents may detect what the child is thinking from some superficially unrelated questions that he may later ask: "Mommy and Daddy, where did the light go when I blew out the candles on my cake?"

What Are Some of the Distorted Mourning Reactions?

The inability to mourn leads to personal disintegration, leading to mental illness. The line of demarcation between "normal psychological aspects of bereavement" and "distorted mourning reactions" is thin indeed, just as is the division between "normality" and "neurosis." The difference is not in symptom but in intensity. It is a *continued* denial of reality even many months after the funeral, or a *prolonged* bodily distress, or a *persistent* panic, or an *extended* guilt, or an *unceasing* idealization, or an *enduring* apathy and anxiety, or an *unceasing* hostile reaction to the deceased and to others. Each manifestation does not in itself determine a distorted grief reaction; it is only as it is viewed by the professional in the composition of the total formulation. The psychiatrist may detect its pathology in the hallucinations of his patients. He may observe its appearance in the obsessive–compulsive behavior, where the bereaved attempts to relieve guilt in a variety of ways, such as extreme cleanliness. He may discover its overtones in psychosomatic diseases, such as the morbid anxiety as to one's own health, with conjuring up of imaginary ailments. In psychotic patients, there is seen the stupor of the catatonic sometimes likened to the death state. Delusions of immortality are found in certain schizophrenics. If there are any doubts, the parent should seek special assistance for the child from the psychologist or child guidance clinic. Without such outside help, few parents, in the midst of their own grief, can give proper support for that child facing a severe mourning reaction.

DOES A CHILD REALLY FEEL GUILT?

There is a degree of guilt involved in every death. It is human to blame oneself for the person's death. Even if the adult knows he did everything in his power to prevent the death and to make the loved one happy, he is still apt to search his mind for ways that he could have done more. After the Cocoanut Grove fire in Boston, one woman could not stop blaming herself for having quarreled with her husband just before his death. Often, the recrimination is an attempt to turn the clock back, undo the quarrel, and magically prevent the loss.

Children more than adults are apt to feel guilty since, in their experience, bad things happen to them because they were naughty. The desertion of the parent must be a retribution for their wrongdoing. Therefore, they search their minds for the "bad deed" that caused it.

Often guilt is induced by the child's misconceptions of reality. One youngster was told that to live you must eat. Since she did not eat her cereal the morning her father died, she concluded that it was she who brought about his death.

Young children believe in magic. That is, if one wishes someone harm, the belief will bring results. When the little boy said in anger, "I hope you die, Mom," and the mother did die, the lad felt responsible and guilty.

Painful thoughts are recalled. "I was terrible to her." "I kicked her." "Why did I call her those awful names?" Any normal child has feelings of intense hostility toward another child or an adult. If this person dies, the youngster may feel that in some way his thoughts contributed to the person's death.

The living sometimes feel guilt simply because they are alive. They may feel they should be censured for wishing that the sick person hurry up and die. Or they feel they should be blamed because they secretly hope that their friends' mommies would die, so they won't be the only one without a mother.

The resultant behavior varies. There may be aggressiveness, with or without excessive excitability. There may be unsociability and obvious despondence. There may be a lack of interest and attention in class, or a degree of forgetfulness of ordinary concerns. It must be underscored that guilt is a normal reaction to the experience of death.

HELPING THE CHILD TO RELIEVE THE GUILT

From a commonsense point of view, the child's guilt is usually unreasonable. Therefore, it is important for the parents to help the child give vent to his anxieties. In his mind he may remember times when he may not have been as good to this person as he should have been. Let him know that all people try to be good and loving but do not always succeed. Nor does one have to. One does the best one can. Tell the child: "You did the best you could. You had nothing to do with his death. All people die." By all means possible, parents should avoid linking suffering and death with sin and punishment.

Explain to him that "wishing does *not* make it so." Try to recall those happy moments when the child did make the deceased very happy. For the youngster who is too young to give shape to his thoughts or to find the words which might relieve his guilt, the best therapy is through relationships with other people. Children learn self-acceptance by being accepted by others. They learn to trust through living with trustworthy parents and teachers. They learn to love by being loved.

THE FUNERAL

SHOULD A CHILD GO TO THE FUNERAL OF A LOVED ONE?

Children cannot and should not be spared knowledge about death. When death occurs within or close to a family, no amount of caution and secrecy can hide from the child the feeling that something important and threatening has happened. He cannot avoid being affected by the overtones of grief and solemnity. All the emotional reactions a child is likely to have to a death in the family—sorrow and loneliness, anger and rejection, guilt, anxiety about the future, and the conviction that nothing is certain or stable anymore—can be considerably lessened if the child feels that he knows what is going on and that adults are not trying to hide things from him.

Often a parent will ask: "Do you think the boys should attend the funeral service of their grandfather? They loved him very dearly. They were very close to him. I'm afraid that if they go they will become too highly disturbed. Don't you think it would be much better if they would stay in the home of some friend or some neighbor during the day of the funeral?"

The parent is usually expecting an affirmative reply to this last question. Parents intend it as a kindness when they shield the child from death, as they send him away to stay with a friend or relative and permit him to return only after the funeral. They are dismayed by the suggestion that the child share in the service honoring the life and memory of someone close to him. Yet recognized child authorities have come to the conclusion that not only is it correct to permit a child to attend a funeral but, from approximately the age of seven, a child should be encouraged to attend. A child is an integral part of the family unit and should participate with the family on every important occasion. The funeral is an important occasion in the life of the family. It may be sad— but it still is a crucial occasion in the life of every family. A youngster should have the same right as any other member of the family to attend the funeral and to offer his last respects and to express his own love and devotion. To shut a child out of this experience of sorrow might be quite costly and damaging to his personality. To deprive him of a sense of belonging at this very emotional moment is to shake his security.

If your child goes to the services at the chapel and cemetery, explain in advance the details of the funeral. Tell him people are buried in places called cemeteries with stones placed on each grave to tell the names of people who rest there, and the place is kept beautiful with flowers and trees for remembrance. Children are more relaxed and less disturbed if they observe the funeral than by the fantasies conjured up by fertile young minds.

If the child does not attend the interment service, he may come to the cemetery later with his family. This is advisable when the child cannot accept the reality of death. There is the case of a boy who was told his mother went on a journey. He became sullen and unmanageable. Whenever the mother's name was mentioned, the child would speak of her in the vilest language. Then finally one day the child was told the truth. He was then taken to the cemetery where he could visit the grave. The child was heartbroken for he now realized that she had really died.

Yet he felt a lot better for the experience. At least now he knew what had happened. And most important he knew that his mother had not run out on him, had not abandoned him or deserted him.

It is difficult to determine whether an apprehensive youngster should be encouraged to attend the funeral. The parents' own judgment should be the best guide. Of course, the child is frightened and may need a little support. Yet, he should never be forced to go. If one can anticipate in advance that there will be hysterical outbursts, it might be wise to keep the sensitive child at home.

Some enlightened adults have helped a youngster feel that he still has an important role to play by asking him to stay home with someone he knows well to answer doorbells and telephones and run errands. After the funeral, he helps to receive friends. During the mourning period that follows, he should not be sent away. He should be given the chance to feel grief and mingle with the family.

The important fact to always remember is that just as children cannot be spared knowledge about death, they cannot and should not be excluded from the grief and mourning following death. With as much calmness and assurance as possible, the parent must give the child calm reassurance of love, not only by tone of voice but by warmth of arms. The parent must keep in mind that children will exhibit their grief differently from adults. A child who is deeply affected by a sense of loss may try to find solace in active play or give expression to his confused feelings by behaving in a noisy and—by adult standards—"improper" manner. It is more important that the child be permitted to express his grief in his own way at a time like this than that he be reprimanded and his burden of guilt increased for the sake of convention and propriety.

How Do We Help the Child Who Has Lost a Loved One?

The adult must understand the youngster's emotional needs. This is accomplished not by cross-questioning but by empathy, understanding, and love. Love contributes to the child's security and gives him the feeling that he is valued. He is then able to love in return.

A parent should encourage what Sigmund Freud calls the "ties of dissolution." That is, reviewing with the youngster pleasant and unpleasant memories of the deceased. As each event is reviewed, a pang of pain is felt at the thought that the experience will never be repeated. As pain is experienced, the youngster is able to dissolve himself of his emotional ties with the dead person. A gradual working over of such old thoughts and feelings is a necessary part of the mourning at any age, and a prelude to acceptance of the death as a real fact.

Assist the child to unburden his feelings through catharsis, confession, remembrance, and release. The child needs to talk, not just to be talked to. He should be given every opportunity to discuss the person who has died and be permitted to feel that, if he wishes to do so, he may even express antipathy as well as affection for the deceased.

Adults should encourage the child to accept the reality of the death. The approach should include a sympathetic, nonshaming attitude toward the youngster's age-appropriate inability to sustain tearfulness and sorrow.

Aid the child to get out of himself. Perhaps he might become more active in a youth group, a club, or a special hobby, such as modeling clay and finger painting.

Respect the youngster's own personality. Avoid efforts to have him become an emotional replacement for the deceased. For example, increased physical intimacy such as sharing a bedroom should be tactfully avoided to prevent an exaggeration of the child's already present sense of guilt at having a dead parent's mate more to himself.

Demonstrate in word and touch how much he is truly loved. A stable and emotionally mature adult who accepts the fact of death with courage and wisdom will bring the truth to the youngster that the business of life is life. Emotional energy formerly directed toward the absent person must now be directed toward the living. This does not mean wiping out the memories of the deceased. Even in death, the absent member can and should remain a constructive force in family life and be remembered in love without constant bitterness or morbidity.

The necessity for carrying on day-to-day routines will aid the process of adjustment, and, in time, special interests and pleasures will again assume their normal place in the scheme of things—for both parent and child. The relationship between parent and child will be essentially the same as before death. When parents accept the reality of death and the need for life to go on, they and their children can maintain their own healthy relationships and find new ones to give meaning and purpose to life, permitting young personalities to develop and mature.

8

J. WILLIAM WORDEN is an Assistant Professor of Psychology at Harvard Medical School and Research Director of Massachusetts General Hospital's Omega Project, an interdisciplinary study of life-threatening illness and life-threatening behavior, supported by the National Institute of Mental Health and the National Cancer Institute.

He received his undergraduate education at Pomona College in California. He holds graduate degrees from Eastern Seminary, Boston University, and Harvard University in the fields of theology, education, and clinical psychology. In 1965 he was a postdoctoral fellow at Stanford University.

In 1968 he and Dr. Avery Weisman began a longitudinal study of terminal illness and suicide at Harvard. Since that time the study has broadened into an investigation of coping and vulnerability among cancer patients.

Dr. Worden has published a number of articles on death and grief in various medical and psychiatric journals. His book, *Personal Death Awareness* (Prentice-Hall), has recently appeared in a translated German edition.

Coping with Suicide in the Family

J. WILLIAM WORDEN

Since my topic is a very broad one, I have decided to address my comments to two issues concerning children. The first has to do with children who are *victims of suicide*, that is, children who take their own lives. The second issue has to do with children who are *survivors of suicide*, children whose parents or siblings kill themselves and with whom it is important to do adequate grief counseling.

Let us first look at the issue of children who take their own lives. From time to time one picks up a copy of a current newspaper and reads that the incidence of suicide among children is on the rise. Certainly if one looks at the public health statistics, the rate of completed suicides among young people seems to be increasing. One of the difficulties in interpreting these suicide statistics is that such information is not recorded for children who are under ten years of age. The theory seems to be that children under ten do not commit suicide because: (1) it is a rare event, and (2) they are unable to grasp the consequences of their actions. This could be debated, but the fact is that statistics on suicide are only recorded for ten years of age and above. Another difficulty in interpreting suicide statistics on children is the way that the figures are grouped. Data on persons from 15 to 24 years of age are lumped together. It is obvious that the developmental issues of a 15-year-old are quite different than those of a 24-year-old. However, when one evaluates the frequency of suicide among children, there seems to be a threefold in-

J. WILLIAM WORDEN ● Harvard Medical School and Massachusetts General Hospital, Boston, Massachusetts 02114.

crease over the last twenty years, no matter how one divides the subgroups.

When considering young people who commit suicide, there are three high risk groups that are of particular interest. The first are the young marrieds. Teenage marriages not only have a very high incidence of divorce but they also tend to spawn an unusually high number of suicide attempts and completed suicides. The second high risk group are young blacks. Young blacks are at high risk for suicidal behavior and often get involved in victim-precipitated homicide, which can be seen as a type of suicide. The third high risk group is college and university students. Suicide is the second leading cause of death among college students, surpassed only by cancer. Death by suicide of college students is greater than for their noncollege counterparts of the same age.

In order to understand suicidal behavior in children and young adolescents, one has to look at *motivation*. Our suicide research at Massachusetts General Hospital corroborates the research of our other colleagues that suicide is not a single phenomenon. Causes and precipitants are multiple. To assume a common motivation in children and adolescents who take their own lives leads to the development of misconceptions.

Some frequent motivations that we find among children and adolescents who attempt suicide are as follows: first, an inability to deal with emerging impulses of adolescence. As is well known, the adolescent has to grapple with emerging impulses of sex, aggression, and the need for intimacy. If these impulses cannot be tolerated or handled well, the adolescent may become involved in some type of suicidal behavior. A child who cannot handle his aggression may retroflect that aggression back on himself. A child who cannot handle emerging sexuality may then take some kind of self-destructive action in order to kill the impulse toward sexuality. There is a certain group of young males who hang themselves as kind of a sexual experience. There is also a group of young women who tend to flirt with death as a type of sensual thing.

A second commonly seen motivation among adolescent suicides is the need to escape from some intolerable situation. Failure in school, a bad home situation, or some other intolerable difficulty may lead one to attempt to escape from the unbearable through killing oneself.

Punishment is another commonly seen motivation in adolescent suicides. The desire to get revenge on someone or to punish someone may lead to this type of behavior. Frequently children say, "If I kill myself, my parents will be sorry and will never forget it."

Another motivation frequently seen at university health services is drug-induced suicide. Students may make self-destructive gestures under the influence of certain drugs, such as a hallucinogenic drug like LSD.

Some children who make suicide attempts have learned this method of coping from other family members or through some form of pathological identification with a family member who has made a suicide attempt. I am following a group of young men, all of whose fathers killed themselves when these young men were approaching adolescence. Each of these young men believes that suicide is eventually going to be his own fate.

Still other children make suicide attempts out of a fear of punishment. Strange as it may sound, fearing failure in school or doing poorly in school may lead one to hurt oneself rather than accept punishment for failure. This can happen on the elementary school level as well as in college.

And then there is the classic motivation for suicide—the "cry for help"—an inability to communicate with others in any other way than to participate in self-destructive behavior.

These are but some of the motivations, and there are others. Generally in our work, particularly in the clinic, we like to think that each suicide has a titre of payback (getting even) and escape. Both of these motivations are present to some degree in every suicide, and it is helpful for the clinician to try and assess how much of each is present in the attempt.

In order to understand better some of the *family dynamics* that lie behind suicidal behavior in children, I will present a recent case: L. was an 11-year-old girl born out of wedlock. Her mother did not want her and so she put her into a series of foster homes until the girl was about seven years of age. At that time, her real father came into her life and invited her to live with him and his second wife. After a period of three years, which was a relatively unsettling time for her, her father and stepmother began to have serious marital difficulties and threatened to split up. About this time, her natural mother came back into her life, adding to her difficulties. The hostility and anxiety she felt toward the situation led her to run away from home. Finally, she was rejected by her real father and sent to live with her natural mother. One of the conflicting difficulties was that her real mother became overly generous and effusive with gifts and affection, a confusing thing to L. A month after she had gone to live with her real mother, a school chum found a note in her pocket saying, "I don't want to live." But nobody did anything about the note. The following month she took an overdose of pills in a shopping center and was rushed to a hospital, given a psy-

chiatric evaluation, and put into therapy. Several weeks later she went
to school one morning and complained of a stomachache. The teacher
dismissed her. Instead of going home, she went to a church and sat for
several hours in a front pew, then later took a .38 calibre pistol belonging
to her mother and killed herself. One of the saddest things about this
event is that several people had seen her at school with the gun tucked
into her jeans, but nobody had done anything about it.

This case illustrates several of the background factors associated
with childhood and adolescent suicides. In the first place, L. was a loner.
She had few, if any, friends. Second, she had a very active schizoid
fantasy life. That is, her fantasy life was more appealing to her than
reality, and there was a very solid barrier between reality and fantasy,
as one finds in the classic schizoid-type person. Third, she had a suicide
history. Our research suggests that a person is a higher risk for completing suicide if they have any history of attempts. Fourth, she had the
availability of a high-risk method—the gun, and fifth, she was in mental
health treatment at the time of the suicide. Not every adolescent who
takes his own life comes from a disrupted family; however, many do.
Other family dimensions frequently associated with suicidal behavior
in children are high levels of fighting within the family, either between
the mother and father or between parents and child. It is also very
common to find the absence of one or both parents. And in cases where
there is no parental absence one may find tremendous cruelty, rejection,
abandonment, scapegoating, and abuse. Institutionalization of a family
member, suicidal behavior by one of the parents, family history of alcoholism, and illegitimacy are other correlates of childhood and adolescent suicides. No one of these variables or even a combination would
be useful in trying to predict suicidal behavior. You would end up with
considerable numbers of false positives. However, these correlates appear in the background of many young people who cope by suicide, and
it gives one a clue as to the etiology behind some adolescent suicidal
behavior.

What should one do if, while working with children, one becomes
aware of the danger signs that the person with whom one is working
may be a suicide risk? These danger signals run anywhere from behavioral clues such as severe depression to an overt history of suicidal
behavior or talking about suicide. I think the major guideline is to ask
the person what is happening. If someone is looking depressed, ask,
"Are you so down that you are thinking about hurting yourself?" I often
use the term "hurting yourself" because it seems to be more acceptable
than "killing yourself." Many people, especially if they have had limited
experience working with suicidal behavior, are afraid they might put the

idea in someone's mind if they ask regarding suicide. This does not seem to be the case, and most people who are in a suicidal crisis appreciate being asked. It gives them permission to share their distress.

The second thing one wants to discover is how the person is thinking about hurting himself. If I am working at the University Health Service and a student at my office appears to be suicidal, I might say, "Are you thinking of hurting yourself?" I would say, "What are you planning on doing?" The person may answer, "Well, I don't know. I hadn't really gone that far in my thinking." I would be less concerned than if the person were to say to me, "I'm going to jump off the Boston University Bridge." I would be much more concerned if the person were to say, "I'm going to jump off the bridge, and I have been walking back and forth looking down at the water." The degree to which the person has a well-worked-out suicide plan gives us a clue as to how immediate the crisis is.

It is important when working with suicidal children not to dispense well-meaning but ill-informed platitudes. Occasionally a child is told, "Why would you want to kill yourself? You're so lucky—you have so much." This type of comment only tends to exacerbate the problem. The main thing is to get the child into appropriate resources for evaluation and treatment.

What about the child who is the *survivor* of a suicide—the child whose parents or siblings have killed themselves? I think the first thing that is important in doing grief work with survivors of suicide is to help the person talk about it. This is true for any grief work, but I think it is particularly true in the case of suicide. If you think of bereavement as a social phenomenon which is facilitated by interpersonal interaction, then it is important after any loss to talk about it. However, there are certain losses which are difficult to speak about, and I think that suicide is one of these. People do not want to talk about it, particularly if the death is an equivocal situation which might have been an accident, but everybody suspects was a suicide. However, reality is facilitated by talking about the situation. I am seeing one young man now whose mother killed herself when he was five years of age. No one ever said anything to him—they shipped him off to a relative, and the father took off for a far distant state. Now that he is a young adult, I have to deal with him and with much of the material that was buried, simply because no one bothered to talk to him at the time of his mother's death. It is especially important to talk to the child about his or her sense of culpability and responsibility.

The second thing to do when doing grief work is to help survivors deal with their feelings. There are various feelings that come as a result of any loss by death—guilt, anger, anxiety, and so forth. I think in the case of suicidal death there is an exacerbation of these feelings, particularly feelings of anger and of guilt. To the extent that the survivor had an ambivalent relationship with the person who died, then one is going to see considerable anger and guilt. In the case of a self-inflicted death there is always anger focused around the fact that the person did not have to do it. There is also a certain high degree of guilt—did I miss something which might have prevented this person from taking his own life? Much of the guilt which is associated with death is irrational guilt and will not stand up to reality testing. However, some people really are culpable and they need help to deal with this real guilt.

The third thing one needs to do with children who are survivors of suicidal death in the family is to help them with their fantasies. There is a sense in the survivor that this might happen to him. If my mother or father can kill themselves, maybe I can kill myself. This type of death raises personal death awareness, which may lead to the experience of anxiety. Encouraging the expression of these fantasies and discussing them with the child is a very important aspect of grief counseling.

I think it is important to help these survivors of suicidal death over time. It takes most people a considerable length of time to work through any loss. I think that working through the loss by suicide may require even more time. The counselor should be aware of important dates such as anniversaries, birthdays, and holidays, and should make counseling available to the child at these particular critical junctures.

All of the things discussed so far in this volume that apply to grieving certainly apply to death by suicide. However, I think a death by suicide exacerbates these dimensions and makes it extremely difficult for one to do the work of grief.

DISCUSSION

MODERATOR: IDA MARTINSON

SPEAKER: Can any of the pharmaceutical tranquilizers contribute to suicides?

J.W. WORDEN: Yes, certainly. For example the potentiating effect of alcohol and drugs. We have a chart that we've drawn up in terms of minimal lethal dosages of the drugs that we most commonly see at the Massachusetts General Hospital. One of the drugs that we've found, and tried to discourage physicians from prescribing, was Doridan. It is difficult but not impossible to lavage it.

B. LAMPERT (Night Supervisor, Quincy City Hospital): I sometimes am the first person to see the attempted suicidal victim awake. I'm always hesitant to say anything in particular except to just show a caring for them as a person, and I don't know if you have any help on that as to what approach to make. It seems senseless to me as one who doesn't know them to say, I'm glad you are alive. Although I am, because the person is a human being. You must have had some experience with this.

J.W. WORDEN: I think it's always better to at least offer to talk about it. To try and find some interjected point or comment that will stimulate discussion. If the person does not want to talk about it, then respect that. I've often found it very helpful just to say to them, "Things must have been pretty rough?" Just some little thing that will then enable them to say, "You don't know how rough it was," or whatever. The unfortunate thing is that when you see them in the Emergency Room they very often will say, "Why did you save me?" "Why didn't you let me die?" Then you see them 10 days later for psychotherapy and they say, "Die, who me?" "No, I was just ingesting a few seconals and I got up to 30 before I realized how many I had taken." So there's a lot of craziness involved too. I don't think its bad procedure to give an entree to the possible conversation, and if the person doesn't want to they will bypass it. I wouldn't feel uncomfortable. I would try to work through your feelings of discomfort. I wouldn't say, "I'm glad you are alive."

B. LAMPERT: I've never said that.

J.W. WORDEN: Maybe you can do it on one level, but if you don't know the patient it may not come through quite the way you might want it to come through. Then you learn about them and you wish they weren't alive.

K. SCHIEL (Pediatric Social Worker, Houston, Texas): I would like to ask you if you could comment a little on helping children to deal with unsuccessful suicide attempts, particularly repeated attempts of their parents, or a suicide attempt that leaves the parent very debilitated.

J.W. WORDEN: Again, it may sound simplistic, but encourage them to talk about it, particularly to talk about any fantasies of culpability on their part or feelings of anxiety (that's another whole area we haven't dealt with). Children particularly feel anxious in the throes of a parent who is wanting to take his own life. Even this 30-year-old man I was telling you about is very anxious when he thinks about the death of his father. He's 30, he's self-supporting, he's married, he's doing well. But because of all the anxiety that he bound up at the time of his mother's death when he was five, he's having some real difficulties with his father's illness and thinking about his father's death. I think encouraging him to talk with someone during psychotherapy or getting at some things indirectly is an easier way to facilitate resolution of these problems. But to encourage expression in whatever media possible is the basic procedure. But not to do anything is least good.

SPEAKER: I know of a very tragic incident where the father had died the year before and the mother was unable to cope without him and she committed suicide in the next year, leaving five children. I don't know of anyone who has addressed the problem for the children in the sense of the man of 30 you mentioned. Each child now has the potential to visualize him/herself as committing suicide. What preventive medicine can be offered to a family such as this? What do we do with the children?

J.W. WORDEN: Are the children farmed out or are they together?

SPEAKER: They are attempting to stay together. I'm not quite sure how they stand at this moment. There are extended family members who are working with them. At the time I knew this was going to be a long-range problem with all of them. Each in their own way. But there was nothing that you could offer really specific. Do you recommend preventive therapy now? Then the next questions are where and how much?

J.W. WORDEN: I guess, without knowing more details, I would try to do some kind of group therapy with them as a family. It needn't take a long time. You can expedite this treatment if you really focus in on the target and know how to facilitate some of these dimensions. I wouldn't necessarily want to do anything more with them unless I identified some real serious problem in that group of five children. But I would certainly want to follow their behavior and their academic progress if that were possible. I think we ought not underestimate how much can be done in a relatively short time. It seems overwhelming. Two deaths, a death by natural causes and one by suicide. But there also could be a tremendous amount of resiliency in these kids. Just really targeting in on the angers, the guilts, the anxieties, the articulations, and the realities can be very, very helpful in a short period of time.

SPEAKER: Whom would you identify as the health team member to be in that role? I feel like there is a void there. I wouldn't even know who to tell this family to go to. Who takes on that role—the short-term family counselor? All of us are doing our bits and pieces as it occurs in our environment.

Where do we have our immediate crisis intervention groups that deal with families who have this problem?

J.W. WORDEN: In Somerville there is a group run by Father Tom Welch for survivors of suicide. Short-term treatment has to be very focused and specific and it can be extremely effective.

9

NED H. CASSEM first obtained his licentiate degree in philosophy at St. Louis University in 1960. He graduated in 1966 as a physician from Harvard Medical School and in 1970 with a bachelor's degree in Divinity from Weston College in Cambridge, Massachusetts. He is a diplomate of the American Board of Psychiatry and Neurology. He has an in-depth interest in such subjects as death and dying, coronary heart disease, and chronic disease and handicap. He trained as a clinical and research fellow in psychiatry at the Massachusetts General Hospital and as a teaching fellow in psychiatry at Harvard Medical School. He is Associate Professor of Psychiatry, Harvard Medical School and Massachusetts General Hospital, and Chief of the Psychiatric Consultation–Liaison Service at that hospital.

Dr. Cassem is a member of many committees and boards. He served for ten years on the Advisory Board of the Foundation of Thanatology in New York. For over a decade he has been on the Executive Board of Directors of Creighton University, Omaha, Nebraska. For five years he was Vice-Chairman and Chairman of the Jesuit Council for Theological Reflection, and he also served for four years on the National Advisory Council of Hospice, Inc. He is a member of the Academy of Psychosomatic Medicine, a fellow of the American Psychiatric Association, and an advisor to the Cancer Counseling Institute of New England. He is a member of the Editorial Board of *Heart and Lung*, a member of the Society of Critical Care Medicine, and a fellow of the American College of Physicians.

Treating the Person Confronting Death

Ned H. Cassem

Help the dying patient? Can one realistically hope to improve the lot of a 35-year-old mother dying of cancer by making her feel better? Feel better about what? At the bedside of the dying, the professional may feel overwhelmed by dread of the encounter or by the presumptuousness of his expectation to help. Yet because the dying have no less right to help than the living, their difficulties and needs require specific attention.

Thanatology is the science or study of death. Broadly conceived, it includes the study of dying as a psychophysiological process, the care of the dying person, including both adults and children, and the care of persons who seek death by suicide. The nondying directly affected by death—the bereaved and the care-giving personnel—are included, as are healthy persons whose lives are disrupted by irrational fears of death.

Psychiatric interest in the concrete problems of the dying is a recent phenomenon. Freud's agonizing seventeen-year struggle against oral cancer (Jones, 1957) is never reflected in his writings, even in those about death. Gifford (1969) has provided a historical review of psychoanalytic theories about death and also reports on the few individuals who became involved in empirical studies of the subject. In 1915 Freud, introducing

Reprinted with permission from *The Harvard Guide to Modern Psychiatry*, edited by Armand M. Nicholi, Jr., The Belknap Press of Harvard University Press, Cambridge, Massachusetts, 1978.

Ned H. Cassem • Harvard Medical School and Psychiatric Consultation–Liaison Service, Massachusetts General Hospital, Boston, Massachusetts 02114.

his notion that the unconscious has no representation of death, explored unconscious convictions of immortality. In 1923, before he linked death and aggression in theory, Freud presented evidence that conscious fears of death represent underlying fears of helplessness, physical injury, or abandonment. In depressive states, on the other hand, fears of punishment, castration, rejection, or desertion may represent fears of death.

The first analyst to investigate the dying themselves was Deutsch in 1933. Little further work was done in this field until Eissler's publication in 1955 signaled an era of empirical interest in dying patients. Beginning with the work of Feifel (1959) and Saunders (1959), significant contributions—from Weisman and Hackett (1961; Hackett and Weisman, 1962), LeShan and LeShan (1961), Hinton (1967), Glaser and Strauss (1965), and several others—multiplied until 1969, when Kübler-Ross's now classic work evoked worldwide interest in the emotional concerns of the dying. The observations and recommendations that follow are based on the work of these investigators as well as my own (Cassem and Stewart, 1975).

THE PSYCHOPHYSIOLOGICAL PROCESS OF DYING

Dying is a process that keeps the body near the forefront of the mind. Having contracted a disease that may eventually cut him down (like heart disease) or devour him (like cancer), the patient interprets or even anticipates bodily changes as ominous. Symptoms are likely to produce fear when present, but fear is often present long before their arrival. The body, once regarded as a friend, may seem more like a dormant adversary, programmed for betrayal. Dying persons, even before disintegration begins, fear many things; loss of autonomy, disfigurement, being a burden, becoming physically repulsive, letting the family down, facing the unknown, and many other concerns are commonly expressed. When all fears were compared for frequency in one sample of cancer patients (Saunders, 1959), the three that topped the list were abandonment by others, pain, and shortness of breath. These fears were expressed before the patients were symptomatic; they felt that as their illness progressed, family and hospital staff would gradually avoid them, their conditions would become increasingly painful, or the illness would encroach on breathing capacity and suffocate them.

Because physicians and families also worry about what will happen to the dying person, it is helpful to know which difficulties are, in fact, the most distressing when the patient is terminal. When Saunders (1959) documented the exact incidence of practical problems in terminal cancer at St. Joseph's Hospital in London, she found that the three most common complaints were nausea and vomiting, shortness of breath, and dysphagia. It was striking that pain did not appear high on the problem list; with proper medication, about 90% of her patients remained pain-free. Largely because of her efforts, avoidance of dying patients decreased at St. Joseph's, but one can expect to find it in most chronic hospitals or nursing homes. Nausea and dyspnea are psychophysiological experiences in which the patient is usually miserable. Relief of these and other troublesome symptoms helps restore peace of mind.

Just as there are familiar physical reactions, like nausea, to a fatal illness, familiar reactions of the mind also occur. To describe the psychological process that begins with getting the news about illness and ends in death, Kübler-Ross (1969) presented a framework of five stages experienced by the dying person. In the first stage, *shock and denial*, the patient says, "No, not me." He may simply be numb and appear completely unaware of the bad news, disagree with the diagnosis, or remain oblivious to its implications. Denial may persist for moments or pervade the rest of the person's life. More characteristically, it fluctuates, waxing and waning over the course of the illness. In the stage characterized by anger, the patient says, "Why me?" Angry outbursts may be directed toward hospital menus, treatment regimens, physicians, families, life in general, or God. Past life may be reviewed, usually with inability to find oneself at fault for hygienic or moral reasons. A sense of unfairness, frustration, and helplessness generate a normal sense of outrage that may turn into periods of bitterness. In the stage termed *bargaining*, the patient says, "Yes, it is me, but . . ." and puts a condition on acceptance, such that he hopes or plans for something that can mitigate disappointment. A characteristic hope would be that life be extended until a special event occurs, such as a wedding, birth, or graduation. In the stage referred to as depression, the patient confronts the sadness of the reality: "Yes, it is me." He may do a good deal of weeping, be intermittently withdrawn and uncommunicative, brood, and be weighed down by the full impact of his condition. Despair and suicide are common preoccupations during this time. The final stage, *acceptance*, is conceptualized as a state reached through emotional work, so that losses are mourned and the end is anticipated with a degree of quiet expectation. It is seen as a restful, albeit weary time, almost devoid of feeling.

The notion of stages is by no means new and is not restricted to dying. As a dynamic process, dying is a special case of loss and the stages represent a dynamic model of emotional adaptation to any physical or emotional loss. Despite its theoretical validity, the concept of the stages is often misused. A common misapplication of the concept in caring for dying patients is the attempt to help an individual through the stages one after the other. It is more accurate and therapeutically more practical to regard them as normal reactions to any loss. They may be present simultaneously, disappear and reappear, or occur in any order. Responding to the emotional distress of the dying often reduces their physical stress. Pain in particular diminishes when a patient is helped to feel understood, less anxious, and less alone.

In fact, some investigators have introduced experimental psychological measures to combat illness directly. In work as yet unpublished, Simonton and Simonton (1974) use relaxation techniques patterned after Jacobson (1938) to induce remission or even cure in cancer patients. Once the patient is relaxed or in a modified trance, suggestions are made that he picture his white cells and immune mechanisms mobilized in concentrated attack upon his malignancy. The results are anecdotal and await controlled replication. The rationale is much the same as that suggested by Surman *et al.* (1973), who demonstrated the effectiveness of hypnosis in the removal of warts—that is, that a person may be able to influence his immune defense systems.

CARE AND MANAGEMENT OF THE DYING ADULT

Questions far outnumber answers wherever the dying are cared for. Many controversies remain unresolved. The investigators mentioned earlier have outlined objectives in the care of the terminally ill that the physician can consider from the time treatment begins. Because the patient's bedside can be an uncomfortable place for the physician, several practical considerations for management are added in support of the goals.

GOALS OF TREATMENT

Deutsch (1933) observed in his clinical sample of dying patients that the decline in vital processes is accompanied by a parallel decline in the intensity of instinctual aggressive-erotic drives. Fear of dying is reduced as the pressures of inner instinctual demands decline. The illness can be viewed by the patient as a hostile attack from an outside enemy or as a punishment for being bad (interpreted by Deutsch as inflicted from within by a harsh superego). Since the patient can worsen his predicament by reacting with increasing hostility toward outside objects or with self-punitive actions to offset experienced guilt, Deutsch's therapeutic objective is a "settlement of differences." The ideal stage is reached when all guilt and aggression are balanced, permitting the patient a guilt-free "regression" to the love relationships of childhood and infancy. Deutsch judges this regression impossible without conflict under other circumstances because of the incestuous nature of the earlier relationships, but he infers that guilt is atoned for by the knowledge of imminent death. One could also infer that the patient's relationship to the therapist would fall under the same protective mantle. For Eissler (1955), therapeutic success depends on the psychiatrist's ability to share the patient's primitive beliefs in immortality and indestructibility. In addition, sharing the dying patient's defenses and developing intense admiration for his inner strength, beauty, intelligence, courage, and honesty are the main forces in the psychiatrist's supportive relationship with the patient. Kübler-Ross (1969) speaks of the unfinished business of the dying—reconciliations, resolution of conflicts, and pursuit of specific remaining hopes. For Saunders (1969), the aim is to keep the person feeling like himself as long as possible. In her view, dying is also a "coming together time" when family and staff are encouraged to help one another share the burden of the terminal illness. LeShan and LeShan (1959), choosing deliberately not to emphasize dying (the minor problem in their views), explore aggressively with patients what they wish to accomplish in living (the major problem). Weisman and Hackett (1961) have coined the term *appropriate death,* for which Weisman (1972) delineates the following conditions: the person should be relatively pain-free; should operate on as effective a level as possible within the limits of his disability; should recognize and resolve residual conflicts; should satisfy those remaining wishes consistent with his present plight and ego ideal; and should be able to yield control to others in whom he has confidence.

Perhaps more important than any other principle in caring for the dying is that the treatment be unique and individualized. This goal can only be accomplished by getting to know the patient, responding to his needs and interests, proceeding at his pace, and allowing him to shape the manner in which those in attendance behave. There is no one "best" way to die.

Treatment Recommendations

Most of what is known about dying patients comes from them. All investigators have emphasized the reversal of expertise manifest in care of the dying. In this field patients are the teachers; those who take care of them can only try to learn. Over the years, observations made by patients on various aspects of their management have helped in the recognition of the following eight essential features in the care and management of the dying patient (Cassem and Stewart, 1975):

Competence. In an era when some discussions of the dying patient seem to suggest that "love" excuses most other faults in the therapeutic relationship, encouraging the misconception that competence in physicians and nurses is of secondary importance for dying patients would be unfortunate. Competence is reassuring, and when one's life or comfort depends on it, personality considerations become secondary. Being good at what one does brings emotional as well as scientific benefits to the patient. No matter how charming physicians, nurses, or I.V. technicians may be, for example, the approach of the person who is most skillful at venipuncture brings the greatest relief to an anxious patient.

Concern. Of all attributes in physicians and nurses, none is more highly valued by terminal patients than compassion. Although they may never convey it precisely by words, some physicians and nurses impart to the patient that they are genuinely touched by his predicament. A striking example came from a mother's description of her dying son's pediatrician: "You know, that doctor loves Michael." Compassion is a quality that cannot be feigned. Although universally praised as a quality for a health professional, compassion extracts a cost usually overlooked in his training. The price of compassion is conveyed by the two Latin roots, *con* and *passio*, to "suffer with" another person. One must be touched by the tragedy of the patient in a literal way, a process that occurs through experiential identification with the dying person. This process of empathy, when evoked by a person facing death or tragic disability, ordinarily produces uncomfortable, burdensome feelings, and internal resistance to it can arise defensively. Who can bear the thought of dying at 20? It is therefore understandable that professionals do not encourage discussion of such a topic by the individual facing it. In addition to guarding against the development of compassion, students are sometimes advised to avoid "involvement" with a patient. When a patient gets upset, hasty exits or evasions are thus more likely. On the other hand, few things infuriate patients more than contrived involvement; even an inability to answer direct questions may be excused when it stems from genuine discomfort on the physician's part. One woman, even though she wanted more information about how much longer she could expect to survive with stage IV Hodgkin's disease, preferred to ask her physician as few questions as possible. "Whenever I try to ask him about this, he looks very pained and becomes very hesitant. I don't want to rub it in. After all, he really likes me."

What are the emotional traumas a compassionate nurse or physician is required to sustain? They can be summarized by the "stages" the terminal patient goes through. Since the stages describe the emotional process in the terminal patient, it is only logical that they call forth similar reactions in a sympathetic observer. As physician and patient view the patient's predicament together, the physician, depending on his sensitivity, is likely to experience shock, denial, outrage, hope, and devastation. Involvement—"real" involvement—is not only unavoidable but necessary in the therapeutic encounter. Patients recognize it instantly. As a hematologist percussed the right side of his 29-year-old patient's chest, his discovery of dullness and the recurrent pleural effusion it signaled brought the realization that a remission had come to an abrupt end. "Oh, shit," he muttered. Then, realizing what he had said, he added hastily, "Oh, excuse me, Bill." "That's all right," the young man replied. "It's nice to know you care."

Comfort. With the terminal patient, comfort has a technology all its own. "Comfort measures" should not indicate that less attention is paid to the patient's needs. In fact, comforting a terminally ill person requires meticulous devotion to a myriad of details. Certain things are basic, like narcotic medication to relieve pain. The most common mistake is inadequate dosage or frequency of medication. For example, even though the duration of action of meperidine hydrochloride (Demerol) is 3 hours, it is common to find orders written for administration every 4 hours. Pain medication orders for patients with continuous pain are too often written "p.r.n.," which can demean the patient by forcing him to beg for medication. The goal of narcotic administration is pain relief with a dosage and frequency adequate to dispel pain and prevent its return. The fear, often covert, of addiction is not justified by experience. Fewer than 1% of hospitalized medical patients treated for pain with narcotics develop a serious problem of addiction (Marks and Sachar, 1973); actual addiction in terminal patients has not been described. Some comfort measures are dramatic, such as pressor infusions to ward off pulmonary edema, while others are simple, like providing fresh air and light in the room.

The ingenuity of skillful, thoughtful caregivers can be taxed in bringing comfort to the terminal patient. Attention to detail demands systematic vigilance by physicians and nurses. An excellent example is mouth care for terminal patients. While most people appreciate the comfort of fresh taste and breath, few could imagine that an entire book, *The Terminal Patient: Oral Care,* could be written on this topic (Kutscher *et al.,* 1973). Yet the mouth is the instrument by which one speaks, tastes, chews, drinks, sucks, bites, and grins. Its care involves the face, teeth, dentures, tongue, gums, lips, and larynx. Pain, dryness, infection, odor, secretions, drooling, hemorrhage, nutrition—these problems can be complicated and cause much discomfort. This book details myriad technical aspects of comfort and suggestions for its realization, such as isotonic saline and 20% Karo mouth washes for stomatitis and potassium chloride rinses for dry mouth, a reminder that dumping syndrome may result because the tip of the nasogastric tube has slipped into the duodenum, and a note that pain in the jaw may be associated with vincristine.

The principles of drug use apply to patients living on borrowed time as they do to patients living under more fortunate circumstances. Alcohol can be used to great advantage for those patients who want it and can profit from its effects. Tranquilizers and antidepressants should be used when symptoms of anxiety and vegetative symptoms of depression distress the dying patient.

Psychedelic drugs were first used for dying patients by Kast (1966) and have been studied extensively at the Maryland State Psychiatric Research Center since 1955 (Kurland et al., 1973). In the hands of the latter investigators, lysergic acid diethylamide (LSD) or dipropyltryptamine (DPT) was an adjunct to brief courses of intense psychotherapy. Cancer patients were selected who were experiencing pain, depression, anxiety, and psychological isolation associated with staff feelings of frustration and inadequacy. After an average interval of 10 hours of psychotherapy, a single psychedelic session was introduced whose full duration averaged 20.2 hours. Flowers and stereophonic classical music were provided on the day of treatment, and after the patient's return to his usual state of consciousness, significant persons (spouse, children, parents, friends) were invited to share the final period of the psychedelic session. All but 6 of 41 ratings for depression, anxiety, pain, fear of death, isolation, and difficulty of management showed significant changes. According to global index ratings, 36% were dramatically improved; 36%, moderately improved; and 19%, unimproved. Only 8.3% were worse. No changes were made in the amount of narcotics consumed before and after psychedelic therapy (Kurland et al., 1973). In 1973, Goldberg, Malitz, and Kutscher edited a book on the use of psychopharmacological agents for the terminally ill. Of great promise is the recent demonstration by Sallen et al. (1975) of the antiemetic efficacy of oral cannabis for patients receiving cancer chemotherapy. Relief of nausea induced by the agents has hitherto been extremely difficult.

Communication.   Talking with the dying is a highly overrated skill. The wish to find the "right thing" to say is a well-meaning but misguided hope among persons who actually do or want to work with terminal patients. Practically every empirical study has emphasized the ability to listen over the ability to say something. Saunders summed it up best when she said, "The real question is not 'What do you tell your patients?' but rather 'What do you let your patients tell you?' " (1969, p. 59). Most people have a strong inner resistance to letting dying patients speak their minds. If a patient presumed to be three months from death says, "My plan was to buy a new [automobile] in six months, but I guess I won't have to worry about that now," a poor listener will say nothing or "Right. Don't worry about it." A better listener might say, "Why do you say that?" or "What do you mean?"

Communication is more than listening. Getting to know the patient as a person is essential. Learning about significant areas of the patient's life, such as family, work, or schooling, and chatting about common interests represent the most natural if not the only ways the patient has of coming to feel known. After a 79-year-old man of keen intellect and wit had been interviewed before a group of hospital staff, one of the staff said, "Before the interview tonight I just thought of him as another old man on the ward in pain." Nothing esoteric is necessary to talk to a dying person. Like anybody else, he gets his sense of self-respect in the presence of others from a feeling that they value him for what he has done and for his personal qualities. Allowing the dying person to tell his own story helps build a balanced relationship. The effort spent getting to know him does him more psychological good than trying to guess how he will cope with death.

The physician can help dissolve communication barriers for other staff members by showing them the uniqueness of each patient. Comments like "This man has thirty-four grandchildren" or "This woman was an RAF fighter pilot" (both describing actual patients) convey information that can help staff find something to talk about with the patients. Awkwardness subsides when patients seem like real people and not merely "a breast CA" or some other disease. This rescue from anonymity is essential to prevent a sense of isolation. Communication is more than verbal. A pat on the arm, a wave, a wink, or a grin communicates important reassurances, as do careful backrubs and physical examinations.

Patients occasionally complain about professional and lay visitors who appear more interested in the phenomenon of dying than in the patients as individuals. A woman in her early fifties, with breast cancer metastatic to bone, brain, lungs, and liver, entered the hospital for a course of chemotherapy. During her entire six-week stay she was irascible, argumentative, and even abusive to the staff. She responded extremely well to treatment, experienced a substantial remission, and left the hospital. She later told her oncologist, apologizing for her behavior, "I know that I was impossible. But every single nurse who came into my room wanted to talk to me about death. I came there to get help, not to die, and it drove me up a wall." A wise caution is to take conversational cues from the patient whenever possible.

Children. Investigators have unanimously concluded that the visits of children are likely to bring as much consolation and relief to the terminally ill as any other intervention. A useful rule of thumb in determining whether a particular child should visit a dying patient is to ask the child whether he wants to visit. No better criterion has been found.

Family Cohesion and Integration. A burden shared is a burden made lighter. Families must be assisted in supporting one another, although this process requires the effort of getting to know family members as well as the patient. Conversely, when the patient is permitted to support his family, the feeling of being a burden is mitigated. The often difficult work of bringing the family together for support, reconciliations, and improved relations can prevent disruption when death of the patient initiates the work of bereavement. The opportunity to be present at death should be offered to family members, as well as the alternative of being informed about it while waiting for the news at home. Flexibility is the rule, with the wishes of the family and patient paramount. After death, family members should be offered the chance (but never pressured) to see the body before it is taken to the morgue. Parkes (1972) has documented the critical importance to grief work of seeing the body of the dead person.

Cheerfulness. Dying people have no more relish for sour and somber faces than anybody else. Anyone with a gentle and appropriate sense of humor can bring considerable relief to all parties involved. "What do they think this is?" said one patient to his visitors. "They file past here with flowers and long faces like they were coming to my wake." Patients with a good sense of humor do not enjoy unresponsive audiences either. It is their wit that softens many a difficult incident. Said one elderly man with a tremor, after an embarrassing loss of sphincter control, "This is enough to give anybody Parkinson's disease!" Wit is not an end in itself. As in all forms of conversation, the listener should take the cue from the patient. Forced or inappropriate mirth can increase a sick person's feelings of distance and isolation.

Consistency and Perseverance. Progressive isolation is a realistic fear of the dying person. A physician or nurse who regularly visits the sickroom provides tangible proof of continued support and concern. Saunders (1969) emphasizes that the quality of time is far more important than the quantity. A brief visit is far better than no visit at all, and patients may not be able to tolerate prolonged visiting. Patients are quick to identify those who show interest at first but gradually disappear from the scene. Staying power requires hearing out complaints. Praising one of her nurses, one 69-year-old woman with advanced cancer said, "She takes all my guff, and I give her plenty. Most people just pass my room, but if she has even a couple of minutes, she'll stop and actually listen to what I have to say. Some days I couldn't get through without her."

BREAKING BAD NEWS

Because so many reactions to the news of diagnosis are possible, having some plan of action in mind ahead of time that will permit the greatest variation and freedom of response is helpful. The following approach is suggested. Begin by sitting down with the patient in a private place. Standing while conveying bad news is regarded by patients as unkind and an expression of wanting to leave as quickly as possible. Inform him that when all the tests are completed, the physician will sit down with him again. Spouse and family can be included in the discussion of findings and treatment. As that day approaches, the patient should again be warned. This warning permits those patients who wish no information, or minimal information, to say so.

If the findings are unpleasant—as in the case of a biopsy positive for malignancy—how can they best be conveyed? A good opening statement is one that is (1) rehearsed so that it can be delivered calmly; (2) brief—three sentences or less; (3) designed to encourage further dialogue; and (4) reassuring of continued attention and care. A typical delivery might go as follows: "The tests confirmed that your tumor is malignant [the bad news]. I have therefore asked the surgeon [or radiotherapist or oncologist] to come by to speak with you, examine you, and make his recommendations for treatment [we will do something about it]. As things proceed, I will be by to talk with you about them and about how we should proceed [I will stand by you]." Silence and quiet observation for a few moments will yield valuable information about the patient, his emotional reactions, how he deals with the facts from the start. While observing, the doctor can decide how best to continue the discussion, but sitting with the patient for a period of time is an essential part of this initial encounter with a grim reality that both patient and physician will continue to confront together, possibly for a very long time.

TELLING THE TRUTH

Without honesty human relationships are destined for shipwreck. If truthfulness and trust are so obviously interdependent, why does so much conspiracy exist to avoid truth with the dying? The paradoxical fact is that for the terminally ill, the need for both honesty and avoidance of the truth can be intense. Sir William Osler is reputed to have said, "A patient has no more right to all the facts in my head than he does to all the medications in my bag." Perhaps a routine blood smear has just revealed that its owner has acute myelogenous leukemia. If he is 25, married, and the father of two small children, should he be told the diagnosis? Is the answer obvious? What if he had sustained two prior psychotic breaks with less serious illnesses? What if his wife says he once said that he never wanted to know if he had a malignancy?

Most empirical studies in which patients were asked whether or not they should be told the truth about malignancy indicated overwhelming desire for the truth. When 740 patients in a cancer detection clinic were asked (prior to diagnosis) if they should be told their diagnosis, 99% said they should be told (Kelly and Friesen, 1950). Another group in this same clinic was asked after the diagnosis was established and 89% of them replied affirmatively, as did 82% of another group who had been examined and found free of malignancy. Gilbertsen and Wangensteen (1962) asked the same questions of 298 survivors of surgery for gastric, colon, and rectal cancers and found that 82% said they should be told the truth. The same authors approached 92 patients with advanced cancer, judged by their physicians to be preterminal, and were told by 79% that they should be told their diagnosis.

How many do not want the truth or regard it as harmful? Effects of blunt truth telling have been empirically studied in both England and the United States. Aitken-Swan and Easson (1959) were told by 7% of 231 patients explicitly informed of their diagnoses that the frankness of the consultant was resented. Gilbertsen and Wangensteen (1962) observed that 4% of a sample of surgical patients became emotionally upset at the time they were told and appeared to remain so throughout the course of their illness. Gerlé et al. (1960) studied 101 patients; members of one group were told, along with their families, the frank truth about their diagnoses, while with the other group an effort was made to maintain a conspiracy of silence between family and physician, excluding the patient from discussion of the diagnosis. Initially greater emotional upset appeared in the group in which patient and family were told together, but the authors observed in follow-up that the emotional difficulties in the families of those patients "shielded" from the truth far outweighed those that occurred when patient and family were told the diagnosis simultaneously. In general, empirical studies do not support the myth that truth is not desired by the terminally ill or harms those to whom it is given. Honesty sustains the relationship with a dying person rather than retarding it. The following example is drawn from Hackett and Weisman (1962):

> A housewife of 57 with metastatic breast cancer, now far advanced, was seen in consultation. She reported a persistent headache, which she attributed to nervous tension, and asked why she should be nervous. Turning the question back to her, the physician was told, "I am nervous because I have lost sixty pounds in a year, the priest comes to see me twice a week, which he never did before, and my mother-in-law is nicer to me even though I am meaner to her. Wouldn't this make you nervous?" The physician replied, "You mean you think you're dying." "That's right, I do," she answered. He paused and said quietly, "You are." She smiled and said, "Well, I've finally broken the sound barrier; someone's finally told me the truth."

Not all patients can be dealt with so directly. A nuclear physicist greeted his surgeon on the day following exploratory laparotomy with the words, "Lie to me, Steve." Individual variations in willingness to hear the initial diagnosis are extreme. And diagnosis is entirely different from prognosis. Many patients have said they were grateful to their physician for telling them they had a malignancy. Very few, however, react positively to being told they are dying. My own experience indicates that "Do I have cancer?" is a not uncommon question, while "Am I dying?" is a rare one. The latter question is more common among patients who are dying rapidly, such as those in cardiogenic shock.

Honest communication of the diagnosis (or of any truth) by no means precludes later avoidance or even denial of the truth. In two studies cited above in which patients had been told their diagnoses outright (including the words *cancer* or malignancy), they were asked three weeks later what they had been told. Nineteen percent of one sample (Aitken-Swan and Easson, 1959) and 20% of the other (Gilbertsen and Wangensteen, 1962) denied that their condition was cancer or malignant. Likewise, Croog *et al.* (1971) interviewed 345 men three weeks after myocardial infarction and were told by 20% that they had not had a heart attack; all had been explicitly told their diagnoses. For a person to function effectively, truth's piercing voice must occasionally be muted or even excluded from awareness. I once spoke on four successive days with a man who had a widely spread bone cancer. On the first day, he said he didn't know what he had and didn't like to ask questions; on the second, that he was "riddled with cancer"; on the third, that he didn't really know what ailed him; and on the fourth, that even though nobody likes to die, that was now the lot that fell to him.

Truth telling is no panacea. Communicating a diagnosis honestly, though difficult, is easier than the labors that lie ahead. telling the truth is merely a way to begin, but since it is an open and honest way, it provides a firm basis on which to build a relationship of trust.

THE ROLE OF RELIGIOUS FAITH AND VALUE SYSTEMS

Investigation of the relationship between religious faith and atti- tudes toward death has been hampered by differences in methodology. Lester (1972b) and Feifel (1974) have reviewed much of the conflicting literature on the relation between religious faith and fear of death. Other research has tried to clarify the way belief systems function within the individual. Allport (1958) contrasted an "extrinsic" religious orientation, in which religion is mainly a means to social status, security, or relief from guilt, with an "intrinsic" religious orientation, where values appear to be internalized and subscribed to as ends in themselves. Feagin (1964) provides a useful 21-item questionnaire for distinguishing the two types of believers. Experimental work (Magni, 1972) and clinical experience indicate that an extrinsic value system, without internalization, seems to offer no assistance in coping with a fatal illness. A religious commit- ment that is intrinsic, on the other hand, appears to offer a good deal of stability and strength to those who possess it.

Many patients are grateful for the chance to express their own thoughts about their faith. If a patient has religious convictions, a useful question can be asked during discussion of the illness: "Where do you think God stands in all this?" followed by "Do you see your illness as imposed on you by God? Why? What sort of a being do you picture God to be?" Answers can be scrutinized for feelings of guilt and for the quality of relationship the individual describes with God.

Belief in an afterlife is another useful area for questioning and helps in assessing tolerance of doubt, an important quality of mature belief. In general, those persons who possess a sense of the presence of God, of being cared for or watched over, are more likely to manifest tranquillity in their struggle with terminal illness. In 1974 Kübler-Ross wrote: "Before I started working with dying patients, I did not believe in a life after death. I now do believe in a life after death without a shadow of a doubt" (p. 167). When M. G. Michaelson reviewed Kübler-Ross's book along with several others on death and dying, he quoted the above passage and added, "Damned if I know what's going on here, but it does seem that everyone who's gone into the subject comes out believing in something crazy. Life after death, transformation, destiny, belief . . . These are the words of hard-nosed scientists fresh from their investigations" (1974, p. 6). Careful investigation of religious faith in the dying needs to be pursued.

CARE OF THE DYING CHILD AND HIS PARENTS

Fatally ill children pose poignant and difficult problems for their parents and caregivers. Easson (1970) and Spinetta (1974) have reviewed contributions to the understanding of these issues in a field that needs more empirical study. Any approach to helping a dying child and his family must be based on the understanding of the child's attitudes toward death.

The most common adult misconception is that children cannot comprehend the meaning of death. Even teenagers are often treated as though they did not understand mortality. Anthony (1940) and Nagy (1948) were the first to study empirically the development of the death concept in children. Table I summarizes the development of this com-

Table I. Concepts of Death in Healthy Children

Concept	Age	Investigator
Reversible	3–5	Anthony (1940), Nagy (1948)
Similar to sleep, journey, departure	3–5	Anthony (1940), Nagy (1948), and most others
Due to violence	3–5	Rochlin (1965), Natterson and Knudson (1960)
Personified	6–10	Nagy (1948)
Externally caused	5–9	Anthony (1940), Nagy (1948), Natterson and Knudson (1960), Safier (1964)
Fact of life	5–9	Most investigators
Inevitable	5–9	Most investigators
	3–5	Rochlin (1965)
Irreversible	9+	Nagy (1948)
	8	Portz (1972)
	7	Steiner (1965)
Universal	9+	Nagy (1948)
	8+	Anthony (1940)
	11	Steiner (1965)
	8½	Peck (1966)

prehension in healthy children. Three phases characterize the gradual understanding of the death concept. Children five or younger regard death as reversible, comparable to a journey or sleep. Six months after the death of her father, a 3-year-old girl asked, "Mommy, will Daddy be home for Christmas?" Because of the inability of children in this age group to grasp the finality of death, I recommend that parents not compare death to sleep, natural as that is, because of the likelihood that a sick child with this association will be afraid to go to sleep at night. (This fear is a common feature of very sick adults troubled by specifically nocturnal insomnia. The fear is often unconscious and its treatment is difficult.) Between the ages of five or six and nine or ten, the child personifies death as a ghost or bogyman who comes to transport the dying person away. The causation is seen as external ("Who killed him?"). At this stage death becomes a definite fact of life, although it is common for the child to regard it as remote—as something that happens only to the very old. The two most abstract elements in the concept of death, irreversibility and universality, are incorporated later. Comprehension of these elements is regarded as marking a complete grasp of the essential notions of the death concept. Acquisition of the concept of irreversibility marks the third and final phase in concept development and appeared after age nine in Nagy's (1948) study. Steiner (1965) and Portz (1972) discovered its appearance at ages seven and eight, respectively. Grasp of death's universality has been found as early as age eight and a half by Peck (1966), though other studies set it at nine or later, and Steiner (1965) found its mean age of onset in her sample to be eleven. The significance of these studies is great: the average child has a complete understanding of the essential notion of death between the ages of eight and eleven. Of course, a child's comprehension can be quite threatening to adults. Michele, age eleven and dying of a brain tumor, said, "My parents won't tell me anything, but I know. I've got a tumor, people die of it . . . Of course I know that I won't get better. Children do sometimes die; I'm going to die too" (Raimbault, 1972).

What fears accompany the growth of this comprehension of death in children? Table II presents a summary of fears specific to the phases of conceptual development outlined above. Children five or younger dread separation from parents. Minimizing the threat of death for them requires specific attention to maintaining contact with parents and reassuring the child about the return of a parent who must leave for short periods. One important application of this principle is the inclusion of children in all death rituals, such as wakes, funerals, *shivah*, and burial. Parents who fail to do so risk frightening the child far more by separating him from his family at a time when death has already caused one sep-

Table II. Specific Fears Associated
with Death in Children and
Adolescents

Age	Fears
Up to 4–5	Separation
6–10	Aggression; mutilation; guilt; loneliness
10+	Own death; abandonment
Adolescence	Shame

aration in the family. No child is too small to be included. As with visitation of the sick or dying, the child should be allowed to go unless he requests not to; he is usually a better judge of the situation than the parent.

Between the ages of six and ten, the child normally develops fears of bodily injury and mutilation, particularly as the victim of aggressive actions. The use of dolls in describing surgical procedures to children in this age group is an effective method of neutralizing some of their fears. This is also the age in which parental discipline becomes formalized in the child's primitive conscience, where "good" acts deserve reward and "bad ones," punishment. The child, now very vulnerable to feelings of guilt, is likely to regard death or illness as a punishment for being "bad." Accordingly, children of this age should be reminded explicitly that illness does not befall them or their siblings because they were bad children. Similarly, all exhortations that imply guilt should be avoided, such as "If you are a good boy and take your medicine, nothing will happen to you." Finally, loneliness is listed here because it is, unfortunately, often experienced by fatally ill children. From around the age of ten on, the concerns of children can be strikingly adult (as noted in the case of Michele, above). They wonder what their own death will be like, particularly whether it will be painful, and, even more, whether they will be left to face it alone when it comes.

Shame is listed as a special concern of the adolescent to emphasize how embarrassed and self-conscious the pubescent youngster can be with a new and rapidly changing body. To have this body threatened or disfigured by a fatal illness can result in feelings of intense shame, even repulsiveness, for the adolescent. All of these age-specific concerns should be assumed to be present to some degree in ill children and great care given to their alleviation.

Like healthy children, fatally ill children understand the meaning of death and are usually aware of their own predicament. Waechter (1972) studied four groups of sixteen children, aged six to ten: those with a fatal illness, those with a chronic but nonfatal illness, those with an acute brief illness, and those with no illness (healthy group). The anxiety scale scores, as well as the derived-fear scores, from the Thematic Apperception Test (TAT) of the fatally ill children were significantly higher than those of all other groups and demonstrated that the most anxiety was expressed by those children for whom death was the predicted outcome. The projective tests revealed significantly more specific concern about death among the fatally ill children. Moreover, every child in this latter diagnostic category (100%) used death imagery at least once in his responses to the protocol. Images of loneliness in this group also significantly exceeded those expressed by the other three groups. The main character in the stories of the fatally ill children died or was assigned a negative future significantly more often than in the other children's stories. Many parents had gone to great lengths to ensure that the diagnosis was never mentioned to the child and the illness never discussed with him. Waechter's study also demonstrated that the less opportunity the child had to discuss the illness at home, the higher his score on the general anxiety scale was. Avoiding talk about the illness appeared to indicate to the child that it was too terrible to discuss.

Even more difficult than discussing the illness with a dying child is the problem of remaining in his presence. What truth is there in the accusation that we avoid and isolate the dying child? What is the child's perception of this? To study these questions, Spinetta, Rigler, and Karon (1974) compared 25 children hospitalized for the first time with a diagnosis of leukemia and 25 hospitalized for a chronic, nonfatal illness; all children were aged six to ten. In an ingenious design, the experimenters used a three-dimensional hospital-room replica, scaled so that 1 inch equaled 1 foot. Using a child-patient doll matched for sex and race with the child, they presented sequentially father, mother, nurse, and doctor dolls with the instruction, "Here comes the nurse; put her where she usually stands." Distance from the child-patient doll was recorded in centimeters in each case. On the first admission, the 25 leukemic children placed the figures significantly farther away than did the control group of chronically ill children. The experiment was repeated with subsequent admissions. While the distance of placement increased for both groups in the subsequent admission, the leukemic children increased the distance significantly more than did the chronically ill, lending strong support to the hypothesis that the child's sense of isolation grows stronger as he nears death. The implication of these empirical studies is clear. Effective care of the dying child must focus both on helping the child communicate his concerns and on helping the parents (and staff) deal with the personal fears that lead them to avoid or isolate the child.

Management of the dying child is an individualized matter, understood best by the parents. The days that can be lived normally will be most enjoyable to the child, and for these days overindulgence may be just as damaging as neglect. Although professionals should not tell parents how to take care of their children, supplying them with a list of specific concerns, like the one in Table II, is helpful. Discovering the child's specific fear is a basic objective. One 10-year-old girl with leukemia, who had done well emotionally up to the time her chemotherapy began to produce hair loss, went to bed in her room and did not come out. Promises of getting a wig proved of no help, but when she learned that the hair loss was an effect of medication, she promptly resumed her normal activities. She had associated loss of hair with death, and when her hair began to fall out, she thought she was dying. Children need encouragement to express and discuss specific fears of this kind (Toch, 1964; Yudkin, 1967; Evans, 1968; Easson, 1970).

What does one tell a child about death or prognosis? Experienced pediatricians are too wise to answer this in any simple way, although most tend to shield the child from labels like "leukemia" or "cancer," and all are wisely careful to avoid predicting when the child will die. However, the discomfort over what to say to the child about illness and death is often a primary obstacle for the parents and the medical caregivers. This uncertainty, coupled with the strong desire not to harm the child or make his predicament any worse, is a major obstacle to allowing the child to express specific fears. Children's questions are often blunt and alarming. "What is death like?" "Am I going to have a coffin?" "Are you and Daddy going to come to visit me when I'm in the grave?" are all questions that have been asked by children under ten. The most helpful technique in dealing with these confrontations is to ask for the child's own answer to the question. Thus "What do you think it's like?" "What do you want done?" "What would you like us to do?" would be helpful initial responses to the three sample questions. Only in this manner can the adult learn more specifically what is on the child's mind. The boy who asked the first question said he thought that in death a bogyman would come to take him away. Later he said he thought death would hurt. His mother was able to comfort him on both points. The 7-year-old girl who asked the second question said she wanted to choose the colors of the casket, while the 7-year-old girl who asked the third girl who asked the third question said that she wanted visitors at the grave and promptly changed the subject to what kind of company she might have in heaven. Although it is painful and difficult to take children seriously, letting them say what is on their minds best guarantees understanding them and finding ways to comfort them. Imagined fears are almost always worse than reality, and astonishing equanimity has been seen in children who are confident that they will not be left alone, attacked, or regarded as bad.

Kübler-Ross has enlightened many about the symbolic language in a child's drawings. The same symbolism can be seen in dolls or other play objects. Sometimes children will talk about drawings, especially if the drawings, not the child, are made the focus of discussion by asking what the child in the drawing thinks, not what the artist thinks. One child who had undergone six major abdominal operations in five months introduced me to a grape named Joe. When I asked where his head was, she told me one pole of the grape was his navel, and in response to the repeated question said that the other pole represented his anus. When I asked what it was like for Joe to be contained between navel and anus, she replied with a smile, "Oh, it's not so bad, most of the time." She then declined to discuss Joe further. Some children will not discuss their drawings or play objects at all, and some are so eloquent that little need be added. Nonverbal communication can be more important at tense moments for the child. Children are reassured by being touched, hugged, or rocked (Kennell, Slyter, and Klaus, 1970), especially if the person doing it is relaxed and accepting. Various authors have pointed out the child's interest in toys, dolls, furry animals, and television; these absorbing objects do not lose their appeal when a child is sick.

Any support given to the parents will greatly benefit the child. While parents will find some understanding of the material presented above very helpful, they have their own needs. From the time they learn of the diagnosis, each may assume that the other will not be able to tolerate the tragedy of their child's illness and both will attempt to minimize any display of weakness and sometimes of any emotion at all. Often numb at first, parents may at times be filled with a sense of injustice or outrage. Repeatedly asking themselves whether they should have noted the illness earlier or acted on the symptoms sooner, and aware that they may have minimized the symptoms even to the extent of regarding them as trivial, emotional, or psychosomatic, the parents may have considerable feelings of guilt. Allowing them to explore and state these concerns is essential. Suicidal thoughts are probably the rule rather than the exception during the child's illness, particularly in the mind of the mother. These thoughts ("When Julie goes, I'm next") can be alarming but are far less threatening if expressed.

While helping parents to maintain as normal a life routine for the sick child's siblings as possible, it is necessary to remember that the well children in the family may become objects of unwitting resentment, another feeling whose emergence can cause considerable distress even in a parent who in no way acts it out. "Why couldn't it be Joe instead of Julie?" is a spontaneous feeling that may cause a parent to experience acute guilt and shame, sometimes for years. If allowed to verbalize such feelings, parents should be reminded that feelings are completely different from actions. Another source of parental discomfort is the need to seek further consultation about treatment for their fatally ill child. The death of another child on the same hospital floor, particularly if afflicted by the same illness, is likely to devastate the parents of a child who is enjoying a remission.

If parents request reading material to help deal with their healthy children, Jackson (1965) and Grollman (1967) are recommended; three useful texts have been designed to be read directly to children after a death has occurred: Grollman (1970), Harris (1965), and Stein (1974). In some cities, organizations exist in which parents of children who are dying or have died support one another; these include the Society for the Compassionate Friends (Stevens, 1973), Candlelighters, and One Day at a Time.

Who cares for the caregiver? Physicians, nurses, social workers, and chaplains who work with dying children have all been asked how they tolerate it. Two major difficulties must be overcome. First, how does one steer a course between incapacitating emotional overinvolvement and callous, detached withdrawal? Schowalter (1970) observes that pediatric house officers who are parents themselves find the horror of a dying child more real than do those who are not parents, but they also tend to have more empathy and work better with the parents than their single peers. He finds a tendency in all house officers studied to avoid getting emotionally involved with dying children. The second dilemma of health workers in this setting is how to obtain gratification from fighting a losing battle. Although they may actually deny that it is a losing battle for some period of time (and this denial can be useful), ultimate reality remains. An interest in and respect for the child and his parents can be developed through encounters with them that provide gratification. This process, of course, implies emotional investment in the child and his parents. Parents of dying children frequently admitted, although with embarrassment, that the desire to see a physician, nurse, or other member of the hospital personnel motivates their hospital visiting more than just seeing their own child. These encounters with parents as well as with the children provide the staff with specific feedback about their importance, effectiveness, and helpfulness in the situation. Flexibility is important. In a survey of pediatricians' attitudes toward the care of fatally ill children, Weiner (1970) notes that as pediatricians get older they develop more flexible attitudes toward parental visiting hours and parental participation in actual child care. Another issue that parents need to discuss with physicians and staff is whether the child should die in the hospital or at home.

Mechanisms to ensure and promote growth in the ability of hospital personnel to take care of dying children and their parents should include formal training. As helpful as that may be, however, other mechanisms are necessary to optimize staff performance and morale in the treatment setting itself, whether inpatient or outpatient. Schowalter (1970) advocates a multidisciplinary team meeting, including pediatric house officers, child psychiatry fellows, nurses, social workers, chaplains, medical and nursing students, and others who work in the setting. Such meetings are conducted by a staff child psychiatrist and senior pediatrician. In similar situations, nurses and house officers have stressed the need for a liaison person to deal with the parents and children directly because both are more likely to feel uncomfortable with the parents. Inviting a parent to a group meeting, either a multidisciplinary or a nursing group, can provide a model for allowing parents to discuss their concerns. Group members can also be invited to express their own feelings about children and parents and to arrive at specific management plans through sharing reactions and observations (Cassem and Hackett, 1975). Where parents' groups exist, the opportunity for staff members to visit maintains contact and provides a chance to see how parents cope with often incomprehensible tragedy.

Care and Management of the Suicidal Person

Death casts its shadow not only on the terminally ill but also on individuals determined or tempted to hasten its arrival. Care of the suicidal person represents a frequent health challenge. The tenth leading cause of death, suicide is known to claim from 22,000 to 25,000 lives annually, although the actual death toll (with many suicides disguised as accidents) may be about twice this number (Schneidman, 1975). For white males aged fifteen to nineteen it ranks second, and for physicians under forty it ranks first among causes of death. In the United States the recorded overall suicide rate is 10 to 12 per 100,000, placing this country midway in international suicide rates listed by the World Health Organization (1968). Japan, Austria, Denmark, Sweden, West Germany, and Hungary have rates higher than 20 per 100,000, while Italy, Spain, New Zealand, and Ireland have rates of 6 per 100,000 or less. For every completed suicide about eight attempts occur. Those who are widowed, divorced, or single kill themselves significantly more often than married people. In the United States, urban and rural dwellers are at equal risk, in contrast to Europe, where risk is greater in urban areas. The suicide rate among blacks, one-third that of whites two decades ago, is now more nearly equal, especially in urban areas. Protestants outrank Jews and Catholics among completed suicides. Half of American male suicides accomplish the act by firearms, a means used by one-fourth of women. More than one-third of female and about one-fifth of male victims use poisonous agents (including gas) to bring death. Hanging or strangulation is chosen by about one-fifth of both men and women.

THE PSYCHODYNAMICS OF THE SUICIDAL PERSON

Self-destruction was seen by Freud (1917, 1920) as murder of an introjected love object toward whom the victim felt ambivalent. Schneidman (1975) reports that this classic and controversial theory stemmed from a statement by Wilhelm Stekel at a Vienna meeting in 1910 that no one killed himself unless he had either wanted to kill another person or wished another's death. Freud's initial conclusions stemmed from his work with depression, in which his discovery of hostility directed inward led to the prominence of aggression in his theory of the dynamics of melancholia. Not all depressed patients, however, engage in suicidal behavior, nor is the suicide attempter always depressed. Zilboorg (1938) points out that some suicidal patients have the unconscious conviction that they are immortal and will not die. In addition to hostility, he found an absence of the capacity to love others in suicidal patients (1937). Rado (1956) has added the concept of expiation: suicide atones for past wrongs and becomes a way of recapturing the love of a lost or estranged object. Menninger's *Man Against Himself* (1938) extends the traditional psychoanalytic concept of self-directed aggression to include other self-destructive behaviors that stop short of suicide, classifying them as chronic (asceticism, martyrdom, neurotic invalidism, alcohol addiction, antisocial behavior, psychosis); focal (mutilation, multiple surgical procedures, malingering, accident-proneness); and organic or psychological (masochism, sadism—with both self-punitive and erotic features). Despite the prominence of self-destructive behavior in everyday life, the relationships between such behavior and depression are not uncontroversial. In a series of careful empirical studies, for example, Beck (1967) was not able to confirm the presence of internalized aggression in depressed patients.

What most appropriately characterizes the state of mind of the person most likely to complete suicide? In addition to hostility, the withdrawal of social support—typified by sudden or intensified estrangement—places a person at much greater risk. Anniversaries of losses and holidays may rewaken all the original, keen feelings of abandonment, loneliness, fury, guilt, defectiveness, isolation, and hopelessness in the individual and increase his proneness to take his life. Beck (1963) stresses that suicidal behavior is most likely when a person perceives his predicament as untenable or hopeless. As psychoanalytic theories of narcissism and ego development expand, self-destruction is coming to be viewed as an effort to restore the balance of wounded self-esteem. Weisman (1971) points out that death itself may not be the wish of the completed suicide victim but that his wish to die signals his conviction that "his potential for being someone who matters has been exhausted" (p. 230). Rochlin (1973), in a creative replacement of the "death instinct" theory, points out that suicide itself, like other forms of aggressive action, serves to restore bruised or shattered self-esteem. Self-inflicted death is, then, a preferred solution to and an honorable way out of the person's crisis of self-esteem, in which hopelessness and helplessness are not only the key subjective features for the individual but also the chief clues for the observer that suicide is about to occur.

CLINICAL EVALUATION OF SUICIDE RISK

In a thorough review of the literature of suicide prevention, Litman (1966) concluded that, despite decades of interest and prolific writing on the topic, scientific study had barely begun. Six years later, Lester (1972a) could find no evidence that suicide prevention affected the suicide rate. Yet prevention depends on the accurate identification of the individual most likely to kill himself. Litman et al. (1974) suggest that the criticism of intervention programs may in part reflect a failure to differentiate between the acutely suicidal person, who is perhaps actually being rescued, and the chronically suicidal person, who, because of a somewhat lower profile, eludes detection and becomes the program "failure."

Depression should always alert the clinician to his duty to assess suicide risk (Fawcett, 1972; Murphy, 1975b). A mental checklist should include age (men's suicide rates peak in the eighties; women's, between 55 and 65); sex (men complete suicides about three times as often as women, while women attempt it two to three times as often as men); alcoholism; absence or recent disappearance of "fight," distress, or strong feeling (sudden calm in a previously suicidal person may stem from a definite resolve to die); refusal to accept help immediately; and prior experience as a psychiatric inpatient.

If the person has already made an attempt, its degree of lethality predicts the danger of a subsequent attempt. To assess this factor, the "risk–rescue" criteria of Weisman and Worden (1972) should be used. Information regarding the degree of risk (method used, degree of impairment of consciousness when rescued, extent of injury, time required in hospital to reverse the effects, intensity of treatment required) and the rescue circumstances (remoteness from help, type of rescuer, probability of discovery, accessibility to rescue, and delay between attempt and rescue) can be established in a careful history. Serious physical illness can increase suicide risk but does not diminish the importance of the foregoing points of assessment; it is the attitude toward the illness that requires explanation. Low tolerance for pain, excessive demands or complaints, and a perception of lack of attention and support from the hospital staff are danger signals in the assessment of suicide potential (Farberow et al., 1963).

Recently Reich and Kelly (1976) analyzed seventeen suicides in 70,404 consecutive general hospital admissions; the only two without a psychiatric diagnosis were cancer patients. The most significant warning of impending suicide was some sort of rupture in the patient's relationship with hospital staff, ranging from angry accusations and complaints to the perception (in the two cancer patients) that the staff was giving up on them.

Psychometric scales for assessment of suicide risk are available (Beck et al., 1974; Zung, 1974). Beck et al. (1975) provide a promising beacon to guide the physician in his search among so many variables. In a rigorous investigation of 384 suicide attempters, employing scales to measure hopelessness, suicide intention, and depression, this group demonstrates that hopelessness is the key variable linking depression to suicidal behavior. Therefore, encouraging physicians to focus on this particular clue may lead to new gains in suicide prevention.

Finally, one should always ask directly whether the person is considering suicide. Excellent evidence exists that a suicidal patient will admit his intentions to a physician if asked (Delong and Robins, 1961). Contrary to popular belief, the physician is more likely than any other individual or agency to encounter the completed suicide victim shortly before his death. Analyzing all successful suicides in St. Louis County in 1968–1969, Murphy (1975a,b) discovered that 81% had been under the care of a physician within six months prior to suicide. By contrast, suicide prevention centers had contact with no more than 2–6% of the suicidal patients in their localities (Weiner, 1969; Barraclough and Shea, 1970; Sawyer *et al.*, 1972). More soberingly, Murphy (1975a) documents that 91% of overdose suicide victims had been under the recent care of a physician (compared with 71% of controls), and in over half the cases, the physician had supplied by prescription the complete means for suicide.

TREATING THE SUICIDAL PERSON

Self-esteem or self-respect is the most basic psychic condition to be guarded if life is to continue. Therefore the restoration and maintenance of the individual's narcissistic equilibrium is the aim of the therapist confronted by a person tempted to destroy himself. Certain practical features of this relationship can be summarized from Schneidman (1975) and others:

1. The therapist should stand as an ally for the life of the individual and in a calm and gentle but firm and unequivocal way make clear that his role is direct and active, including intervention to prevent suicidal behavior when the person makes this known. This intervention may include sending the police to the patient's home, depriving him of the means of suicide (pills or weapons), or hospitalizing him by involuntary commitment. This approach is justified by those situations where suicidal communication is a cry for help (Farberow and Schneidman, 1965) and must be accompanied by equally forthright reminders to the patient of the therapist's drastic limitations—that he is only rarely able to control another's behavior; that responsibility for behavior belongs to its initiator; that unconscious hopes for someone else to stop the execution of a plan are very often disappointed; and that without the patient's actually cooperating (at least unconsciously), the therapist could not prevent or reverse a single suicide attempt.

2. This active relationship requires a delicate but deliberate violation of the usual confidences of the therapeutic relationship. Self-destructive plans are never to be held in confidence when their

disclosure may prevent the death of the patient. It is not the plan that prompts action from the therapist (he can be content to hear in confidence innumerable variations) but the emergence of indications that execution of plans is likely.

3. Repeated monitoring of suicidal potential is an important feature of working with an actively suicidal patient. Suicidal intent, the extent of hopelessness, and events on the patient's "calendar," such as anniversaries of losses and most holiday seasons, require direct questioning for proper assessment.

4. Despite his status as an ally for life, the therapist must also have the capacity to hear out carefully and to tolerate the feelings of despair, desperation, anguish, rage, loneliness, emptiness, and meaninglessness articulated by the suicidal person. The patient needs to know that the therapist takes him seriously and understands. This understanding may require the therapist to explore the patient's darkest feelings of despair—a taxing empathetic task.

5. Reduction of social isolation and withdrawal is essential. Active advice and encouragement may be required, as well as family, couple, or group treatment. Self-esteem develops through relationships with others, and introjection of a benign and supportive love object is an essential feature of ego development. A shortage of such love objects is often so prominent in the lives of suicidal patients that replacement of bad introjects may seem (and some say theoretically is) impossible. These persons are skilled at distorting interpersonal relationships and evoking rejection. For this reason the therapist's investment in and respect for them may mark the first and only relationship in which they can begin to eliminate distorting perceptions of all relationships.

6. Relationships, work, hobbies, and other individualistic activities enhance and maintain self-esteem. Community agencies may help the patient renew or initiate such activities.

7. Coexisting psychiatric disorders must be treated. Suicide does not respect diagnostic categories. Potential for completed suicide, while representing an escape from an intolerable, hopeless crisis of self-esteem, must be dealt with in a manner appropriate to the psychopathology of the individual. Thus with the suicidal schizophrenic, reality testing, increased medication, and even ECT may be essential to prevent death. With the borderline patient or narcissistic personality, one is likely to spend even more time than usual interpreting negative transference or rehearsing the typical sequence of behaviors outlined by Adler (1973): highly

idealized expectation, followed by inevitable disappointment and reactive rage, culminating in self-destructive behavior. Similar considerations for depression, anxiety, and other conditions must be made. If the patient has no psychiatric illness, as in the case of patients with cancer or dependence on chronic hemodialysis, attention to factors that can restore self-esteem and respect are also paramount.

8. Finally, the therapist himself must at times seek support or consultation when treating a patient with potential for completing suicide. Of all the difficulties encountered, the most troublesome is almost surely the countertransference hatred that suicidal patients often evoke in their caregivers (Maltzberger and Buie, 1974). When hostility is prominent in the patient, hypercritical, devaluating, scornful rage may be directed at the therapist for long periods of time. Then the patient's feelings as well as the therapist's own reactions can make treatment seem like an ordeal.

Treating a suicidal person can be quite anxiety-provoking for the therapist. The need to balance consideration for the patient's safety with the goal that he live his life independently in his own world reminds us how limited the therapist's powers are—that is, they are no stronger than the patient's desire to make use of help. The therapist who appreciates his ultimate inability to stop the person who really wants to kill himself is far more likely to be effective in restoring the person's sense of self-esteem and wholeness. Respect for independence, like investment in a patient's well-being, is itself therapeutic. Clarifying these limitations with the patient helps convey respect for his autonomy and reminds the therapist that a completed suicide can occur despite complete fulfillment of his responsibility. Both are thereby better enabled to see that the risks of their mutual encounter are worth taking.

CARE OF THE BEREAVED

Heberden, listing the causes of death in London in 1657, assigned tenth rank to "griefe." Three centuries later, Engel (1961) argued eloquently that grief should be classified as a disease because of its massive impact on normal function, the suffering it involves, and the predictable symptomatology associated with it. Parkes's *Bereavement* (1972) represents the most comprehensive contemporary summary of the empirical factors associated with the entire process. *Bereavement* refers to the process of accommodation to a specific loss, including anticipatory grief and mourning (Weisman, 1974); *grief* denotes the conscious impact of loss on an individual; *mourning* is the reactive process of coping with the loss.

How can one distinguish normal from abnormal grief? In his pioneer work with survivors of the Cocoanut Grove fire, Lindemann (1944) describes pathognomonic features of normal grief: (1) somatic distress, marked by sighing respiration, exhaustion, and digestive symptoms of all kinds (C. S. Lewis wrote in 1961, after the death of his wife: "No one ever told me that grief felt so much like fear. I am not afraid, but the sensation is like being afraid. The same fluttering in the stomach, the same restlessness, the yawning. I keep on swallowing," p. 7); (2) preoccupation with the image of the deceased, a slight sense of unreality, a feeling of increased emotional distance from others, and such an intense focus on the deceased person that the grief-stricken may believe themselves in danger of insanity; (3) preoccupation with feelings of guilt; (4) hostile reactions, irritability, and a disconcerting loss of warm feelings toward other people; and (5) a disruption or disorganization of normal patterns of conduct.

In abnormal grief reactions, Lindemann found that the postponement of the experience or the absence of grief is usually associated with much more difficulty during bereavement. Parkes (1972) observes that hysterical or extreme reactions to loss (such as prolonged screaming and shouting, repeated fainting, and conversion paralysis) carry a prognostic significance as bad as the absence of grief. Cultural norms must of course be consulted to determine what is extreme in expressing grief. Lindemann (1944) lists nine signs of abnormality, which he calls "distorted reactions": (1) overactivity without a sense of loss; (2) the acquisition of symptoms belonging to the last illness of the deceased, presenting as conversion (hysterical) or hypochondriacal complaints; (3) a recognized medical disease, such as ulcerative colitis, rheumatoid arthritis, or asthma (although Lindemann linked these diseases to the group of psychosomatic conditions recognized at the time, subsequent investigations have shown that after the death of a loved one, survivors are at a significantly increased risk of death and illness); (4) alteration in relationships to friends and relatives, with progressive social isolation; (5) furious hostility against specific persons, resembling a truly paranoid reaction; (6) such suppression of hostility that affectivity and conduct resemble a schizophrenic picture, with masklike appearance, formal, stilted, robotlike movements, and no emotional expressiveness; (7) lasting loss of patterns of social interaction, with absence of decisiveness and initiative; (8) behavior that is socially and economically destructive, such as giving away belongings, making foolish business deals, or performing other self-punitive actions with no realization of internal feelings of guilt; and (9) overt agitated depression. It is important to note that Lindemann was impressed by the rarity of these manifestations of abnormal grief among the families of the Cocoanut Grove victims. Clayton *et al.* (1968) and Clayton (1974), in extensive investigations of bereaved persons, have firmly established in follow-up how well most bereaved persons recover. At three months after the loss, Clayton found that four out of five patients were improved, and only 4% were worse. Rees and Lutkins (1967) have shown that widows and widowers have a significantly increased death rate for their age category in the first year following the death of their spouses. Holmes and Rahe (1967) have demonstrated that loss in many forms increases the risk of illness. Contrary to Rees and Lutkins's data, Clayton failed to find an increase of morbidity among a sample of bereaved persons in the St. Louis area.

How long does bereavement take? A parent who recently joined a parents' group after the death of his child asked this question. Another parent replied, "My son died thirty years ago, so I don't know yet," eloquently expressing the way grief can remain while behavior returns to normal. Treatment of the abnormal reactions is of primary importance.

The Process of Mourning

While many schemes are useful in conceptualizing mourning, that of Bowlby (1961) is classic.

1. In the first phase, the survivor is preoccupied with the lost person. In accord with animal models of loss, this time is characterized by searching and protest. The behaviors may seem bizarre yet are entirely normal. One boy of fourteen lost his best friend from nephritis. For three months thereafter he carried on prolonged conversations with his departed friend, with the conviction not only that his questions and discussion of future plans were heard but that his friend was present. Vivid hallucinatory experiences of the deceased person occur in about 50% of bereavement reactions (Rees, 1970). A mother who lost her 20-year-old daughter repeatedly found herself in the girl's room. She wore the girl's clothes and on occasion put on the girl's nightclothes and lay in her bed striving for some sort of meeting with her. The German artist Kollwitz described working on a monu-

ment for her younger son who was killed in October 1914. Two years later she noted in her diary, "There's a drawing made, a mother letting her dead son slide into her arms. I could do a hundred similar drawings but still can't seem to come any closer to him. *I'm still searching for him,* as if it were in the very work itself that I had to find him" (Parkes, 1972, p. 39; italics added). Children of Israeli soldiers missing in action have been observed repeatedly to lose objects, complain loudly and tearfully about the losses, and, suddenly finding the object, carry it about on display, urging adults in the family to rejoice with them in the discovery; in one case the object was a picture of the missing father.* It is important for an observer to recall that even apparently bizarre behavior in a bereaved person may represent the searching or protest phase of the mourning process.

2. The second phase described by Bowlby is one of disorganization, a preoccupation with the pain of the experience, characterized at times by turmoil or even despair. One father, two years after the loss of his 10-year-old son, told a group: "It hurts. I could tell you all about how you'd feel going through the whole thing, but I could never tell you about the emptiness that comes . . . afterward. That hurts more than anything—at times it aches." C. S. Lewis noted after the death of his wife, "Her absence is like the sky, spread over everything" (1961, p. 13). The pointlessness of life and the reasonableness of suicide are common in the minds of the bereaved in this period. Social interaction seems impossible, and yet solitude is intolerable. "I want others to be about me," Lewis observed. "I dread the moments when the house is empty. If only they would talk to one another and not to me" (p. 7).

3. Finally, a phase of reorganization occurs in which normal functioning and behavior are restored. Reversals during this time are the rule, and the reappearance of the earlier two phases should be expected. During the reorganization process, the bereaved person is caught off guard by memories and repeated realizations of the pain of loss. A widow, at the ringing of a phone, may be devastated by the sudden realization that her husband will never phone her again. Returning to scenes or sights shared with the deceased is a common occasion of grief and an opportunity to

---

*Reported by M. Rosenbaum at Psychiatric Grand Rounds, Massachusetts General Hospital, Boston, Massachusetts, in 1975.

further the process of mourning. Death after a long illness, for example, may seem acceptable to a survivor because it terminates the pain and debilitation of the lost person. But return to a scene that recalls the lost person in a healthy state may precipitate new feelings and make the death entirely unacceptable. One widow used to carry the picture of her husband as he looked at his worst. When she felt sad she would take the picture out and look at it, "to remind myself that he couldn't be expected to go on."

Psychologically, death represents the rupture of the survivor's attachment to the lost person—a bond heavily laden with feelings, conveniently oversimplified as those of love and those of hate. For example, a wife dies. Because of her the husband felt loved, better, stronger, whole; her death comes as a personal blow, even to his own self-esteem ("narcissistic injury"). He now feels smaller, weaker, fragmented. He may feel cheated, angry, and desirous of lashing out. For all his hostile feelings toward her he now feels guilt, probably in proportion to his negative feelings ("I might have been angry, but I wouldn't have wished this for her in a million years"). In the classic view of Freud (1917), the libido invested in the lost object, through the process of mourning, was freed to allow the survivor to invest feeling in new love objects. Despite impressions given by some theorists, investment in the lost person is not forsaken. Rather, a progressive clarification of the attachment seems to occur, until both positive and negative feelings toward the dead person are restored to a realistic balance and the lost one is viewed with more perspective. In this clarification process, the survivor puts in perspective his own role and value in the relationship, thereby restoring his own damaged self-esteem.

BEING HELPFUL TO THOSE IN MOURNING

No less than the dying, mourners are outcasts of society. Their presence is painful to many who surround them, and efforts to silence, impede, or stop mourning are common. Most societies and subcultures have rituals for beginning the process of mourning, such as wakes or *shivah*, funerals, and burial services. These rituals have significant values (see Cassem, 1976): (1) They provide an occasion for the gathering of the social network of support that surrounds the family unit. (2) Those gathered activate important memories about the deceased, sometimes providing the bereaved with information and relationships not previously known. (3) Tributes paid to the dead person emphasize both his worth and the fact that he is worth the pain and stress of grieving. (4) Family units may be drawn closer together. (5) Rituals permit expression of sorrow, set some limits on grieving, and provide legitimate outlets for expressing positive feelings. (6) Rituals reemphasize the reality that the deceased person is dead and gone, often by providing a view of the body (failure to see the dead body generally retards grieving). (7) The funeral permits other people in the community to pay their respects and initiate their own grieving. (8) Death rituals for the community provide occasions to grieve for losses unrelated to the deceased, a chance to continue unfinished mourning. (9) Rosenblatt (1975) presents cross-cultural evidence from seventy-eight societies that overt expressions of anger following a death are less common where ritualists deal with the body up to and during burial. (10) Because the bereaved generally remember nonverbal expressions of concern and affection more clearly than verbal expressions, rituals provide the formal occasion to bring others into the mourners' presence.

The task of the mourner is threefold: first, to experience and reflect upon his feelings toward the lost object during life and the feelings evoked by death; second, to review the history of the attachment; and third, to examine his own wounds, attend to their healing, and confront the task of continuing without the lost object. To help the bereaved, one need only facilitate the three parts in this process.

Allowing the bereaved to express feelings is essential. The most important part of this process is to avoid the maneuvers that nullify grieving. Clichés ("It's God's will"), self-evident but irrelevant reassurances ("After all, you've got three other children"), and outright exhortations to stop grieving ("Life must go on") should be carefully avoided. Some acquaintance with the nature of mourning is helpful, because many of the expressions of pain or seemingly aberrant behavior are necessary elements in the searching–disorganization–reorganization process. Tolerance of reversals from cheerful to grief-stricken feelings during the reorganization phase is essential.

Presence means more than words. Above all, the friend who can remain calm and quiet in the presence of a weeping or angry or bitter mourner is highly valued. Embarrassment only increases the discomfort of the grieving person. A touch can be worth a thousand words, as can moist eyes or a quiet tear. Friends who arrive at the home of the bereaved with food, or offer to take the children on an outing, or make some other tangible gesture do more than those who ask to be called if anything is needed.

Sharing memories may help complete the memories of the deceased.

The bereaved may be helped by looking through old photograph albums or collections of letters as they try to put the lost person into perspective. Persevering patience over the long haul is essential. Many acquaintances disappear after the funeral and burial rituals. Those who can continue to visit make a great difference. Letters or notes are valued by mourners. Anniversaries are key foci in the grieving trajectory. Special attention to the bereaved on that day is part of the most basic care of mourners.

Return to activity is usually an essential feature of the recovery process, both because it brings the mourner back into contact with concerned fellow workers and because the therapeutic effects of work on self-esteem help repair the narcissistic injury of the loss. Most bereaved persons benefit from returning to work within three to six weeks after the death of a loved one. Self-help groups continue to be among the most effective modalities for permitting expression of emotion, demonstrating that bereavement and its feelings are universal and supplying the compassion and respect necessary for rebuilding self-esteem. The prototype for these groups is the widow-to-widow program (Silverman et al., 1974).

Reading can sometimes help bereaved persons. Often the most consoling books are stories of persons sustaining tragic losses, such as Anne Morrow Lindbergh's (1973) or C. S. Lewis's (1961). Edgar Jackson's *You and Your Grief* (1966), written for the bereaved family, provides helpful directives.

SUDDEN DEATH

Death with time to prepare is difficult enough, but sudden death inflicts a unique trauma on the survivors. Death on arrival at the hospital, still-birth, sudden infant death, accidental or traumatic death, sudden cardiac arrest, death during or after surgery, murder, suicide—each carries its own set of horrors for the survivors, and in them the shock of death is dramatically intensified. Guilt is likely to be a much more serious problem because of the total absence of any preparation or opportunity to "do things right." Where violence and disfigurement are present, these feelings will be even further intensified.

General rules for dealing with the bereaved apply here, with certain specific emphases.

1. The chance to view the body, even when mutilated, should be

offered to the family members. If severe mutilation is present, the family should be very carefully warned; they should be told that the body is disfigured and that they may wish not to view it but may do so if they desire. In addition, a chance to do something for the body can at times be very helpful to a survivor who has had no chance to do anything. Some funeral directors offer parents of children who have died suddenly the opportunity to wash and prepare the child's body for burial or to help in this procedure; most parents accept and benefit greatly from this opportunity. The need to view the body is greater when death is sudden.

2. Patience with the prolonged numbness or shock of the family is an essential feature of their care.

3. Physical needs can be attended to and may be the only avenue of communication. Leading the family to a quiet room, providing comfortable seats, bringing coffee, and other little acts of kindness performed in a compassionate and quiet way are helpful.

4. The numb survivor may also benefit from very gentle questions that help to review the last hours of the deceased. This review starts the searching process and leads to areas where guilt is present. What happened? Were there any prodromal symptoms? Any premonitions? Who saw him last? Families who do not wish to explore these crucial questions at the time should not be pushed.

Later issues are very similar to those mentioned above. The questions of guilt are usually more agonizing. The day may be relived thousands of times. In certain cases of sudden death, specific conflicts should be kept in mind. Parents stricken by a sudden infant death should receive all the available standard information on this syndrome, in which careful explanations emphasize that the causes are unknown and the usual self-accusations about what could have been done are groundless. Survivors of a murder victim often experience retaliatory murderous fury (sometimes unconsciously). Fear of not being able to control vindictive aggression can amount to panic in some individuals, with an inability to identify the feeling. Survivors of a suicide victim are often the victims of an angry gesture and frequently must cope with their own hatred of the victim.

The main handicap for professionals is lack of an opportunity to develop a relationship with the survivors prior to the death of the victim. It is a time when close friends and relatives need to be mobilized.

ANTICIPATORY GRIEF

Lindemann (1944) uses the term *anticipatory grief* to refer to a special case of separation in which death is assumed and later found to be untrue—a husband missing in action in World War II, for example. Having gone through the phases of grief, the mourner in this situation can be at a disadvantage if the lost person returns, for the grief work may have been done so well that she finds herself freed of her attachment to the love object and unready to receive him back. Since that time, the term has come to signify the reaction experienced in the face of an impending loss. Sudden death provides no opportunity for anticipatory grief. On the other hand, Parkes (1970) points out that most of the widows he has studied knew in advance that they would lose a husband. Aldrich (1974) notes several differences between anticipatory and conventional grief. First, anticipatory grief is shared by both the patient or victim and the family, whereas conventional grief is experienced by the bereaved alone. Second, anticipatory grief has a distinct end point in the physical loss of death, whereas conventional grief may be prolonged for years. Third, the pattern of the two types may be different, with anticipatory grief often accelerating as death nears and conventional grief beginning at a high level of intensity and gradually decelerating. Fourth, hostile feelings toward the (about-to-be) lost person differ in the two states. If the lost person is still alive, as in anticipatory grief, the survivor's negative feelings tend to be experienced as more dangerous. Fifth, hope can accompany anticipatory grief in a way that it cannot once death has occurred. Finally, the question arises whether anticipatory grief can make conventional grief unnecessary or whether the work of mourning can be accomplished before death occurs. It is a fact that where anticipatory grief is possible, the intensity of the feelings of loss may actually decelerate as death draws nearer. In this case, death itself is viewed as increasingly acceptable. On the other hand, Aldrich notes that the sample of Cocoanut Grove bereaved appeared to accomplish their mourning more quickly than did the sample of Parkes's widows (1970). No evidence can as yet satisfactorily answer the question of whether unexpected loss requires longer grieving than an anticipated loss.

Weisman (1974) reminds us of the importance of viewing grief and love as mirror images of each other. Each enhances the other. Rochlin (1965) has argued in detail that the powerful dynamic impetus produced by our losses is indispensable to emotional maturation. Mourning can lead to some of man's highest achievements as well as result in many pathological states when delayed or unresolved.

DIFFICULTIES OF CAREGIVERS

Cultural taboos spare no one from their subtle pressures. The "American way of not dying" (Paul Ramsey's phrase) pervades the population. Even in the major arenas where death now occurs—general hospitals, nursing homes, and chronic-care facilities—staff members may become uncomfortable at the mere mention of death. Health professionals who attend the dying are not immune to bedside discomfort. Glaser and Strauss (1965), in a careful study of nurses on a general hospital floor, documented that the sicker the patient was and the closer he was to death, the less time the staff spent with him. This relationship was true even though the more seriously ill patients were located closer to the nursing station. Despite a consensus that dying persons fear abandonment, avoidance of them is the rule rather than the exception.

Why does the doctor tend to avoid some dying patients or, if he does not avoid them, remain quite ill at ease in their presence? Encounters with dying persons generally represent either a personal threat to the professional or a threat to his relationships. A personal threat can arise from exposure to serious disability or dying. "There is to be no mention of the word *death* here" can represent a denial that death will ever claim the doctor. The surgeon who says, "There can be no such order as 'DNR' [Do Not Resuscitate]," may be denying that failure is possible in the struggle against death. Avoidance of paralyzed or disfigured patients may protect the physician from the realization that the same fate could befall him. As mentioned earlier, absorbing and responding to a dying person's plight is very difficult; compassion can impose a heavy burden on the sympathetic professional. "I don't think I can take another story like the one I just heard" expresses a familiar and realistic worry.

A dying person poses a threat to the professional's own human attachments. First of all, the doctor is reminded that death means loss of the relationship with the patient himself, and of all the investment and caring that has gone into it. Secondly, imminent loss of a patient reminds the professional of his own past losses and of threatened ones. Care of a dying woman, for example, can recall the death (real or feared) of a mother, sister, or spouse. Wounds of an incompletely grieved loss are reopened. Losses in one's life are not discrete; they are cumulative. Those who care for the dying sustain repeated bereavement.

Because of the repeated stresses posed by these threats, professionals develop defenses. One course is to avoid involvement with patients in order to minimize personal loss or discomfort. This is the probable source of fixed "styles" of relating to patients. For example, the physician who follows the initial communication of the diagnosis to the patient with a standardized, nonstop monologue detailing diagnosis, treatment, and course of the disease is not trying to baffle, overwhelm, or lose the patient altogether but rather to minimize his own anxiety by handling every initial "bad news" session the same way. Intellectualization, discussing the dying person mainly as an interesting "problem" or "case," functions protectively as well. Some professionals say of the patient that he "can't take it," which is often a projection of their own fears. Further examples would illustrate other defense mechanisms; what is important is that physician discomfort and avoidance arise mainly from instincts of self-defense rather than from any desire to alienate or harm the patient.

Criticism of physicians tends to be widespread and can generate unfortunate hostilities. A common accusation is that doctors are more afraid of death than other people and entered medicine for that reason. Sophisticated critics refer to Feifel's 1965 study of physician attitudes toward death, which shows that physicians spent significantly less time in conscious thought about death than did two control groups of other professionals. Feifel *et al.* (1967) have also demonstrated more fearfulness of death in medical students than in two control samples. Patient dissatisfaction with doctors' communication patterns, their avoidance mechanisms, and their manner in the sickroom compounds the alienation of dying patient and physician.

In response to such criticism some physicians have become defensive or even hostile themselves. When forced to assume the adversary role, the doctor is at a distinct disadvantage. The physician as well as those who work with him might be helped by keeping certain reflections in mind. Although everyone is fascinated with death, physicians grapple with it more frequently than others and are sought out by those who wish to prevent or postpone it. Conflict or discomfort in the presence of death is intensified by the combination of two factors: frequent exposure and high responsibility. The findings of Feifel *et al.* (1967) may also be accounted for by the increased exposure of medical students to death and to the widespread belief of the population that the physician is the last barrier against it. Natural disappointment that a terminally ill person cannot be saved can lead to resentment. As one who could not reverse the illness, the doctor may become the object of some or all of the family's resentment and blame. Finally, those who accuse doctors of inability to communicate with dying patients seem to forget that everyone has difficulties in this regard, including the patients' families.

What can be done to decrease the defensiveness of the physician?

In addition to enjoying the exercise of greater responsibility in helping prevent or postpone death, doctors pay for this privilege by stronger feelings of failure and guilt when their hopes and the patient's are disappointed. Colleagues or coworkers of an oncologist who is particularly gruff will accomplish much more with a sympathetic approach than with a hostile one—"It must be difficult to work with patients whose illnesses always have such poor odds for cure or recovery." Or one might say to the surgeon after an untimely failure, "It's been a tough day, but you did everything you could." Such sympathetic comments have the greatest likelihood of turning tense situations into mutually supportive ones. Finally, more systematic efforts need be undertaken to help health personnel who work on the front lines deal with their own feelings about suffering, disappointment, failure, and death. Ideally this process should begin early in training, so that the repeated exposure to traumatic experiences can be accompanied by emotional growth with its major features, compassion and equanimity.

Fear of Death in Nondying Persons

In 1915, after Freud's original statement that death has no representation in the unconscious, an oversimplified interpretation of psychoanalytic theory tended to translate fear of death as fear of castration. In fact Fenichel, interpreting Freud, stated, "It is questionable whether there is any such thing as a normal fear of death; actually the idea of one's own death is subjectively inconceivable, and therefore probably every fear of death covers over unconscious ideas" (1945, pp. 208–209). Contemporary literature emphasizes the opposite—that fear of death motivates a large portion of every human's behavior, generating multiple activities aimed at avoiding or denying mortality (Becker, 1973). Meyer (1975) has explored the relationship of many neurotic anxieties to underlying fears of death. The tendency to diagnose the neurosis of our era as thanatophobia conveys erroneously that no one faces death with equanimity. Sheps (1957) has actually asserted that a terminal patient who failed to show marked anxiety was denying fear of death. Weisman (1972) points out the "tendentious fallacy" involved here, reminding us that acceptance of death without denial is not only a possible but a frequent occurrence. When suspecting a neurotic fear of death, the clinician should explore for guilt, masochism, and dependency, with close attention to the social function of the symptom (death fear).

Death is often viewed as the ultimate punishment. Hostile and erotic feelings or deeds are usually what the patient believes have earned this fate. One year after marriage, for instance, a young woman experienced the onset of panic states in which she thought death was imminent. Work with her revealed these attacks were immediately preceded by an impulse to seek a sexual relationship with a man other than her husband. When panic-stricken, however, she was unable to leave the house. Her panic state actually protected her from acting out her impulse, serving a useful psychic function.

Masochistically inclined persons may also speak of a horror of death. Prolonged, detailed, florid, or melodramatic descriptions of pain and suffering are clues to the masochistic trait. For example, a middle-aged woman with a diagnosis of terminal carcinoma was admitted to a chronic hospital to die. She spoke at great length of her imminent demise, the repeated and terrible tragedies throughout her entire life, and her fear of death. One physician, skeptical of the diagnosis, investigated further. Thorough workup revealed no trace of malignancy; on receiving the news of misdiagnosis, the patient became profoundly depressed at the prospect of discharge. Although her voiced fears also brought her attention (secondary gain), she clearly talked of misery with enthusiasm (primary gain).

For other individuals, a specific death neurosis, such as cardiac neurosis, which preoccupies them with the possibility of sudden death, becomes the justification for a regressive withdrawal to infantile dependency. Cardiac disease (or a heart attack) can be an honorable excuse for abandoning adult responsibilities. These patients' frequent verbal reminders that sudden death is only hours or moments away permit them to tyrannize their families—"Don't stay home on my account; of course, I may not be here when you return." Here the expression of fear functions to justify the patient's helplessness, while maintaining nearly complete control of his family's services. The problems of living are so complicated and overwhelming for some persons that preoccupation with death becomes a welcome diversion. Ruminating about death is easier than shouldering the burdens of life.

When the shadow of death indicates that life is now measured, life and time become more precious. Deeper meanings and clearer priorities often emerge for the dying person, who can no longer afford to postpone opportunities. He must make the most of his time. Those who care for persons confronting death or its consequences share the opportunity to examine what makes a life meaningful. Sharing burdens, insights, and support, patient and professional are given a chance to absorb and pursue these discoveries together.

REFERENCES

Adler, G.: Hospital treatment of borderline patients. *Am. J. Psychiatry* **130**:32, 1973.

Aitken-Swan, J., and Easson, E. C.: Reactions of cancer patients on being told their diagnosis. *Br. Med. J.* **1**:779, 1959.

Aldrich, C. K.: Some dynamics of anticipatory grief, in *Anticipatory Grief*, B. Schoenberg, A. C. Carr, A. H. Kutscher, D. Peretz, and I. Goldberg (Eds.). New York: Columbia University Press, 1974.

Allport, G.: *The Nature of Prejudice*. New York: Doubleday, 1958.

Anthony, S.: *The Child's Discovery of Death*. New York: Harcourt, Brace, 1940.

Barraclough, B. M., and Shea, M.: Suicide and samaritan clients. *Lancet* **2**:68, 1970.

Beck, A. T.: Thinking and depression: I. Idiosyncratic content and cognitive distortions. *Arch. Gen. Psychiatry* **9**:324, 1963.

Beck, A. T.: *Depression: Clinical Experimental, and Theoretical Aspects*. New York: Harper & Row, Hoeber Medical Division, 1967.

Beck, A. T., Kovacs, M., and Weisman, A.: Hopelessness and suicidal behavior. *J.A.M.A.* **234**:1146, 1975.

Beck, A. T., Schuyler, D., and Herman, I.: Development of suicidal intent scales, in *The Prediction of Suicide*, A. T. Beck, H. L. P. Resnik, and D. J. Lettieri (Eds.). Bowie, Maryland: Charles Press, 1974.

Becker, E.: *The Denial of Death*. New York: Free Press, 1973.

Bowlby, J.: Processes of mourning. *Int. J. Psychoanal.* **42**:317, 1961.

Cassem, N. H.: The first three steps beyond the grave, in *Acute Grief and the Funeral*, V. R. Pine, A. H. Kutscher, D. Peretz, R. C. Slater, R. DeBellis, R. J. Volk, and D. J. Cherico (Eds.). Springfield, Illinois: Thomas, 1976.

Cassem, N. H., and Hackett, T. P.: Stress on the nurse and therapist in the intensive care unit and the coronary care unit. *Heart Lung* **4**:252, 1975.

Cassem, N. H., and Stewart, R. S.: Management and care of the dying patient. *Int. J. Psychiatry Med.* **6**:293, 1975.

Clayton, P. J.: Mortality and morbidity in the first year of widowhood. *Arch. Gen. Psychiatry* **30**:747, 1974.

Clayton, P. J., Desmarais, L., and Winokur, G.: A study of normal bereavement. *Am. J. Psychiatry* **125**:168, 1968.

Croog, S. H., Shapiro, D. S., and Levine, S.: Denial among male heart patients. *Psychosom. Med.* **33**:385, 1971.

Delong, W., and Robins, E.: The communication of suicidal intent prior to psychiatric hospitalization: A study of 87 patients. *Am. J. Psychiatry* **117**:695, 1961.

Deutsch, F.: Euthanasia, a clinical study. *Psychoanal. Q.* **5**:347, 1933.

Easson, W. M.: *The Dying Child*. Springfield, Illinois: Thomas, 1970.

Eissler, K.: *The Psychiatrist and the Dying Patient*. New York: International Universities Press, 1955.

Engel, G. L.: Is grief a disease? *Psychosom. Med.* **23**:18, 1961.

Evans, A. E.: If a child must die. *N. Engl. J. Med.* **278**:138, 1968.

Farberow, N. L., and Schneidman, E. S. (Eds.).: *The Cry for Help*. New York: McGraw-Hill, 1965.

Farberow, N. L., Schneidman, E. S., and Léonard, C. V.: Suicide among general medical and surgical hospital patients with malignant neoplasms. *Med. Bull. Veterans Admin.* **MB-9**:1, 1963.

Fawcett, J.: Suicidal depression and physical illness. *J.A.M.A.* **219**:1303, 1972.

Feagin, J. R.: Prejudice and religious types: A focused study of Southern fundamentalists. *J. Sci. Study Religion* **4**:3, 1964.

Feifel, H.: The function of attitudes toward death, in *Death and Dying: Attitudes of Patient and Doctor. Group Adv. Psychiatry* Symp. 11, **5**:33, 1965.

Feifel, H.: Religious conviction and fear of death among the healthy and the terminally ill. *J. Sci. Study Religion* **31**:353, 1974.

Feifel, H. (Ed.): *The Meaning of Death*. New York: McGraw-Hill, 1959.

Feifel, H., Hanson, S., Jones, R., and Edwards, L.: Physicians consider death, in *Proceedings of the 75th Annual Convention of the American Psychological Association*, E. Vinacke *et al.* (Eds.), Vol. 2.

Fenichel, O.: *The Psychoanalytic Theory of Neuroses*. New York: Norton, 1945.

Freud, S. 1915: Thoughts for the times on war and death, in *Standard Edition*, J. Strachey (Ed.), Vol. 14. London: Hogarth Press, 1957.

Freud, S. 1917: Mourning and melancholia, in *Standard Edition*, J. Strachey (Ed.), Vol. 14. London: Hogarth Press, 1949.

Freud, S. 1920: Beyond the pleasure principle, in *Standard Edition*, J. Strachey (Ed.), Vol. 18. London: Hogarth Press, 1955.

Freud, S. 1923: The ego and the id, in *Standard Edition*, J. Strachey (Ed.), Vol. 19. London: Hogarth Press, 1961.

Gerlé, B., Lunden, G., and Sandblom, P.: The patient with inoperable cancer from the psychiatric and social standpoints. *Cancer* **13**:1206, 1960.

Gifford, S.: Some psychoanalytic theories about death: A selective historical review. *Ann. N.Y. Acad. Sci.* **164**:638, 1969.

Gilbertsen, V. A., and Wangensteen, O. H.: Should the doctor tell the patient that the disease is cancer? in *The Physician and the Total Care of the Cancer Patient*. New York: American Cancer Society, 1962.

Glaser, B. G., and Strauss, A. L.: *Awareness of Dying*. Chicago: Aldine, 1965.

Goldberg, I. K., Malitz, S., and Kutscher, A. H. (Eds.): *Psychopharmacologic Agents for the Terminally Ill and Bereaved*. New York: Columbia University Press, 1973.

Grollman, E. A.: *Talking About Death*. Boston: Beacon Press, 1970.

Grollman, E. A. (Ed.): *Explaining Death to Children*. Boston: Beacon Press, 1967.

Hackett, T. P., and Weisman, A. D.: The treatment of the dying. *Curr. Psychiatr. Ther.* **2**:121, 1962.

Harris, A.: *Why Did He Die?* Minneapolis: Lerner, 1965.

Hinton, J.: *Dying*. Baltimore: Penguin Books, 1967.

Holmes, T. H., and Rahe, R. H.: The social readjustment rating scale. *J. Psychosom. Res.* **11**:213, 1967.

Jackson, E. N.: *Telling a Child about Death*. New York: Channel Press, 1965.

Jackson, E. N.: *You and Your Grief*. New York: Channel Press, 1966.

Jacobson, E.: *Progressive Relaxation*. Chicago: University of Chicago Press, 1938.

Jones, E.: *The Life and Work of Sigmund Freud*. Vol. 3. New York: Basic Books, 1957.

Kast, E. C.: LSD and the dying patient. *Chicago Med. Sch. Q.* **26**:80, 1966.

Kelly, W. D., and Friesen, S. R.: Do cancer patients want to be told? *Surgery* **27**:822, 1950.

Kennell, J. H., Slyter, H., and Klaus, M. H.: The mourning response of parents to the death of a new-born infant. *N. Engl. J. Med.* **283**:344, 1970.

Kübler-Ross, E.: *On Death and Dying*. New York: Macmillan, 1969.

Kübler-Ross, E.: *Questions and Answers on Death and Dying*. New York: Macmillan, 1974.

Kurland, A. A., Grof, S., Pahnke, W. N., and Goodman, L. E.: Psychedelic drug-assisted psychotherapy in patients with terminal cancer, in *Psychopharmacologic Agents for the*

Terminally Ill and Bereaved, I. K. Goldberg, S. Malitz, and A. H. Kutscher (Eds.). New York: Columbia University Press, 1973.

Kutscher, A. H., Schoenberg, B., and Carr, A. C.: *The Terminal Patient: Oral Care.* New York: Columbia University Press, 1973.

LeShan, L. and E.: Psychotherapy and the patient with a limited life span. *Psychiatry* **24:**318, 1961.

Lester, D.: The myth of suicide prevention. *Compr. Psychiatry* **13:**555, 1972a.

Lester, D.: Religious behaviors and attitudes toward death, in *Death and Presence*, A. Godin (Ed.). Brussels: Lumen Vitae Press, 1972b.

Lewis, C. S.: *A Grief Observed.* Greenwich, Conn.: Seabury Press, 1961.

Lindbergh, A. M.: *Hour of Gold, Hour of Lead.* New York: Harcourt, Brace, 1973.

Lindemann, E.: Symptomatology and management of acute grief. *Am. J. Psychiatry* **101:**141, 1944.

Litman, R. E.: The prevention of suicide, in *Current Psychiatric Therapies*, J. H. Masserman (Ed.), Vol. 6. New York: Grune & Stratton, 1966.

Litman, R. E., Farberow, N. L., Wold, C. I., and Brown, T. R.: Prediction models of suicidal behaviors, in *The Prediction of Suicide,* A. T. Beck, H. L. P. Resnik, and D. J. Lettieri (Eds.). Bowie, Md.: Charles Press, 1974.

Magni, K. G.: The fear of death, in *Death and Presence*, A. Godin (Ed.). Brussels: Lumen Vitae Press, 1972.

Maltzberger, J. T., and Buie, D. H.: Countertransference hate in treatment of suicidal patients. *Arch. Gen. Psychiatry* **30:**625, 1974.

Marks, R. M., and Sachar, E. J.: Undertreatment of medical inpatients with narcotic analgesics. *Ann. Intern. Med.* **78:**173, 1973.

Menninger, K. A.: *Man Against Himself.* New York: Harcourt, Brace, 1938.

Meyer, J. E.: *Death and Neurosis.* New York: International Universities Press, 1975.

Michaelson, M. G.: Death as a friendly onion. *New York Times Book Review* 21 July 1974, pp. 6–8, 1974.

Murphy, G. E.: The physician's responsibility for suicide. I. An error of commission. *Ann. Intern. Med.* **82:**301, 1975a.

Murphy, G. E.: The physician's responsibility for suicide. II. Errors of omission. *Ann. Intern. Med.* **82:**305, 1975b.

Nagy, M. H.: The child's view of death. *J. Genet. Psychol.* **73:**3, 1948.

Natterson, J. M., and Knudson, A. G.: Observations concerning fear of death in fatally ill children and their mothers. *Psychosom. Med.* **22:**456, 1960.

Parkes, C. M.: The first year of bereavement: A longitudinal study of the reaction of London widows to the death of their husbands. *Psychiatry* **33:**444, 1970.

Parkes, C. M.: *Bereavement: Studies of Grief in Adult Life.* New York: International Universities Press, 1972.

Peck, R.: The development of the concept of death in selected male children. Doctoral dissertation, New York University, 1966.

Portz, A.: The child's sense of death, in *Death and Presence*, A. Godin (Ed.). Brussels: Lumen Vitae Press, 1972.

Rado, S.: *Psychoanalysis of Behavior.* New York: Grune & Stratton, 1956.

Raimbault, G.: Listening to sick children, in *Death and Presence*, A. Godin (Ed.). Brussels: Lumen Vitae Press, 1972.

Rees, W. D.: The hallucinatory and paranormal reactions of bereavement. M.D. thesis, 1970.

Rees, W. D., and Lutkins, S. G.: Mortality of bereavement. *Br. Med. J.* **4:**13, 1967.

Reich, P., and Kelly, M. J.: Suicide attempts by hospitalized medical and surgical patients. *N. Engl. J. Med.* **294:**298, 1976.

Rochlin, G.: *Griefs and Discontents: The Forces of Change.* Boston: Little, Brown, 1965.

Rochlin, G.: How younger children view death in themselves, in *Explaining Death to Children*, E. A. Grollman (Ed.). Boston: Beacon Press, 1967.

Rochlin, G.: *Man's Aggression.* Boston: Gambit, 1973.

Rosenblatt, P. C.: Use of ethnography in understanding grief and mourning, in *Bereavement: Its Psychosocial Aspects*, B. Schoenberg, I. Gerber, A. Wiener, A. H. Kutscher, D. Peretz, and A. C. Carr (Eds.). New York: Columbia University Press, 1975.

Safier, G.: A study in relationships between the life and death concepts in children. *J. Genet. Psychol.* **105:**283, 1964.

Sallen, S. E., Zinberg, N. E., and Frei, E., III.: Antiemetic effect of delta-9-tetrahydrocannabinol in patients receiving cancer chemotherapy. *N. Engl. J. Med.* **293:**795, 1975.

Saunders, C.: Care of the dying: The problem of euthanasia, 1–6. *Nurs. Times* **55:**960,994,1031,1067,1091,1129, 1959.

Saunders, C.: The moment of truth: Care of the dying person, in *Death and Dying*, L. Pearson (Ed.). Cleveland: Case Western Reserve University Press, 1969.

Sawyer, J. B., Sudak, H. S., and Hall, S. R.: A follow-up study of 53 suicides known to a suicide prevention center. *Life-Threatening Behav.* **2:**227, 1972.

Schneidman, E. S.: Suicide, in *Comprehensive Textbook of Psychiatry*, A. M. Freedman, H. I. Kaplan, and B. J. Sadock (Eds.). Vol. 2, 2nd ed. Baltimore: Williams & Wilkins, 1975.

Schowalter, J. E.: Deatb and the pediatric house officer. *J. Pediatr.* **76:**706, 1970.

Sheps, J.: Management of fear in chronic disease. *J. Am. Geriatr. Soc.* **5:**793, 1957.

Silverman, P., MacKenzie, D., Pettipas, M., and Wilson, E. (Eds.): *Helping Each Other in Widowhood.* New York: Health Sciences, 1974.

Simonton, O. C. and S.: Management of the emotional aspects of malignancy. Paper presented at the symposium New Dimensions of Habilitation for the Handicapped, held by the Florida Department of Health and Rehabilitative Services, 14–15 June 1974, at the University of Florida—Gainsville, 1974.

Spinetta, J. J.: The dying child's awareness of death: A review. *Psychol. Bull.* **81:**256, 1974.

Spinetta, J. J., Rigler, D., and Karon, M.: Personal space as a measure of a dying child's sense of isolation. *J. Consult. Clin. Psychol.* **42:**751, 1974.

Stein, S. B.: *About Dying: A Book for Parents and Children.* New York: Walker, 1974.

Steiner, G.: Children's concepts of life and death: A developmental study. Doctoral dissertation, Columbia University, 1965.

Stevens, S.: *Death Comes Home.* New York: Morehouse-Barlow, 1973.

Surman, O. S., Gottlieb, J. K., Hackett, T. P., and Silverberg, E. L.: Hypnosis in the treatment of warts. *Arch. Gen. Psychiatry* **28:**439, 1973.

Toch, R.: Management of the child with a fatal disease. *Clin. Pediatr.* **3:**418, 1964.

Waechter, E.: Children's reaction to fatal illness, in *Death and Presence*, A. Godin (Ed.). Brussels: Lumen Vitae Press, 1972.

Weiner, I. W.: The effectiveness of a suicide prevention program. *M.H.* **53:**357, 1969.

Weiner, J. M.: Attitudes of pediatricians toward the care of fatally ill children. *J. Pediatr.* **76:**700, 1970.

Weisman, A. D.: Is suicide a disease? *Life-Threatening Behav.* **1:**219, 1971.

Weisman, A. D.: *On Dying and Denying.* New York: Behavioral Publications, 1972.

Weisman, A. D.: Is mourning necessary? in *Anticipatory Grief*, B. Schoenberg, A. C. Carr, A. H. Kutscher, D. Peretz, and I. K. Goldberg (Eds.). New York: Columbia University Press, 1974.

Weisman, A. D., and Hackett, T. P.: Predilection to death: Death and dying as a psychiatric problem. *Psychosom. Med.* **23**:232, 1961.
Weisman, A. D., and Worden, W. J.: Risk–rescue rating in suicide assessment. *Arch. Gen. Psychiatry* **26**:553, 1972.
World Health Organization: *The Prevention of Suicide.* Geneva, 1968.
Yudkin, S.: Children and death. *Lancet* **1**:37, 1967.
Zilboorg, G.: Considerations on suicide, with particular reference to that of the young. *Am. J. Orthopsychiatry* **7**:15, 1937.
Zilboorg, G.: The sense of immortality. *Psychoanal. Q.* **7**:171, 1938.
Zung, W. W. K.: Index of potential suicide (IPS): A rating scale for suicide prevention, in *The Prediction of Suicide*, A. T. Beck, H. L. P. Resnick, and D. J. Lettieri (Eds.). Bowie, Md.: Charles Press, 1974.

ROBERT M. VEATCH obtained his bachelor's degree in Divinity (magna cum laude) from the Harvard Divinity School and his M.A. and Ph.D. in medical ethics from Harvard University. Dr. Veatch is currently Professor of Medical Ethics at The Kennedy Institute of Ethics, Georgetown University. From 1970 to 1979 he was Senior Associate and Staff Director of the Research Group on Death and Dying and the Research Group on Ethics and Health Policy at the Hastings Center, Institute of Society, Ethics and the Life Sciences. At present he is a member of the Editorial Board of the *Journal of the American Medical Association*, a member of the Subcommittee on the Protection of Human Subjects at Montefiore Hospital, New York, contributing editor to *Hospital Physician*, and associate editor of the *Encyclopedia of Bioethics*.

His work and writing on death and dying and various aspects of medical ethics and health are widely recognized. In addition to scores of published papers on medical ethics he is the author of three books: *Death, Dying, and the Biological Revolution; Value-Freedom in Science and Technology;* and *Case Studies in Medical Ethics*. In addition he is editor or co-editor of seven other texts, including *The Teaching of Medical Ethics, Death Inside Out, Ethics and Health Policy,* and *Population Policy and Ethics: The American Experience*.

Ethics and the Care of the Child with Terminal Illness

ROBERT M. VEATCH

Health care is overwhelmingly a profession of pathos. The emotional strain of coping with crisis tells on nurses, physicians, and other health care providers just as it shatters the stability of the sick one and those close to him. Nothing, however, exceeds the agony of coping with the child in desperate and acute crisis—where those standing at the bedside know that the outcome may be severe handicap or even death. We lash out at the world demanding explanation, grasping for guidance. It is appropriate that a volume be published that deals with these most tragic cases. It is fitting that we have addressed the social, psychological, and religious dimensions of the terminally ill child.

But it is also appropriate to pause to explore the ethical quandaries of deciding about the care of the terminally ill child. Our agony is at least in part the ethical agony of not knowing what to do to show our care for the terminally ill child.

The general themes of the ethics of the care of the terminally ill are now debated regularly in a large and important literature by philosophers, theologians, nurses, physicians, and lay people who have known the burden of such decisions.[1] The debate has become increasingly sophisticated. The defensiveness, the fear of open discussion, the paternalism of only a decade ago is giving way to increasing openness and even moral clarity in all but the most difficult cases.

I would like to explore two remaining ethical problems in the debate about the care of the terminally ill. These are areas of controversy in-

---

ROBERT M. VEATCH ● The Kennedy Institute of Ethics, Georgetown University, Washington, D. C. 20057.

creasingly dominating the scholarly debate as well as legal and clinical discussion. We shall not be able to make ethically responsible decisions at the bedside until we have much more clear information about these problems than we do now. The first problem is whether it is possible or helpful to make a sharp ethical separation between the child who is inevitably dying and the one who could be saved by medical intervention but will die unless treated. The second is what standard to use in deciding what is the ethically appropriate treatment of a terminally ill child.

THE DYING CHILD WHO COULD LIVE

In that earlier era of defensiveness about death and dying some of us were lulled into ethical insensitivity by the confused and mistaken belief that ethically the obligation of the health care professional was to preserve life at all costs and that this justified providing every treatment that could possibly extend life. We now realize that that position is untenable for many reasons. First, the vast majority of medical professionals—nurses, physicians, and other health workers—never have held that the health professional's primary duty was to preserve life. It is not in the Hippocratic Oath, nor in the medieval codes, in Percival's Code, the AMA Principles of Ethics, the ANA's Codes for Nurses, or any of the international codes of ethics for health professionals. Health professionals have traditionally held that their primary moral obligation was to serve the interests of the patient, to avoid harm, or to give care in a way that produces the best balance of benefit over harm for the patient. That commitment itself is remarkably controversial and is rejected by many outside of the health professions but is at least a far cry from the simplistic notion that the health professional's duty is only to preserve life.[2] The simpler position is that preserving life is always the dominant duty and has been held by only a small minority both inside and outside of medicine. Even if the health professional ethic did require preserving life at all costs, there is no reason that should be binding on the rest of society. Most realize that there are often interventions possible that will pump a few more breaths into the dying child that are not morally required because they are not fitting. They turn the child into an object to be manipulated by a skilled technician. But children are more than objects; caring professionals are more than technicians. If the child gives us enough time, there are always interventions that are possible that we choose not to do, because they are no longer fitting.

Recently there has been some tendency to determine whether an intervention is appropriate by asking whether a patient is terminally ill.[3]

This leads to the first currently unresolved problem in the care of the terminally ill child. In the now famous case of Joseph Saikewicz, the terminally ill and severely retarded patient dying of leukemia, the court seemed to make a great deal of the fact that Mr. Saikewicz was dying and dying inevitably.[4] After affirming that the state has an interest in the preservation of human life, the court went on to argue that "there is a substantial distinction in the state's insistence that human life be saved where the affliction is curable, as opposed to the state interests where, as here, the issue is not whether, but when, for how long, and of what cost to the individual that life may be briefly extended."[5]

The California Natural Death Act, the bill passed in 1976 giving statutory authority for the first time to citizens to have some limited control of the way in which they die, relies on the same distinction between those inevitably dying and those who are still capable of living if treated.[6] Although the bill is seen by some as a useful first step in giving a greater voice to patients, it limits the legal authority of patients to have treatment stopped under the bill to cases where the patient is certified as terminally ill—when death will result regardless of the application of life-sustaining measures.

On reflection we discover that "terminal illness" can mean two quite different things. A report by the Subcommittee on Terminal Illness of the National Cancer Institute (NCI) defines terminal illness in such a way that it limits the term to conditions not amenable to cure or long term control.[7] Terminal illness is, according to the report, "characterized by a progressive deterioration with impairment of function and survival limited in time, usually from several days to a few months." In the definition itself the report fails to address whether survival is necessarily limited in time regardless of any possible intervention that may be attempted. An infant with spina bifida and myelomeningocele has impairment and limited survival time unless he is operated upon, but, with surgery, he may survive indefinitely (although he may be severely handicapped, physically and mentally).

Elsewhere in the NCI report, the committee does make clear that it also limits terminal illness to "disease entities not amenable to cure or long-term control." Once again a radical separation is made between the inevitably dying terminally ill and those who could live if treated.

The moral reasons given to justify nontreatment of the inevitably dying child, however, challenge the legitimacy of this sharp separation. A wide variety of moral traditions have offered two reasons for making treatment of the terminally ill patient expendable.[8] First, the treatment is expendable, or unreasonable, or elective, or heroic, or extraordinary (it is crucial to realize that these terms mean about the same thing in this context) when it is useless given the patient's condition. This means that surgery that will not correct the child's problem is expendable. It also means that an IV drip may be expendable as extraordinary given some patient's condition in spite of the usual but confusing common sense meaning of the word extraordinary. Treatments are expendable if they are useless given the patient's condition.

But if that is a criterion, it is foolish to add the requirement that the child must also be dying inevitably. To be sure the fact that a child is inevitably dying may change the evaluation of whether the treatment is useful, but a treatment that is useless given the patient's condition must certainly be expendable even if the patient is not inevitably dying. To hold otherwise would be to inflict all sorts of useless treatments on patients willy-nilly.

Second, the treatment is expendable if it will inflict a grave burden on the patient. It is conceivable that an intervention could be classified as useful for a particular patient but still inflict such agony on the patient that morally it should not be required of him. The excruciating pain of treatment of some burn patients could be waived by the patient or his agent on this basis.[9] The treatment of a patient in chronic renal failure by hemodialysis is apparently unbearably burdensome for some patients. American law, Catholic moral theology, and most secular commentators on the problem have held that overwhelming burden to the patient, whether the burden be physical or psychological, is enough to justify nontreatment decisions by the patient or his agent. In the case of the dialysis patient and some of the burn patients this is seen as justified even though the patient could live indefinitely if he continues to receive the treatment. There is an agony which no human should be obligated to bear. This moral justification of the nontreatment decision must hold whether the patient is inevitably dying or not. The decision-maker for the patient should not be constrained by the fact that the patient is only temporarily in a terminal state. These same justifications for letting the patient die—uselessness and grave burden—must apply whether the child is inevitably dying or will live if treated. Of course, the fact that some children will live if treated must be taken into account by the decisionmaker in evaluating questions of uselessness and grave burden, but the radical separation of the two kinds of cases is a serious mistake. If the decisionmaker has an obligation to serve the interests of the patient he or she must be permitted to make both treatment and nontreatment decisions based on these criteria of uselessness and grave burden regardless of whether the child is inevitably terminal. The issue is particularly critical with the child because many of the most difficult cases are those where the child is dying but could live indefinitely if treated: the spina bifida case, the child with Down syndrome and an intestinal atresia, the infant with a congenital cardiac malformation—they all could live if treated but still force the ethical question upon us of whether that treatment is fitting or appropriate or ethical.

SUBJECTIVE AND OBJECTIVE STANDARDS

We are still left with a second question of contemporary controversy: In making these decisions about the care of the terminally ill child who ought the decisionmaker to be and what standard ought that person use? I think it is crucial to distinguish two kinds of decisionmakers: primary and secondary decisionmakers. Once one realizes that the crucial questions are ethical ones, this has a great impact on the role of the health professional in making these decisions. The real question, to use the phrase of Pope Pius XII, is whether the treatment is fitting given the circumstances of persons, places, times, and cultures.[10] That is, the decisionmaker must be in a position to assess whether the treatment will be unbearable for the patient and whether the treatment is fitting given the beliefs, values, and commitments of the patient and his subcultural tradition. This means that the technical, medical aspects of the decisions are logically secondary. Of course, the decisionmaker must have accurate and empathetic information about the possibility of various interventions and their impacts. On that basis the decisionmaker will determine whether the proposed interventions are useless or burdensome, given the beliefs, values, and commitments of the patient, but the technical, medical information about diagnosis and prognosis are logically never sufficient to decide whether a treatment is morally appropriate.

This means that an objective basis for decisions about the treatment of terminally ill children will be deceptively difficult to find. In the debate over the treatment of children with spina bifida, physicians with enormous experience have been engaged in a heated controversy over the development of technical criteria for nontreatment, what they call "nonselection" for surgery, to close the opening of the spinal cord knowing that nontreatment means the almost certain death of the infant. John Lorber, for example, has proposed four technical criteria that he claims accurately predict the seriousness of an infant's permanent handicap should that infant have the spinal opening closed with surgery.[11] Lorber states bluntly that infants "who have any one or any combination of the following should not be given active treatment."[12] He goes on to list criteria including "grossly enlarged head, with maximal circumference of 2 cm or more above the 90th percentile related to birth weight."[13] His

criteria apparently are put forth to measure degree of permanent handicap. Head circumference measures extent of hydrocephaly and therefore future retardation. Height of spinal opening measures degree of
paralysis.

The criteria have been attacked by many people including Sherman
Stein, Luis Schut, and Mary Ames.[14] They seem to accept, however, the
idea that a set of technical measures could be found. They simply disagree with Lorber's empirical measures as accurate predictors of outcome. Both Lorber and his critics miss the crucial ethical point. One can
not move from a prediction of outcome to decisions about nontreatment
without making an ethical judgment. One must decide about whether
children with severe mental retardation ought to be allowed to live. To
attempt to move directly from technical measures to treatment decisions
is what I call the "technical criteria fallacy."[15]

Once one realizes that the technical information about diagnosis
and prognosis with and without treatment are merely data upon which
to make the moral decision about the appropriateness of treatment, one
sees why I conclude that health professionals are placed in a very unique
role in the decisionmaking process. They are crucial in many important
capacities, but they are what I call secondary decisionmakers. They provide data about medical possibilities, but as secondary decisionmakers
they do much more and serve many more important roles. They are
advisors to the decisionmakers and provide psychological as well as
moral support. They are companions in the agonizing process of decisionmaking. Occasionally they are called upon to do even more (more
about that later), but if the crucial elements of the decision are funda-
mentally moral judgments about what is fitting given the circumstance
of the patient, his beliefs, values, and his religious, philosophical and
cultural heritage, then health care professionals play a very special role
as counselors and advisors to the decisionmakers about only limited
aspects of the decisions about the care of the terminally ill patient.

The primary decisionmakers must be those with the most intimate
knowledge of the patient and/or those with the perspective to determine
what is most likely to be in the patient's best interest given the patient's
peculiar commitments, values, and circumstances.

Identifying who that person is is a difficult task. In the case of the
competent adult patient we have reached a moral and legal conclusion
that the patient himself should be the primary decisionmaker. He normally is in the best position to judge whether the proposed treatment
would be fitting, given his peculiar beliefs and values, or whether the
treatment would be so burdensome that he ought to refuse it.

Realistically, however, even the competent patient may make a mistake in such judgments. It is possible to say that objectively a treatment may be in a patient's best interest even though he believes it is not. Rejection of the use of technical criteria as the basis for an objective decision about treatment of the terminally ill does not necessarily rule out the possibility that there is some morally objective basis for saying that a treatment is in the patient's best interest.

In the case of the competent patient, however, it may make little policy difference what is objectively in the patient's best interest. We value personal freedom and the principle of self-determination. Even if we had some infallible way of knowing what was objectively in a patient's best interest we would not let that information override the patient's subjective determination of what treatment should be refused. That competent patient still is under a moral obligation to apply the criteria of uselessness and burdensomeness in deciding what treatment he should accept and what he should refuse, but he is the primary decisionmaker and his subjective assessment will carry the day.

With children, however, the picture is much more complicated. Someone must choose, and it must be someone other than the patient. In cases of older children their own views will be taken into account by the primary decisionmaker, but they cannot necessarily be decisive unless the patient is judged mature enough to take on that burden of decisionmaking himself. The courts have found children as young as sixteen capable of making decisions to refuse the amputation of a cancerous leg, but barring such a determination someone else must make the decision for the patient.

The question forced upon us in a fresh and traumatic way recently is: Ought the decisionmaker be obligated to come as close as possible to choosing the course that is objectively in the patient's best interest, or can the decisionmaker be given some latitude to incorporate subjective factors, just as the competent patient would?

The problem is made worse by the realization that there is no agreed-upon method for knowing exactly what is objectively most in a young child's best interest. In a democratic society we acknowledge the fallibility of every human being. We know that each of us could be wrong in assessing the best interests of another (or even our own best interests). We still are convinced that some decisionmaking methods are likely to come closer to objectively right decisions than others. In decisions about treatments of terminally ill children we hold that a court is the best mechanism we have for approaching objectivity. Of course, the judge is also a fallible human, but we believe that due process, the adversarial procedures, and the awesome sense of responsibility that accompanies the role of judge and public scrutiny combine to be the best we have at approaching objectivity. That is why we often turn to the courts to arbitrate such life and death questions. The judge's task is to determine as closely as possible what is objectively in the child's best interest, not what he himself would prefer.

But in decisions about the terminal care of a child, do we always want to follow the course that most closely approximates an objective judgment about what is in the child's interest? Strange as it may seem at first, I think we do not. There *are* limited circumstances where the most reasonable course must be followed for a terminally ill child, where we would turn to the courts to make a decisive judgment about whether treatment should be rendered. These are cases where the patient is incompetent and where he does not stand within a caring familial unit. Joseph Saikewicz had no family members willing to take on the role of responsible primary decisionmakers. It was appropriate that we turned to the court to find out what the most reasonable course to follow was. The court had the awesome task of choosing the one course that was most likely to be in Mr. Saikewicz's best interest.

But what about cases where the child stands within a caring familial unit? We affirm the importance of the family as a basic unit within our society. Familial autonomy is a principle fundamental to our tradition just as individual autonomy and self-determination are.[16] The caring family accepts the social responsibility for the child and makes countless decisions about its nurture throughout its immature life. Parents are not only permitted but expected to incorporate the familial subculture into their decisions. They teach the child a unique set of religious beliefs and ethical values, whether they be atheism, the Catholicism of Karen Quinlan's family that led them to refuse a respirator for their daughter, or the Jehovah's Witness theology that will lead the child to refuse a life-saving blood transfusion as soon as he is able and competent to do so.

We permit parents to teach their children sets of values that are not shared by the rest of society. In doing so, we permit them to constantly make choices that seem not necessarily to be objectively the most reasonable choices about what is in their child's best interests.

To be sure, we place severe limits on parental discretion. Parents may teach their children the Jehovah's Witness theology, but they cannot themselves refuse a life-saving blood transfusion for the child while he is still a minor. The parent may choose deviant school systems, but cannot, as the Amish would, choose no schooling at all. When a family is available and willing to serve as responsible guardians of the interests of their child, then the terminally ill child's care will primarily be in their hands. Our social commitment to the importance of familial autonomy requires this. They are charged with the right and the obligation of making judgments about what they think is in the child's best interest even if they choose a course that seems objectively not to be the most reasonable course. They are—they must be—the primary decisionmakers. They are given this right and responsibility by society until they demonstrate that they have deviated so far from what is the most reasonable course that their judgment must be reviewed.

At this point we discover an additional crucial role for the health professional. He or she must be among those charged with the responsibility of seeking out a review of the parental decision in those rare, tragic cases when it is believed that the parent is acting foolishly, maliciously, or irresponsibly.

Like a relative or concerned neighbor or clergyman, health professionals bear the caring responsibility of protecting the child from parental decisions that exceed the limits of reasonableness. As secondary decisionmakers they have the obligation to counsel the parent, to provide information about alternative courses of action, and finally, if necessary, to seek review of the decision. They do not in themselves have the authority to overturn the parental decision. We entrust judges with that responsibility, but the secondary decisionmakers have the ethical obligation of seeing that the review is undertaken.

In cases where a judge reviews the initial parental decision about the care of a terminally ill child, his task will be to determine if they are acting responsibly as guardians. His role here, it seems to me, must be different from the cases where the child has no responsible family member stepping forward to play the role of primary decisionmaker. In the cases of the incompetent patient without family members willing to serve as guardian, the judge has no alternative but to seek the most reasonable course, the one that objectively serves the patient's interests as well as possible. In the case of the child with family standing by, the judge must ask a somewhat different question. He must ask if the parental decision is sufficiently reasonable that it can be tolerated by the court as sufficiently in the child's best interest that the family should be permitted to make it within the limits of familial autonomy. Family members may make a range of decisions but a range within rather narrow bounds of the limits of reasonableness. When those limits are exceeded, the judge will have to appoint a new guardian to protect the interests of the child. That is why in Karen Quinlan's case the judge never determined that stopping the respirator was the most reasonable course, the one most likely to be in Karen's best interest. He never faced that question. Rather, he had to determine only whether Mr. Quinlan was

acting responsibly as guardian, whether his proposed course was sufficiently within the limits of reasonableness, that the subjective factors
taken into account by Mr. Quinlan were acceptable.

This provides a rather complex answer to the question of who
should make decisions about the care of the terminally ill child and what
standard should be used. It is an answer that sees important but different
roles for health professionals, parents, and the courts because of the
different ethical positions each is in. Parents have the obligation to serve
the interests of their terminally ill child by choosing a course that is in
their child's best interests given the familial beliefs and values although
it is not necessarily the most objectively reasonable course. Judges have
the responsibility of reviewing parental decisions in cases where someone has questioned whether the parent has exceeded the limits of reasonableness. They also have the obligation to serve as primary decisionmakers in those tragic cases where children or other incompetent
patients have no family members willing to serve as their guardians. In
these cases the judge must seek the *most* reasonable course. Health
professionals have a rather different role: They must first provide necessary medical information for making the decision. Then they must be
available as trusted counselors and caregivers for the patient and family.
Finally, in rare cases they must fulfill their obligation to their patients by
seeking court review in cases where they believe the guardian for the
patient has exceeded the bounds of reasonableness in deciding about
the child's terminal care.

REFERENCES AND NOTES

1. Beauchamp, T. L., and Perlin, S. (Eds.): *Ethical Issues in Death and Dying*. Englewood
   Cliffs, New Jersey: Prentice-Hall, 1978.
   Fuchs, V. R.: *Who Shall Live?* New York: Basic Books, 1974.
   Jonsen, A. R., and Garland, M. J. (Eds.): *Ethics of Newborn Intensive Care*. San Francisco:
   Health Policy Program, School of Medicine, University of California and Berkeley:
   Institute of Governmental Studies, University of California, 1976.
   Ramsey, P.: *Ethics at the Edges of Life—Medical and Legal Intersections*. New Haven: Yale
   University Press, 1978.
   Swinyard, C. A. (Ed.): *Decision Making and the Defective Newborn*. Springfield, Illinois:
   Charles C Thomas, 1978.
   Veatch, R. M.: *Death, Dying, and the Biological Revolution—Our Last Quest for Respon-
   sibility*. New Haven: Yale University Press, 1976.
   Weir, R. F. (Ed.): *Ethical Issues in Death and Dying*. New York: Columbia University
   Press, 1977.
2. Veatch, R. M.: The Hippocratic ethic: Consequentialism, individualism, and pater-
   nalism, in *No Rush to Judgment: Essays on Medical Ethics*, David H. Smith and Linda M.

Bernstein (Eds.). Bloomington, Indiana: The Poynter Center, Indiana University, 1978, pp. 238–265.

3. Ramsey: *Ethics at the Edges of Life*, pp. 171–181.
4. *Superintendant of Belchertown State School and Another v. Joseph Saikewicz*. Mass. Adv. Sh. 2461, 2501, 370 N.E. 2d 417. Massachusetts Supreme Judicial Court, November 28, 1977.
5. Ibid., p. 425.
6. The California Natural Death Act. Chapter 3.9, Part 1 of Division of the Health and Safety Code.
7. U.S. Department of Health, Education, and Welfare: *The Interagency Committee on New Therapies for Pain and Discomfort—Report to the White House*. Washington, D.C.: U.S. Government Printing Office, May 1979, pp. IV–3.
8. See Veatch, *Death, Dying, and the Biological Revolution*, pp. 105–10.
9. Imbus, S. H., and Zawacki, B. Z.: Autonomy for burn patients when survival is unprecedented. *N. Engl. J. Med.* **297**:308, 1977.
10. Pope Pius XII: The prolongation of life. *The Pope Speaks* **4**:395, 1958.
11. Lorber, J.: Results of treatment of myelomeningocele. *Dev. Med. Child Neurol.* **13**:290, 1971.
12. Lorber, J.: Selective treatment of myelomeningocele: To treat or not to treat? *Pediatrics* **53**:307, 1974.
13. Lorber, J.: Early results of selective treatment of spina bifida cystica. *Br. Med. J.* **4**:201, 1973.
14. Stein, Sherman C., Schut, Luis, and Ames, Mary D.: Selection for early treatment in myelomeningocele: A retrospective analysis of various selection procedures. *Pediatrics* **54**:556, 1974.
15. Veatch, R. M.: The technical criteria fallacy. *Hastings Center Rep.* **7**:15, 1977.
16. See, *In the Matter of Karen Quinlan*, 70 N.J. 10, 355 A. 2d 647. Supreme Court of New Jersey, March 31, 1976.

DISCUSSION

MODERATOR: IDA MARTINSON

H.S. SCHIFF: We've been talking a lot about moral decisions. I'm very curious whether any follow-up work has been done with Jehovah's Witnesses families where a child has been transfused. Is the child considered of a lower status in the family or considered tainted in some way? What is the quality of that child's life if the court's decision prevails?

R.N. VEATCH: To my knowledge there are no studies with the detail that you suggest. There is a complication in the Jehovah's Witnesses cases. Some Jehovah's Witnesses, but not all, will accept a transfusion when it is required by the court as not interfering with the theological commitments of the sect. I'm sure that that kind of variation in the belief system would have a bearing on the acceptance of the child. Even, however, if we were to conclude by such a study that the child was somehow placed in an inferior status as a result of having the transfusion, I don't see any other alternative for a society but to intervene when the limits of what I call reasonableness have been exceeded so far. We simply can not permit parents to let their children die in such circumstances. We have to intervene as a society and say that someone else takes on the guardian role.

SPEAKER: We took care of a young adolescent girl who had a blood dyscrasia and who was a Jehovah's Witness. Her parents had refused treatment as did the girl. The hospital went to court and obtained permission to treat her. The girl ultimately died. But the last days before she died, she was extremely emotionally distraught because she had to accept blood products. This was also very difficult for the nursing staff. We had to administer the blood. We felt that it was a personal affront to this girl to do it. While she was only 14, she was emotionally mature and very intelligent. I would like people to keep that in mind when we talk about the law and the rights of parents and children.

R.N. VEATCH: The case you described raises a subtlety in both the legal and ethical discussion that we should highlight. Our earlier statements about the rights of minors to refuse treatment generally holds. As we begin to deal with mature minors, of which a 14-year-old might possibly be one, cases arise where, through individual court review, a treatment refusal has been honored. This has occurred not on the basis of the parental refusal (parents don't have that right) but in special situations where the court became convinced that the minor understood the decision and could take responsibility for it. In a situation of a 14-year-old with a long history of Jehovah's Witness beliefs, it's conceivable that a court could take that out. I would not assume immediately that a court would order blood transfusions for any 14-year-old. It didn't in this circumstance. We are getting very close to the borderline where judges must agonize as the rest of us do. But there are certainly cases on record where minors have been permitted to refuse treatment on their own on the basis of being able to persuade the court that they fully understand and are mature enough to take responsibility for the decision. I think that that's an encouraging exception to the general rule that minors will have treatment ordered for them in such cases.

—————————11

GEORGE J. ANNAS is Associate Professor of Law and Medicine at the Boston University School of Medicine and Chief of the Health Law Section at the Boston University School of Public Health. He is the editor of *Medicolegal News* and *Nursing Law and Ethics*, both publications of the American Society of Law and Medicine, and writes a regular column entitled "Law and the Life Sciences" for the *Hastings Center Report*. For the past five years Professor Annas has served as Vice-Chairman of the Massachusetts Board of Registration in Medicine, and he is Chairman of the Board's Complaint Committee. His books include *The Rights of Hospital Patients* (Avon, 1975), *Genetics and the Law* (Plenum Press, 1976), *Genetics and the Law II* (Plenum Press, 1980), and *Informed Consent to Human Experimentation: The Subject's Dilemma* (Ballinger, 1977). With Leonard Glantz and Barbara Katz he has just completed *The Rights of Doctors, Nurses and Allied Health Professionals* (Avon, 1981) and is currently working on the second edition of *The Rights of Hospital Patients*.

11

Kids Are People Too, Sometimes
Parents vs. Children

GEORGE J. ANNAS

Parental authority and children's rights have been the subjects of many recent statutes and lawsuits. In early Greece and Rome, parents had absolute life and death authority over their children. And parents are enjoined in Ecclesiastics to "bow down" the children's necks. Indeed, it was not until the 1960s that states passed child abuse and neglect statutes designed to protect children from their parents. Courts also have routinely required parents to provide lifesaving blood transfusions to their children, even over the religious objections of their parents, and proscribe the use of nonapproved drugs, such as Laetrile, on children. Nor can parents any longer withhold lifesaving medical treatment with impunity.

Nonetheless parental authority to consent to or withhold medical treatment from children remains a murky area of the law. It is replete with tension between protecting the "best interests" of the child and fostering the traditional family by promoting parental authority. The problem is that while usually these interests can be upheld simultaneously, when the "mature" child opposes parental desires, a choice between them must often be made.

The difficulty in making this choice is portrayed by two recent decisions of the United States Supreme Court. While ostensibly based upon

GEORGE J. ANNAS • Boston University School of Medicine and Health Law Section, Boston University School of Public Health, Boston, Massachusetts 02118.

interpretations of the United States Constitution, the primary thrust of both opinions is a determination of whether the state statute under challenge as unconstitutional is likely to strengthen or weaken legitimate parental authority. Since there is no reason to believe the judges on the Supreme Court have any particular expertise on the family, this is a game any number can play. The cases are extreme: a daughter's decision to have an abortion against the wishes of her parents and the parents' decision to commit their child to a mental institution.

THE MINOR'S RIGHT TO AN ABORTION

The abortion case involved an issue the court had dealt with previously. It had decided a state could not give parents an absolute "veto power" over their daughter's decision to have an abortion.[1] A Massachusetts statute, which required that the daughter ask her parents' permission before an abortion and only after they refused would she be given the right to petition a court to decide if she were "mature" enough to make the decision herself or that an abortion would be in her "best interests," was struck down as being inconsistent with the no veto power rule. While eight of the nine justices agreed the statute as written is unconstitutional, four used the occasion to discuss the family and the limit on children's rights.

Conceding that children are "not beyond the protection of the Constitution" and that "neither the Fourteenth Amendment nor the Bill of Rights is for adults alone," the court, however, emphasized the authority of parents and their primary role in society:

> The unique role in our society of the family . . . requires that constitutional principles be applied with sensitivity and flexibility to the special needs of parents and children. . . . The duty to prepare the child for "additional obligations" . . . must be read to include the inculcation of moral standards, religious beliefs, and elements of good citizenship. . . . This affirmative process of teaching, guiding, and inspiring by precept and example is essential to the growth of young people into mature, socially responsible citizens.[2]

The court underlined the point with a line from a previous decision: "It is cardinal with us that the custody, care and nurture of the child reside first in the parents, whose primary function and freedom include *preparation for obligations the state can neither supply nor hinder*" (emphasis added by the court). This is followed by a discussion of the necessity of legal support for parental authority to aid parents in discharging their obligations. While the court concluded that the present Massachusetts statute is unconstitutional, it went out of its way to suggest how these defects could be corrected, (e.g., court access before parental notice), apparently believing that by so doing it was supporting the family.

The law remains, therefore, that the state cannot give parents the authority to absolutely prevent their pregnant daughters from having abortions, as their daughters have a constitutional right to make this decision in consultation with their physician. The daughter's right of self-determination was placed above parental authority.

COMMITMENT OF CHILDREN

Given this precedent, is a law which permits parents to commit their children to state mental hospitals, without any requirement of notice or a right to a hearing by the child, constitutional? The answer here is even more completely based on the Court's view of the family rather than its view of the Constitution than in the abortion decision. Chief Justice Burger wrote the opinion of a five-justice majority:

> Our jurisprudence historically has reflected Western Civilization concepts of the family as a unit with broad parental authority over minor children . . . parents generally have the right, coupled with the high duty, to recognize and prepare [their children] for additional obligations. . . . Surely this includes a "high duty" to recognize symptoms of illness and *to seek and follow medical advice*. The law's concept of the family rests on the presumption that parents possess what a child lacks in maturity, experience, and capacity for judgment required for making life's difficult decisions. [Emphasis supplied][3]

The Chief Justice then noted limits to parental authority and cited cases where parents put the child's physical or mental well-being in jeopardy. But he was unable to distinguish operations which are "not agreeable" to the child and "involve risks," like tonsillectomy or appendectomy, from involuntary commitment to a mental institution for an average term of 100 to 350 days. In his sweeping words, "Most children, *even in adolescence*, simply are not able to make sound judgments concerning many decisions, including their need for medical care or treatment. *Parents can and must make those judgments*" (emphasis supplied).

RECONCILING THE DECISIONS

Why is it that parents cannot decide about abortions for their minor daughters but can decide that their daughters (and sons) should be committed to mental institutions? There are three rationales offered implicitly or explicitly by the court, none of which are convincing. The first is that the decisions are different in nature; the second is that parental authority to commit only extends to "immature" minors; and the third is that physicians can perform the function of safeguarding the rights of minors in cases where parents are not acting in the best interests of their children:

1. Justice Burger suggests the decisions are different because, in the commitment case, parents have no "absolute right" to order commitment; the statute requires that the superintendent of the mental institution make an independent judgment that the child is mentally ill and that he can be helped by treatment. The problem with this explanation is that substantially an identical situation exists in the abortion decision. The court has previously made it clear that there is no "absolute right" to an abortion—the decision must be made in consultation with a physician, and only a physician may perform the abortion. The decisions are not persuasively distinguishable on this ground.

2. Since the plaintiffs in the commitment case, J.L. and J.R., are both very young children, it could be argued that the decision to commit can only be constitutionally made without a hearing if the child is too "immature" to make any intelligent choice himself, since minor women can only exercise their own choice to have an abortion if they are "mature and well-informed enough to make" an intelligent decision on their own. J.L. was admitted to a mental institution at the age of six shortly after his natural parents were divorced and his mother remarried. He was diagnosed as having a "hyperkinetic reaction to childhood." Two years later he was returned to his home for two months, but his parents were unable to control him and eventually relinquished their parental rights to the state altogether. J.R. was removed from his family at the age of three months and declared a neglected child. He was placed in seven foster homes and finally determined to be "borderline retarded," suffering from an "unsocialized, aggressive reaction to childhood."

These two children are obviously neither mature enough to make their own decisions, nor from the type of family the court seems to admire. Nevertheless, while the facts of the case could confine it to immature minors, the statute in question does not so limit parental authority nor does the overbroad language of the Chief Justice. At one point, as already noted, he talks about "most children, even in adolescence" not being able to make sound judgments; at another, of the availability of habeas corpus as a remedy, to be pursued by, among others, "the person detained in a facility."

PHYSICIANS AND COURTS

3. Finally, one could argue that the Court has entrusted physicians with protecting the constitutional rights of children in need of medical care. The language of the commitment case certainly bears this reading. Justice Burger not only uses this as his basis for distinguishing the cases, but also adds considerable language about the respective roles of courts and administrative agencies, as opposed to the role of physicians. His conclusion is that only physicians can decide if a child needs institutionalized care, and therefore this is a decision that courts have no business being routinely involved in:

> Medical diagnostic procedure is not the business of judges. *What is best for a child is an individual medical decision that must be left to the judgment of physicians in each case.* . . . The questions are essentially medical in character . . . we do not accept the notion that the shortcomings of specialists can always be avoided by shifting the decision from a trained specialist using the traditional tools of medical science to an untrained judge or administrative hearing officer after a judicial-type hearing. Even after a hearing, the nonspecialist decision-maker must make a medical-psychiatric decision. [Emphasis supplied]

The Chief Justice also argues that independent physicians are the child's best defense against getting "dumped" into an institution by parents more concerned with their own best interests than those of the child. Two points about this remarkable denigration of due process deserve mention. First, the hearing would, if it were more than a charade, involve the child's representative getting an expert of his own to examine the child and make a recommendation to the court. If the recommendation was different from that of the committing physician, either one must conclude that the case for commitment is not as strong as the Chief Justice seems to believe it generally will be, or that "experts" either have no "scientific" basis for their opinions in this field or will testify for whatever side pays them. Either way, there seems no reason to prefer an untested expert to one required to face an impartial judge and have his expertise questioned.

Second, while Justice Burger's language reads like a *carte blanche* to physicians to act on what they think is in the best interests of their minor patients (referred to them, of course, by their parents), the law can only be interpreted this way by ignoring the abortion case. In that case the four justices writing the opinion of the court quote Justice Stewart's previous opinion in which he discusses the role of the abortion clinic physician in looking out for the best interests of the pregnant minor with approval: "It seems unlikely that she will obtain counsel and support from the attending physician at an abortion clinic, where abortions for pregnant minors frequently take place." Indeed it does. It also appears unlikely that institutional psychiatrists will not be biased in favor of institutional care, just as surgeons are biased in favor of surgery.

An Alternative: The Court's View of the Family

Therefore, none of these rationales is entirely persuasive. An alternative suggestion is that the decisions are based on the Court's view of the family and how its strength can be maintained. Both decisions focus on the family and both purport to strengthen it. There is a problem with this rationale as well: It is one thing to say that parents should have, at least initially, the right to have their child committed (with the concurrence of a physician), it is quite a different thing to say that the state, standing *in loco parentis*, should have the same rights. Nevertheless, this is precisely the argument Justice Burger makes for the court. He concludes that wards of the state should receive the same constitutional treatment that nonwards receive, and since nonwards do not get a hearing prior to commitment, wards of the state should not get one either. In his words, "we cannot assume that when the State of Georgia has custody of a child it acts so differently from a natural parent in seeking medical assistance for the child." This argument puts theory above fact and also undercuts almost everything the Chief Justice had to say about the importance of the family initially. Justice Brennan, speaking for three dissenters, certainly seems correct when he notes that saying that a child does not have a right to a hearing prior to commitment because social workers are obligated by law to act in the best interests of children is "particularly unpersuasive." "With equal logic it could be argued that criminal trials are unnecessary since prosecutors are not supposed to prosecute innocent persons." And, he might have added, commitment hearings are unnecessary because psychiatrists are not supposed to commit children who are not suffering from mental illness that is treatable in an institutional setting.

I believe the decisions are not reconcilable. But the explanation for this need not be complex. The court is attempting primarily not to interpret the United States Constitution, but to decide what course is "best" for the country by attempting to decide what course is best for the family. The result is unsatisfactory and contradictory because the court does not know—as perhaps none of us know—what is wrong with the modern family and what can be done about it. Disputes over teenage pregnancy and the confinement of children to mental institutions are both often the product of failed families. It is a mistake to attempt to construct a model of an "ideal" family from these two tragic circumstances. The result is destined to be flawed and inconsistent.

THE PROBLEM FACED BY LOWER COURTS

Since the United States Supreme Court has yet to locate a unified theory of the family and parental authority in the area of personal medical decisionmaking, it should not be too surprising if lower court decisions on this matter are not always consistent either. General rules are easy to state. Courts will almost always permit "mature minors," i.e., minors who understand the nature and consequences of their decisions, to consent to treatment or refuse treatment themselves (consistent with the abortion decisions). On the other hand, parents will almost always be permitted to make such decisions, so long as they involve elective, nonemergency treatments, for their children.[4] In the area of emergency medicine, on the other hand, courts will almost always order treatment for children to save their lives, even over the religious objections of their parents.[5] The rationale is analogous to child neglect and has been stated by the United States Supreme Court in another context: "Parents have a right to become martyrs themselves but do not have a right to make martyrs of their children."[6]

Three recent cases illustrate the difficulty courts and society have with these concepts in specific cases. The first involved a young boy, Chad Green, who was dying of leukemia. He had been placed on a chemotherapeutic regime at a major Boston hospital. After a few months, his parents discontinued the medication and began using metabolic therapy. The physicians at the hospital learned of this and brought a "care and protection" petition to court, alleging child neglect and asking the court to order the treatment resumed. The court agreed, upon finding that the child would die without the chemotherapy and that with it he had better than a 50/50 chance of a complete cure.[7] In a subsequent proceeding, the court denied the parents' request to supplement the chemotherapy with Laetrile and other vitamins, and the Greens fled the state with their child.[8] He died a short time later in Mexico, while receiving Laetrile treatments.

A similar case arose in New York in which a seven-year-old was suffering from Hodgkins disease.[9] The parents took the child to a physician who recommended chemotherapy and radiation. They did not like this suggestion, and therefore consulted with another physician who agreed to treat their child with Laetrile, megavitamins, and nutritional therapy. The New York Department of Child Welfare argued in court that this treatment was child neglect. Unlike the Massachusetts courts, the New York courts disagreed. New York's highest court ruled in favor of the family's decision. Nevertheless, the family took the child to Puerto Rico for treatment and he died a short time later.

While these cases reached different outcomes, it is important to note two apparently crucial differences. First, Chad Green's condition was deteriorating and all the medical experts agreed that he would die without chemotherapy. The other child's cancer, on the other hand, was stable and the family agreed that they would use conventional therapy if their child got worse. Second, this child was under the care of a licensed New York physician; no Massachusetts physician would care for Chad without chemotherapy nor testify that it was acceptable medical practice to do so. Accordingly, it seems safe to say that if parental medical decisions go against *unanimous* medical opinion, the parents are not likely to be upheld in the courts. On the other hand, if they can find a licensed physician to treat their child consistent with their wishes, and the child is not in danger of imminent death, courts will be inclined to permit the parents to make the decision.

A third case, which I believe was wrongly decided, illustrates the difficulties of any hard and fast rules in this area. That case involved a 12-year-old Down syndrome child named Phillip Becker. He was diagnosed as having a heart defect, and the physicians agreed that if he did not have open heart surgery to correct it (a routine operation involving less than 10% mortality or morbidity), he would die a slow and lingering death over a number of years. The parents opposed the surgery and would not consent to it because they believed their child's life was not worth prolonging and did not want him to survive them. The court, after hearing from the parents and two physicians, sided with the parents. While admitting the decision was troubling, the trial judge found the procedure "elective" and thought the parents' decision was "within the range of debatable options."[10] His decision not to intervene was affirmed by a California Appeals Court. This case stretches parental autonomy to the limits, as it permitted the parents to make a life and death decision even though the child had *never* lived at home with the parents but had been institutionalized since birth. The case also illustrates how far at least some courts will go to foster what they see as "the family." The judge in the Becker case said he was "worried about 1984" and the death of the family. What he should have been worrying about—and what we all should be worrying about—is the protection of children who are unable to protect themselves. Sometimes this necessitates overruling parental decisions.

Courts are mirrors of society, but sometimes they lose sight of their obligations to protect the helpless and disadvantaged. We can sympathize with their difficult decisions, but we should insist that their decisions be principled and that all errors be made on the side of life. Many decisions that courts have made in the name of the family illustrate Mark Twain's famous remark: "The more you explain it, the more I don't understand it."

References and Notes

1. *Planned Parenthood of Missouri v. Danforth,* 428 U.S. 52 (1976).
2. *Belotti v. Baird,* 443 U.S. 622 (1979).
3. *Parham v. J.L. & J.R.,* 442 U.S. 584 (1979).
4. See generally, Annas, G.J.: *The Rights of Hospital Patients.* New York: Avon Books, 1975, pp. 137–140.
5. Ibid., p. 138. Also see *State v. Perricone,* 37 N.J. 463, 181 A.2d 751 (1962).
6. *Prince v. Massachusetts,* 321 U.S. 158, 170 (1944).
7. *Custody of a Minor,* 1978 Mass. Adv. Sh. 2002, 379 N.E.2d 1053 (1978).
8. *Custody of a Minor,* 1979 Mass. Adv. Sh. 2124, 393 N.E.2d 836 (1979).
9. *In re Hofbauer,* 65 A.D. 2d 108, 411 N.Y.S. 2d 416 (App. Div. 1978), *aff'd,* 47 N.Y. 2d 648, 393 N.E. 2d 1009, 419 N.Y.S. 2d 936 (1979); also see Horwitz, E.T.: Of love and Laetrile: Medical decision making in the child's best interests, *Am. J. Law Med.* 5:271, 1979.
10. *In re Phillip B.,* 92 Cal. App. 3d 796, 156 Cal. Rptr. 48 (1979); also see Annas, G.J.: The courts and Phillip Becker, *Hastings Center Rep.* 9:18, 1979, and Robertson, J.: Involuntary euthanasia of defective newborns: A legal analysis, *Stanford L. Rev.* 27:213, 1975.

Discussion

Moderator: Ida Martinson

Speaker: What are a child's rights to die? I'm a bereaved parent. My child died after 8 years of cardiac disease and he decided himself that he wanted his fourth open heart surgery. We went along with it, the hospital went along with it, everybody went along with it. But it had been his decision that he didn't want to take his digitalis anymore.

G.J. Annas: There are two answers: the real answer and the legal answer. It's unlikely that a child would hire an attorney. If you were against him and didn't want to let him die, and if he hired an attorney to try and go against your consent to treatment, he would undoubtedly lose. The parental authority to consent to standard medical treatment would outweigh a minor child's desire to die. That is the legal interpretation. Whether that's right or not, I don't know. Obviously it's going to depend a little on the child's age. The closer they get to 18 the more the courts are likely to do what the child wants. If there's a question at all that the child lacks maturity to make a judgment, the court is always going to side with the parents. I think you're right to make the decision as a parent as to whether you believe what your child wants. If you believe that's really what he wants, then that's what you should do. But the law looks at that differently. The case would never get to court unless somehow your child hired a lawyer. It's just hard to envision that kind of case.

Speaker: Is there a legal age of consent for youngsters who decide they want no further treatment, no respirator, and so forth?

G.J. Annas: The general age of consent is 18. If it's elective surgery, the courts will allow for delay until they are 18, if no serious side effects or permanent harm would result. If it can't be postponed, in general, the parents will be permitted to make the decision if it's consistent with standard medical care.

Speaker: If the child believes very strongly that he does not want this surgery done, can it still be waived?

G.J. Annas: Yes. The only current exception is abortion. The child has a legal right to make the decision for an abortion if she can understand the nature and consequences of her decision. Parents cannot interfere with that right but no other case has been litigated on something that has to be done and can't wait until the age of 18.

———————————12

THE REVEREND JOHN A. CARR was born in Pennsylvania in 1932. After a few childhood years in Pennsylvania and New York, he moved to, and grew up in, Derby, Connecticut. He was born with the congenital absence of both hands and one foot. After graduating from Ohio Wesleyan University with a B.A. in 1953 and from the Yale Divinity School with a B.D. in 1957, he spent ten years in the pastoral ministry of the New York Annual Conference of the United Methodist Church, serving churches in Westbury and Hauppauge, New York, and North Haven, Connecticut. He entered training for the hospital chaplaincy in 1966 and became a chaplain supervisor (teaching chaplain) in the Association for Clinical Pastoral Education, Inc., in 1970. He received his S.T.M. degree from the Yale Divinity School in 1968. Reverend Carr is currently on the staff of the Department of Religious Ministries at the Yale–New Haven Medical Center and is responsible for chaplaincy coverage at the Yale Psychiatric Institute in New Haven and Gaylord Hospital in Wallingford, where he trains seminarians, clergy, and laity in developing skills of hospital ministry.

He formerly taught a class on Personal Counseling and Crisis Ministry at the Yale Divinity School and served on the boards of directors for the M.S. Society and the Bureau of Homemakers Services. In addition, he has been appointed to the Connecticut Governor's Committee on the Employment of the Handicapped.

Currently, Reverend Carr is Co-Chairperson of the Task Force on Disabled Persons for the Consultation on Church Union (COCU); he is Chairperson of the Eastern Region of the Association for Clinical Pastoral Education and also serves in their National House of Delegates. He is a Certified Chaplain in the College of Chaplains of the American Protestant Hospital Association.

Reverend Carr has been interested in issues of racial and social justice during his parish ministry and chaplaincy. Recently he has been studying and speaking in the areas of healing and holistic health.

He is married to Mary Alice Shuman (Sammy). They have a son, David, and a daughter, Karen.

12

Coping with Handicap
Searching for the Boundaries

JOHN A. CARR

INTRODUCTION AND SCENE SETTING

All of us can truly learn and grow together if we participate in the challenge of entering into a living dialogue with others on the concerns that we are focused on in this volume. They are not empty, dull, or academic concerns (although we can often present them that way), but represent crucial and dynamic issues that have or will directly affect us all. One of the difficulties in joining in such a discussion is the very fact that it is personal. All of us are vulnerable and have loved ones who are vulnerable. We all will have to live with experiences of difficulty, pain, sadness, and disability in our own lives, either from our own life situations or from those close to us. Those close to us are also our own life situations.

I will not hesitate, therefore, to write both personally and autobiographically. I have lived with awareness of some of these concerns for over 45 years, and I am still growing and arriving at new understandings and insights. I have taken, and have been given, various perspectives on coping and living all these years with certain specific limitations or handicaps in my own life. That particular school of hard knocks has taught me, as it has so many others, some lessons that were both difficult and vital. I did not always like the homework assignments, but I expect I will have to meet such continuing assignments for the rest of my life.

For almost 25 years I have been involved in the life-rewarding, if not materially-rewarding, vocation of ministry. In 10 years of parish

JOHN A. CARR • The Yale–New Haven Medical Center, Yale Psychiatric Institute, New Haven, Connecticut 06511, and Gaylord Hospital, Wallingford, Connecticut 06492.

ministry, I was confronted again and again by situations in people's lives that brought them unexpected hardships and unanticipated sorrow. I found myself with families who were overcome by disappointments in plans as well as by termination of plans. I can remember many years ago sharing in a chilling send-off of a husband and father to an extended treatment center after he was diagnosed as having tuberculosis. None of us knew much about what that meant, and it seemed like we were sending someone off to Siberia, perhaps never to return. In those first 10 years of parish involvement, I saw very humble people deal with newly acquired life shocks and threats that they had never thought would happen to them or to their families. In these various situations the church and its pastor did not always provide comfort or healing, but the gradual realization dawned on me that the support and presence of others, the community of caring, was a significant factor in how people were able to pull themselves together and continue to live in spite of the misfortunes in their lives.

My education and experience continued as I changed my vocational focus from parish minister to hospital chaplain. In the last 14 years I have come to know people, patients, and their families who come from their homes to an institution, hoping for care and cure of slight and serious difficulties, with acute and chronic conditions that will temporarily or permanently affect their lives and their understanding of themselves. We who are the staff of such institutions can provide our expertise of knowledge, skill, procedures, and other resources, and all of the above is expected and necessary. We also provide the setting and climate where the important work of treatment, recovery or struggling toward success or failure, goes on. I continue to experience in these specialized settings that the person who understands that he is not alone is more able to take on whatever struggle his life situation has brought to him.

The significance of this recounting for me is that although the expectations, role, and circumstance might change, it is I who must share with and learn with others as we confront life experiences of a variety of degrees of importance, of threat, of surprise, and of challenge. It is in a blending of focus or interaction with others that direction and energy for each of our life struggles or journeys is derived.

GAINING PERSPECTIVE ON THE SCENE

Is it overly dramatic to use such terms as *journey* or *struggle* in developing the theme of coping with handicap? We are only talking about persons we know, and some of these persons are handicapped or limited and therefore not even fully complete persons. Hey! Wait a minute. We are not even talking about full and complete persons. We have just stumbled onto one of the dilemmas of assessment or evaluation, whether it be by oneself or by another, one close or distant in caring, or one trained or untrained in some professional degree. We now have to decide about the identity of the human who is living in the struggle. And in a paradoxical way, each struggle takes place both in isolation and in interaction with the community, be it with one, with a few, or with large numbers. The dimension of the struggle might change depending upon the variety of involvements but the intensity of the struggle does not change. For when a person is confronted with the unexpected that happens in his life to change it, that event marks him and determines him as different from those around him or those who are different from how he himself used to be. That person is truly in a struggle. And if it is a real struggle one cannot predict the outcome. One can assume but can never really know the results of a struggle until its conclusion.

Can we do anything to help someone while they are in their struggle for self-identity and self-realization, which occurs at the crucial moment when they realize or experience their brokenness, their trauma, their incompleteness? We can, but we must be very careful. I have found two important elements related to this period of realization in the struggle. One is the fact or actuality of the brokenness that is assaultive to the person. Our caring and treating, our attention to the wounds of body and mind are given with quality skill and resources in order to effect an increased condition of health or improvement. The other element calls us to involve and sensitize ourselves to the extent or state of the wounding, with a resultant understanding of the true sense of personhood. Does or has the wounding diminished the sense of the person's self worth, his sense of being fully a person? This becomes a struggle for ourselves as well as for the wounded or handicapped person because we have to seriously consider how we feel and think about those who have acquired or demonstrated brokenness. We perhaps even have some inner anxiety or fear that we are not simply doubting the limitations of the wounded other, but also of ourselves. This situation can lead to a whole variety of insights and actions. It can be humorous or tragic; it can be helpful or in itself debilitating. We can ignore or deny our own wounded condition and therefore lose an essential grasp of ourselves. We can become distracted to ourselves at the expense of a lack of care or treatment to the other who needs us there for response to him.

Here are a few examples that might even sound familiar to you. A staff person in a care-giving institution says, "Just leave me alone today. I have a miserable cold and I don't feel like relating to anyone." Yet that same staff person may expect pleasant cooperation from a patient who is still reeling from a loss of an important body part or of a very familiar life function or identity. We might hear, if there is a place for honest confession on the part of staff, that loss of sleep really leaves them insensitive and awkward, but the patient who has been kept from good rest by assaults to his body or mind is expected to respond to staff with proper attitude or decorum. We are really caught in denying to others what is so important to ourselves when we are hurting. We can only learn about life caring and coping if we realize that, in spite of the unevenness of condition or complaint, both persons are humans having human responses. It is not a case of the higher quality person, who is staff, treating a lower quality person, who is a patient. It is, in actuality, two persons who have different official assignments in an interaction but who share a common struggle of grasping the concept and essence of personhood. They find gnawing fears that suggest that limitation means failure at being a person or rejection from the company of other persons. Sydney Jourard, among others, in his collection of essays entitled *The Transparent Self*, reminds us that we can facilitate the reliability of the data coming from a care receiver, if not from any other, by how we are willing to let them see ourselves.[1] The self-sharing must not be more important than the space of the one receiving care, and it must certainly be open and sensitive and in relation to the one receiving care or at least hopeful of receiving it. Over time we develop many defenses to hide the human qualities of the handicapped.

I would like to give two further examples, one taken from inside almost every hospital I have ever visited and one from the public arena. I go to a patient's open door while a staff person is still present. (I might add that sometimes the work going on should have required that the door be closed.) I knock on the door and the staff person tells me to come in. Oh, what a helpful staff person because the patient cannot speak. Well, that is sometimes the case, but more often the patient who can speak for him- or herself is never really given a chance to invite others, or decide to invite others, into his/her space. Maybe there is even a question as to whose space it is.

The public arena example is similar in some respects. A friend of mine and her roommate were taking a vacation trip to Florida. She travels in a wheelchair, while the other woman does not. Although my friend is very articulate and shows that quality in her initiation of conversation and her management of her needs, all responses were made to the other woman. This first surprised and then appalled them both. They had to resort to bitter humor to deal with the repetition of the transactions that were occurring so artificially. "Does she take cream in her coffee?" "Would she like, does she want?" My friend said that by the end of her trip she felt much more debilitated. There was also somewhat of a strain between the two friends, as everyone else was giving their relationship different values and interpretations than they were.

The common factor in each case was that the patient, in the first example an actual patient, while in the second a woman in a wheelchair (who was actually a staff person), was treated as more disabled than they were, as less significant than they were. These are brief examples of how small and subtle or large and obvious the eroding of human dignity and full personhood can be. If a patient or any person aware of his own disability can fall into the trap of diminishing his worth he can certainly find a lot of unaware helpers surrounding him. As the old saying goes, "With friends like that, who needs enemies?"

What Is the Reality?

As we delve deeper into this subject of coping with handicap we are faced with the same questions that any handicapped person must seriously raise for himself if he is going to cope successfully. What is reality anyway? Not just what does it all mean, but what is it? I now want to share a very deep and personal conviction. It certainly is not uniquely mine, but it is essential to my own view of reality and my sense of personhood for myself and for others. Who is handicapped anyway? I was born without hands and without a foot, but I spent the first 30 years of my life proving that I was not handicapped. Then I began to accept the reality that I was and finally grew to the point of allowing myself to associate with others who had obvious physical handicaps.

In the old days, if someone else who was not obviously all there came into the collection of people I was part of, I would quietly resent it and avoid as much as possible any nearness or contact. Once I allowed myself to throw off such hindering or negative attitudes, I was truly less restricted, more free to grow and more fully be. I became less handicapped. I was more able to cope. My boundaries were expanded and I could move in many ways over more territory or space. Was I not therefore more of a person than before, even though I had a greater sense of my limitation?

The kind of personal movement to which I just referred has led me to realize what was reality for me. And this is the knowledge and conviction that I would bring to others and try to put deep inside them. It comes in three parts: (1) We are all handicapped. Everyone. (2) Not all our handicaps are obvious. (3) Many times the most obvious handicap is not the most painful or difficult one. If we can take these three points seriously, and I certainly do, we realize that even as we use labels and categories they are somewhat false distinctions. They are false at the same time that they are true because the issue is one of differences put into different perspectives. Persons who work with those designated as handicapped do not really stand completely outside the category of handicap. I think it could be said that most patients have some additional suffering or anxiety because of the handicaps or limitations or disability found in the makeup of the staff. This, of course, can be compounded if the patient or staff person is unaware of the staff person's truly handicapping condition.

I was in a coffee shop one morning when a woman came in with a small child who was obviously, by her conduct, her child. The child was apparently the product of a biracial sexual union, for while the mother was white, the child's father was probably black, although he was not present to confirm this. After they left, the man next to me, who did not know me, said, "Isn't that too bad." I replied, "What is?" He said, "What they are putting that poor child through," clearly talking about the child being biracial. I replied, with as much calm as I could muster, "Why, I guess that is up to us, isn't it?"

That is the hopeful possibility for any who deal with their own or someone else's limitation or disability. Coping with a handicap will depend upon how human interactions occur, to allow more or less progress toward meaningful life. If someone dropped books or packages, I would not hesitate to help pick them up, even if that person had the obvious use of both hands and all ten fingers. I would not be performing that act because that person was or was not handicapped, or to prove how able I was in spite of appearance, but because another human being had had an unexpected situation happen and because I could provide some of the remedy. Even that simple offer to help could prove to be an affront to him, but then that would be a sharing of information about a different kind of handicap on his part as well as sharing some level of his ability to cope. Our acts of helping to pick up dropped books or packages may be very superficial but they nevertheless become incorporated into the dropper's experience. Edgar Jackson, in a very helpful book, *Coping With Crises in Your Life*, states, "All of life is involved in crisis. The skills that we develop emerge from the total experiences of the past with all of its previous crises. How we have developed tends to determine how we will function in each new set of circumstances."[2] What a person can and will accept will be partly determined by what his experiences have allowed him to become thus far.

I have deliberately chosen an example of the person on the street dropping books because I have often been using institutional and care-giving settings and have used the term *patient*. I want us to be consciously aware of the fact that we are not confining ourselves to categories of patients. In that respect we want to ask ourselves, as I ask other hospital staff and as I ask patients, singularly and in groups, "Where is the magic line? Where does patienthood stop and personhood begin?" Should there and can there be a distinction and what does it mean? As we continue to ponder such questions, we will again realize that we are talking about ourselves as well as others and we are dealing with fluid and, hopefully, growing concepts and not fixed, rigid categories. This leads to what I feel needs significant emphasis and work as we try to understand how one, anyone, copes with crisis and handicap.

Searching for the Boundaries—A Shared Task

I have tried to convey in this material not only the fact that it is inaccurate to think of only special people as handicapped, but also that the process of finding boundaries for identity, behavior, or capability is not sufficiently worked on, or at, in isolation. I need to use with deliberateness the concept of *dialogue*. I am convinced that the only ultimately proven progress made in determining one's growth or potential, one's self-understanding and identity, comes via the experience of dialogue. For me, dialogue means more than words back and forth. It means careful listening, speaking, and attention back and forth. It means sweat and tears and commitment to more of the same in continuing negotiated efforts to come to an understanding and then push beyond to a different and more fulfilling understanding. I am indebted to a person who was a patient and who introduced me to Erving Goffman's book, *Stigma: Notes on the Management of Spoiled Identity*. So much material in this book is helpful in considering the issues before us today. As Goffman very carefully describes, and provides analyses of, the factors pertaining to those who carry some kind of limitation, mark, or stigma, he says, "the term stigma, then, will be used to refer to an attribute that is deeply discrediting, but it should be seen that a language of relationships, not attributes, is really needed."[3]

For me the development of relationships in a creative experience of dialogue is what finally determines where a person stands and the direction in which he will move. It is essentially the same work that any two persons must do, but when one or both have been given, or have accepted, the label of handicapped, that work more explicitly perhaps has for a goal, the sense of what possible reality or meaning in life that handicapped person will have. Again, any interaction developed through dialogue will have that result but may come as an unconscious or implicit goal. I use terms such as *sweat*, *tears*, and *negotiate* because dialogue is a very difficult and halting task. It is filled with terror because each person in the setting of dialogue is being confronted with how he and others see themselves. He is confronted with what he is, what he is not, and what he might become. Dialogue results in change, and the change, even if eventually considered positive, can be painful. If a person has finally had to make a painful adjustment to what he has become as a result of a newly acquired handicap, he is then asked again to move into more new or unknown territory only having the promise that it will not be finding his way back to the way it used to be.

In spite of, as well as because of, all the work of identifying conditions, treating conditions, and relating to persons in and with their given conditions, with labels of identity affixed as a result, the work of going or growing beyond that which is already known to what may be is continued with those chosen for, or choosing, dialogue. That is what I mean by boundaries. The move is toward the horizons of the possible. It may be the one who is handicapped who is pushing or resisting, or it may be the other. Tentative movement and experimentation may be a dance of risk, but also, one hopes, an intentional commitment to find where the possibilities for thinking about, feeling, and living life can be. And, as is true in dialogue, when one who is involved moves, so must the other. I conceive of the boundary that is being searched for as that area that moves one beyond his present given condition toward a greater possibility. It is in that unknown shadowy place where some may say that the handicapped person is denying, but he might say that he is seeking, his fuller potential. That boundary is never firm and sure, but its place is only determined by the quality and intentional commitment to the dialogue. All of us present can, and will in some way, determine the boundaries for those we relate to.

Lawrence LeShan, in an intriguing book called *Alternate Realities: The Search for the Full Human Being*, provocatively suggests that the determining of reality is a combination of discovering and creating reality.[4] We go out and learn what is given but we also bring our creativity to the journey and thus find what is real for us. Many would say that there is much music worth listening to; others would say that there is very little music worth listening to, and these groups would disagree probably on which is which. In our intentional dialogue we will involve what has meaning, what does not have meaning, and why. Only then can one decide if any specific life will have meaning. It is, of course, my own experience that the very entering into dialogue creates and nurtures meaning. I cannot consider quality of life without relationships, and the dialogue entered allows for an increasing potential. So as we search for the boundary we move back and forth across it to lesser and fuller areas that we claim for ourselves and the other in our dialogue claims for himself.

A Gift from the Handicapped Who Struggle at the Boundary

The persons who arrive at the boundary challenge each of us to discover and determine our own. We are shown that movement can be made and what seems difficult can be surpassed. Life that seems very empty might be found to be quite full. We can be called to our own boundary as we are called to move others toward finding their own. Each of these acts are gifts that can make our own life sweeter.

We are also provided with examples of boundary living by the handicapped that can later directly apply to ourselves. I once spoke to a group of people on the general topic of their health and needs, and I asked how many had ever been in a wheelchair. Two people raised their hands. I then asked how many could guarantee that at this time next year they would not be in a wheelchair. That is part of both the actuality and the mystery of life. Our attitudinal barriers that prevent us from entering into possible dialogues may come from our dread that we are also potentially the injured, more broken than now. In denying our efforts to fight for a world more open to the handicapped, whether we refer to architectural or attitudinal barriers, we may be denying ourselves accessible avenues we will need later.

Another gift is just to ponder the meaning of given (by birth) or acquired (because we are born in a finite condition) handicaps among us. As we reach the threshold of knowledge and technology so that we can eliminate the possibility of birth defects and have a less spoiled world, we are confronted by the fact of spoilage possible during life. I use the term *spoilage* because it provides some of the degree of feeling we have about the broken. But gifts are provided to us all by those that different societies would consider broken, and therefore we have a richer world. Who knows what our condition of life would be if every contribution from everyone considered undesirable by some would disappear. The gift from those who live on the boundary is the promise that our own world can be richer in gifts and meaning.

Again, let me use a personal example. I, as a person with an obvious handicap, might not lower the level of fear, intolerance, or anxiety found in a person who cannot accept my personhood because of my handicap. But I might soften him up just a little so that the next person makes a little progress. If, and when, the person who is threatened becomes more allowing, he becomes more a person fully able to relate.

To conclude, I would like to say two things about the use of language. As we hear words such as *disabled, limited, handicapped*, we are being subtly focused on that which is impaired and not on that which is full, has potential, is gift-producing. We say, "Is the glass half empty or half full?" Can we not wonder if the life in a person is emptying or fulfilling?

I also want to acknowledge the difficulty in correcting disagreeable habits and writing in sexist language. I would allow that every *he* and *his* could also be a *she* or *her*. The insensitivity to the mindset supported by sexist language is one more failure to fully enter into the living dialogue that pushes us all to more openly accept others as our equals in life. When we do not make ourselves sensitive to the affront, intended or otherwise, we allow a world of lesser possibility for some and therefore for all.

The churches and faith communities are trying to come to grips with how openness and acceptance are talked about but are not practiced in concrete situations. As a minister, I must see that the church that I am involved in enters the struggle for people's fullest potential. A new book just published as a Faith and Order Paper of the World Council of Churches demonstrates this concern from a variety of authors and places around the world. It is called *Partners in Life*.[5] That, in essence, is what our joining together in this volume seems to be all about; we are seeking to learn how to more effectively be partners in life. I would extend an invitation to each of us to look at our own readiness to participate in the dialogue and join in finding the boundaries for meaningful life.

References

1. Jourard, S.: *The Transparent Self*. Revised Edition. New York: Van Nostrand, 1971.
2. Jackson, E.: *Coping with Crises in Your Life*. New York: Hawthorn Books, 1974, p. 78.

3. Goffman, E.: *Stigma: Notes on the Management of Spoiled Identity*. Englewood Cliffs, New Jersey: Prentice Hall, 1963, p. 3.
4. LeShan, L.: *Alternate Realities: The Search for the Full Human Being*. New York: Ballantine Books, 1976.
5. Muller-Fahrenholz, G. (Ed.): *Partners in Life: The Handicapped and the Church*. Geneva: World Council of Churches, Faith and Order Paper No. 89, 1979.

DISCUSSION

MODERATOR: MARY S. CHALLELA

A. MILUNSKY: Reverend Carr has unique talents and abilities and has triumphed over his congenital handicap. What in your background, Reverend Carr, can you identify as the factors which enabled you ultimately to adapt to and to overcome your handicap? By recognizing those factors for us, we might be able to help others more effectively.

REVEREND J. CARR: Although I did not use many personal references, it was the area that I talked about—dialogue and community and support. My mother and father were very supportive. No one knew I was going to be born with a handicap until I was born. It wasn't expected by anyone including the doctor. There was no reason in my mother's pregnancy to suspect that I would be handicapped. They decided that I would be an only child. Not because they could not have other children, but they did not know how many resources I would need as I grew up. I was able to grow up rather normally without special schooling or aid, but they didn't know that ahead of time. My mother said when I was very young, "He is not going to be spoiled and he is not going to be sheltered." So I was thrust out into the community. As I grew and as I went into different places there was always the trauma of being different and new to somebody. But there was always a support group that would rally around. So when I talk about dialogue and support and coming into a relationship, I'm really talking about what happened in my own life. People were willing to know me as a person including, as was said yesterday so well, all the warts.

A. MILUNSKY: In schools, wherever they may be, there is this natural cruelty of children one to another. You must have experienced these problems. How can we teach our children to be sensitive to those of us with handicaps? And I, of course, know that we all have handicaps.

REVEREND J. CARR: I think in a variety of ways. I remember things in my childhood that made significant impressions. One was part of my own prejudice toward others. I remember as a small boy going to a movie. (Until 10 I grew up in a very small town, so everyone had a chance to get to know me gradually and I got to know everybody else gradually.) I saw a woman who came to town who had a rather disfigured face because of burns, and I can remember the terror and even the nightmares I had because of that. I later wondered what kind of terror and nightmares I might cause in other people. When I moved to Connecticut I was in sixth grade, one month into the school year. We moved on a Saturday morning. School started on Monday morning. I was thrust into a brand new school which had first through seventh grades and a common schoolyard. I'll never forget that morning in that schoolyard when all these kids saw me for the first time, and the comments like "Look at the funny kid" and "Look at the kid with the funny arms" or "broken arms." I found out all of a sudden that I had an instant peer group. The peer group was my other sixth grade classmates. They also probably had those same questions, but I was one of theirs. And they rallied around me and I probably made more friends in my own class as a result of their rallying because they were called forth to take care of one of their own. Not someone who is handicapped but someone who is a sixth grader. So again the whole issue is that no matter in what ways we are different or unique from each other, there are some ways in which we are common to each other. If you can have the courage to find out which ways you have in common with somebody else, whoever that somebody else is, you can provide for that person a support group and that becomes very important. So I guess helping others to find a support group will take care of some of the traumas in life. But there are nightmares. There are nightmares on your part and that of others. Hopefully, as their tolerance level rises as they get to know you as a human being, the nightmares might become dreams.

A. MILUNSKY: What, Reverend Carr, do you feel about having a handicapped person as an important resource person as part of the health care team? For example, I have patients who like yourself have severe physical handicaps. Is it useful or is it counterproductive to have someone like yourself sit down with parents of such a child?

REVEREND J. CARR: That will depend on the work you can do together and how you can try to develop that work. It's not automatically helpful to have a handicapped person participate. I can give two examples. At Gaylord Hospital, which is a physical rehabilitation hospital, I met an ex-patient one day in the hallway with whom I had never worked. He said, "Boy, you just turned my life around." I asked, "What do you mean? We never even worked together." he said, "I was in the cafeteria one day in my wheelchair and I saw you take your wallet in and out of your pocket to pay for your food and that just made me start thinking what will be possible for me, if he can do that." That was one instance. Another was when a person with whom I worked in a psychiatric hospital felt they were even more debilitated, because they saw what I could do and they were just so aware of what they couldn't do. It took a while to let them know that I grew up working at it. One quick example. When I was in high school, I went away to church camp at about 16 or 17 years of age. I tried to play ping-pong, which I did not do very well. A girl beat me at seven straight games of ping-pong. It kind of tells you how stupid and foolish I was to play that many games. I went away to college that fall and must confess that I may have achieved a Ph.D. if I had hit the books rather than the ping-pong table! I became a very good ping-pong player! Age and weight have crept up on me, but I'm still fairly good. When I went back as a counselor to that same church camp, the thing for the high school students to do would be to try to beat me. I still run into adults today who can remember playing ping-pong in the old days. So it depends on the relationships that develop. I would have to say (and this may be future job security) yes, I think it is helpful. But only if the hard work is done in finding out what the symbols and the tokens and the signs of a person mean to the hard work of getting to know each other.

13

MARY S. CHALLELA is Director of Nursing and Training at the Eunice Kennedy Shriver Center for Mental Retardation, a university-affiliated facility that provides interdisciplinary education for students and service to developmentally disabled clients. A graduate of Massachusetts General Hospital's School of Nursing, Dr. Challela received her baccalaureate, masters, and doctor of nursing science degrees from the Boston University School of Nursing. She is also a graduate of the post-master's program at the University of Washington Child Development Center in Seattle. Dr. Challela has conducted a variety of short-term training and continuing education courses for community health nurses throughout New England. She is an Assistant Clinical Professor at the Boston University School of Nursing and has participated in national conferences on the developmentally disabled child.

Helping Parents Cope with a Profoundly Mentally Retarded Child

MARY S. CHALLELA

Helping parents cope with a severely retarded child requires a variety of attitudes and skills, among which are objectivity, knowledge, empathy, and caring. It requires an understanding of the dynamic interplay between feelings and reality by the professional who enters into a helping relationship with the parents.

For the purpose of this chapter, *parental coping* is defined as managing the day-to-day activities of meeting the handicapped child's needs, the parents' own needs, and those of other children in the family, in a realistic manner. It includes continuing with their own lives in a way that is both growth-promoting and comfortable.

There are many tasks parents must learn if they have a severely mentally retarded child. These may range from the general caretaking activities of hygiene and feeding to more specific areas of promoting their child's potential development and independence. Before parents can be expected to assume any of these tasks effectively, they must first be allowed and encouraged to respond emotionally to the crisis of handicap.

MARY S. CHALLELA • Eunice Kennedy Shriver Center and Boston University School of Nursing, Waltham, Massachusetts 02154.

BIRTH AS A CRISIS EVENT

The birth of a first child often arouses many ambivalent feelings in parents, from the joy of having produced a new life to the somber acknowledgment of new responsibilities for which there has been little preparation. It may result in a reaction described by Caplan as a crisis or an upset in a steady state characterized by a period of disorganization.[1] In order to restore equilibrium, the parents must learn new coping methods and establish new patterns of relationships. Successful resolution of the crisis promotes personal growth, equips the individual to cope successfully with subsequent crises, and results in a higher level of personality reorganization.[2]

EFFECT ON THE FAMILY SYSTEM

The family may be conceptualized as a social system in which there are norms prescribing appropriate role behaviors. A crisis event such as the birth of a new member forces the reorganization of the family system. Roles are reassigned, new positions must be learned, and values have to be readjusted.[3]

Viewed within this framework, the birth of a baby generates a change in the system. The previous relationship between husband and wife, in which they were able to devote all their efforts and time to doing for each other, must be realigned. Each must now take on a dual role; wife/mother and husband/father. Now they must incorporate into their daily lives a third person, someone who is not only a stranger, but also completely dependent upon them, making many demands on their time, energy, and resources. The dyad becomes a triad.

The mother, who is usually the primary caretaker, has less time for her husband, who himself is usually eager to learn how to care for his infant. He learns to do many of the caretaking tasks as well as the fun things, such as play, thus freeing his wife to do other activities and enjoy some time for herself. However, even with all of these demands on his parents' time and energy, the infant soon becomes an integral part of the family unit. As the baby grows and develops, he changes in both appearance and behavior, responds to their care and efforts and provides much positive feedback. Eventually, the child learns to do things independently and becomes a source of pride and happy expectations for the future. Furthermore, he is living proof of his parents' self-worth and a measure of their immortality. Although there continue to be periods of stress and crisis, the parents have learned to cope and adapt to these new events as well as to anticipate them. Equilibrium becomes easier to reestablish as time goes on and the system retains its harmony.

EMOTIONAL IMPACT OF AN ABNORMAL CHILD

Consider then the reaction of parents to the birth of a handicapped child. Now they must cope with two crises: one, the birth itself, and two, the fact of not having reproduced a normal child. The abnormal child confronts the parents with their failure to produce a normal infant and is a severe blow to their self-esteem.[4] Society does not accept what is not perfect, particularly if the result cannot be "fixed or mended." More often than not, parents are advised to institutionalize a profoundly retarded child for the "good" of both the child and the family. Feelings of shame and guilt are experienced and often expressed by some parents.

Parents' initial reactions to a handicapped child are influenced by many factors: (1) how and when told of the abnormality, (2) their degree of social isolation, (3) the type and severity of handicap, (4) social class and education, (5) sex of parent, (6) attitudes of family and friends, and (7) information received from and attitudes of professionals. Furthermore, the handicapped child may be perceived as punishment for past transgressions and may cause increased stress on an existing dissatisfying marriage.[5] Parents are concerned with their ability to care for the child both physically and financially, with the effect on other siblings, and with the child's future. If the child is severely handicapped, requiring intervention by many therapists and frequent visits to hospitals or clinics, the parents may feel overwhelmed by the program of care they are expected to assume, the number of different professionals with whom they must endlessly interact, and the financial burden imposed. A handicapped child also presents the parents with the difficult task of having to face friends and relatives with an imperfect child. In addition to these factors, parents often have to live through the nightmare of not knowing whether their child will live or die; this is particularly the case when the newborn infant requires intensive care or immediate surgery to save his life. In one study of 275 families with handicapped infants, the parental adjustment at birth was based on the child's survival. The investigators reported that subsequently everything else was weighed against "he might have died."[6] In these instances, parents may not go through the therapeutic grief and mourning stages. One mother, whose infant required constant attention for management of seizures, was never able to discuss her feelings of disbelief and anger until the child was stabilized medically at 3 years of age. When visited by the nurse prior to a scheduled evaluation, the mother began to verbalize her long-buried feelings of sadness and anger.

With these facts in mind, it is essential to recognize that professionals have major responsibilities to the parents and child. Since severe handicaps can usually be diagnosed at birth or within the first year of life, the task falls on nurses, doctors, and social workers in newborn and pediatric settings. What are the needs of parents?

NEED FOR EMOTIONAL SUPPORT

Primarily, they need to know something about the condition and to talk about their feelings when the diagnosis is first made. Although literature states that all parents go through a mourning period characterized by grief and loss, not all parents go through the process in the same way or in the same time frame.

Parents of retarded children are not all alike in feelings, attitudes, and responses to life situations. Each set of parents is composed of individuals who themselves have had unique experiences and who have come from various backgrounds. They may be normal or maladjusted, just like parents of nonretarded children. To categorize or stereotype them is a disservice to their individuality. Some parents may cry; others may refuse to openly express themselves; still others choose to intellectualize the situation. Professionals need to reassure parents that whatever their reactions are, they are acceptable. It serves no purpose to tell them that all parents go through an adjustment period and explain what this is. Let parents do it in their own way, in their own style! Just as it is impossible to predict outcome for the child, it is also impossible to predict outcome for the parents. Olshansky states that all parents of retarded children have a particular type of reaction that is lifelong, which he calls "chronic sorrow," a pervading feeling of psychological grief.[7] How this is expressed varies from parent to parent. One mother of a Down syndrome child stated, "Sometimes we think she may be normal," because she was developing so well. This is not unusual in the case of a child with Down syndrome during the first year of life. Often it is not until parents have a normal child that they realize the extent of the disabled child's delays and retardation. Some parents seem to resent the progress of the normal younger sibling. They are reminded of the imperfection of the abnormal child, an imperfection they would like to deny, overlook, or rationalize.

NEED FOR INFORMATION

Parents may be afraid to ask the doctor for much information. It often falls on the nurse to help them with this task. How they are told and how the information is interpreted may have a long-lasting effect on their feelings about professionals.[5]

Although they may have been given information about the child's diagnosis and subsequent needs, it is not unusual for parents to seek a second opinion. Barsh found that the reasons for seeking a second opinion were based on needs for (1) an empathetic physician, (2) the authoritarian competency of professionals, (3) diagnostic sensitivity, (4) experienced security and realism, and (5) empathy for themselves and the child.[6] His sample consisted of children with diagnoses of cerebral palsy, organic brain damage, Down syndrome, blindness, and deafness. All of these reasons for seeking a second opinion given by parents point to the need for understanding, knowledgeable, caring, and honest professionals, whether it be at the time of birth or at subsequent periods of crisis and stress. In relation to information concerning the diagnosis, it is essential that parents be given time not only to absorb what is heard but also to ask questions. Professional jargon should be avoided. Information should be couched in familiar words or phrases. It should be repeated periodically using the same familiar terms. When possible, it is helpful to provide parents with written information that is appropriate to a lay person's level of education and understanding. Explanation should be geared to the parents' tolerance to accept and hear what is being said. Although it may not be possible to predict outcome or prognosis, parents can be encouraged to expect some progress, even though it may be limited to the child's learning head control or drinking from a cup. Accomplishment of simple tasks can instill a sense of pride in parents for their severely handicapped child's abilities.

TELLING FAMILIES AND FRIENDS

Suggestions or help on how to tell their families and friends is necessary, in addition to the parents' need for correct information regarding diagnosis and special care. It is advisable that family members be told as soon as possible so that the child's parents can deal with their reactions in a concrete fashion based on reality. Relatives should be encouraged to send the same acknowledgments they would at the birth of a normal child. It is painful for parents not to receive cards and gifts. Unfortunately, society has no built-in supports for parents of abnormal newborns and does not know how to respond to the situation. Parents of abnormal children themselves are reluctant to send out announcements because of their guilt and fear of what friends and relatives might think. As a result of this, parents experience loneliness and isolation. The nurse or other professionals involved with the family can help bridge the gap between the grieving parents and their relatives, particularly the child's grandparents. The way in which this is approached may have lasting results on future relationships. The clergy can have a powerful influence on the immediate and extended family's attitudes toward the birth of an imperfect child. Religious rites, such as the baptism ritual or circumcision, denote this child as worthy of inclusion in the total family life in all of its facets.

Concurrent with the psychological and emotional needs of parents as they cope with a disabled child are the concrete and immediate demands of learning how to care for the physical and emotional needs of the infant. The mentally retarded child differs from the physically handicapped child in that there is no cure or treatment that will alter the severity of his mental limitations. Also, a severely retarded child will often have some physical limitations due to his inability to learn from his environment as a direct result of abnormal neurological development. In this case he will need consistent and special assistance to develop some of the skills of his normal peers. He may also have a medical disability, such as a seizure disorder, which further interferes with development and is the focus of attention until control is effected.

The nurse's knowledge of medical problems and their management, as well as her knowledge of parental and infant psychological and emotional needs, identifies this professional as a critical support person at this time. The nurse also serves as the liaison between the parents and the physician, who, one hopes, is sensitive to the parents' anxieties and concerns.

During the child's first year of life, parents' major concerns include the physical aspects of care, integrating the child into the family routines, and how to interpret his cues. Feeding and sleep problems have been the major areas identified by parents of normal infants; these are also the major concerns of parents of profoundly retarded infants. These babies often have problems with tongue control and swallowing, so that feeding consumes a great deal of time and emotional energy; consequently, they may have nutritional deficiencies. Sleep habits are also often a source of problems, as the baby may be irritable, demanding, and given to sleeping in short naps; therefore, the parents receive very little sleep themselves. The child's lack of physical progress may be discouraging to parents; consequently, they tend to maintain the child's dependency instead of engaging him in developmental activities. Rather than beginning to teach the child how to learn a new skill, it is easier to do all the care themselves. The child may have problems reaching for a toy, which developmentally has relevance eventually to holding a bottle, a cracker, and a spoon. Parents may continue to do all the feeding activities and not spend any time in helping the young child learn how to reach and grasp; therefore, the parents' time will continue to be consumed by an activity the child might have eventually learned to do for himself. Actually, the parents think they are being "good" parents because no one has helped them to recognize the relationship between two different but related activities. To them being good parents is meeting the dependency needs of their child. He may forever remain a baby in their eyes. Realistically, there are many ambiguities involved in caring for the severely handicapped child. It may be difficult for parents to see the difference between overprotectiveness and attending to their child's special needs. It is the responsibility of the professional—whether teacher, nurse, social worker, physician, or therapist—to guide them toward this awareness.

CHILDHOOD AND ADOLESCENCE

As the child proceeds through the chronological stages of childhood, new crises arise for the parents as they realize again how different their child is. This is particularly evident during the early school years and again in adolescence. Young siblings pass him by, as do children of neighbors and friends. The chronic sorrow rises and falls during these times and in between them. If husband and wife have been supportive of each other during the early years, their relationship will continue to strengthen as together they cope with their feelings and the reality of their child's condition. It should be emphasized how important it is to help parents maintain an open communication system, to listen to each other, and to support one another.[8] Sometimes, mothers assume much of the care of the disabled child, which leaves the father on the outside. Most fathers want to be included in their child's program and will respond to the guidance of sensitive and caring professionals.

LONG-TERM PLANNING

The profoundly retarded child is one who will always require supervision and care; he will never be able to assume responsibility for all (or sometimes any) of his own needs. Therefore, parents are faced with a bleak future and often feel trapped with a never-ending sense of responsibility for care and protection. As the child grows in size, his physical care becomes much more burdensome than at a young age; therefore, it is important that parents be involved in setting up a program of activities that will help the child learn self-help skills, however limited, thus relieving themselves of some responsibility. At the same time the parents will receive positive feedback from the child when he responds to their efforts and becomes more independent. A child who can indicate his toileting needs relieves the parents of an enormous burden. Not only is their caretaking made easier, but finding a babysitter is facilitated as well. This will also expedite his fitting into community programs.

COMMUNITY PROGRAMS

In preparation for any program, it is recommended that the handicapped child be evaluated by a variety of specialists, preferably in an interdisciplinary setting. This prevents fragmentation of care, promotes better utilization of disciplines, and saves time for parents. This may be the time when the parents are first confronted with a varied group of professionals with different terminologies. How this situation is approached and handled may affect the parents' acceptance of ideas and their involvement in a program of care.

Coping with a profoundly disabled child takes the strength of the entire family. Families have need of periodic respite from these demands. Community agencies often can assist in providing either home assistance or alternative placement for periods of time. The latter allows the family to take vacations, attend to other family activities, and just live a normal life. Locating and planning for such services may be the responsibility of a nurse or social worker. Helping parents with the financial burden of these and ongoing services is a responsibility of society. Society, however, may be slow to respond if there is no advocate for the family. This role, too, is an essential one and may be assumed by an interested and knowledgeable individual.

Another way to help parents cope with a profoundly retarded child, in addition to guidance in child care, promotion of development, and integration of the child into the family unit, is to encourage them to join a self-help group of other parents of disabled children. This has proven beneficial to most parents and families. Together they resolve many of the problems that arise in parenting the severely handicapped child, become a formidable force for obtaining services through legislation and social pressure, serve as a socializing agency for each other, and reach out to other parents of disabled children.

SUMMARY

Helping parents of a profoundly retarded child is a challenge to professionals, beginning with the physician and nurse in the newborn nursery and continuing throughout the life cycle with the involvement of other disciplines. Parents need emotional support and counseling in dealing with the initial and subsequent crises, education in learning how to care for the child's special needs, guidance in their relationships with other family members, and continued interest and encouragement to pursue their individual growth as important members of society. It is essential, however, that professionals remain cognizant of the fact that families differ in life-style, levels of functioning, and ability to cope with crises. Therefore, ongoing assessment of these factors should serve as the basis for any intervention.

REFERENCES

1. Caplan, G.: *An Approach to Community Mental Health.* New York: Grune and Stratton, 1968, p. 18.
2. Hill, R.: Genergic features of families under stress. *Soc. Casework* **39**(2):139, 1958.
3. LeMasters, E. E: Parenthood as Crises, in *Crisis Intervention: Selected Readings,* Howard J. Parad (Ed.). New York: Family Service Association of America, 1965, p. 111.
4. Solnit, A. J., and Stark, M.: Mourning and the birth of a defective child. *Psychoanal. Study Child* **16**:523, 1961.
5. Cohen, P.: The impact of the handicapped child on the family. *Soc. Casework* **42**:137, 1962.
6. Barsh, R. H.: *The Parent of the Handicapped Child.* Springfield: Charles C Thomas, 1968.
7. Olshansky, S.: Chronic sorrow: A response to having a mentally defective child. *Soc. Casework* **42**:190, 1962.
8. Lepler, M.: Having a handicapped child. *Am. J. Mat. Child Nurs.* **3**(1):32.

BIBLIOGRAPHY

Drotar, D., Baskiewicz, A., Irvin, N., Kennell, J., and Klaus, M.: The adaptation of parents to the birth of an infant with a congenital malformation: A hypothetical model. *Pediatrics* **56**:710–717.

Young, R. K.: Chronic sorrow: Parents' response to the birth of a child with a defect. *Am. J. Mat. Child Nurs.* **2**(1):38.

14

ALLEN C. CROCKER is Director of the Developmental Evaluation Clinic at the Children's Hospital Medical Center in Boston (and Associate Professor of Pediatrics at the Harvard Medical School). His interests were initially in general pediatrics, and particularly in the area of children with inborn errors of metabolism. For the past dozen years he has been concerned with the evaluation of and the planning of programs for children with mental retardation and other developmental disorders, as well as the provision of support services for their families. He is also involved with parent groups and professional organizations that address the needs of exceptional children.

The Involvement of Siblings of Children with Handicaps

ALLEN C. CROCKER

Personal adaptation to the circumstances of human exceptionality is challenging for the primary individual and, in diverse aspects, for all others as well who are in that individual's life system. In this regard, major attention has generally been accorded to responses and adjustments of the first person, and then to the situation of the parents and other care-providers. Less acknowledgment is offered to significant fellow adventurers, the sisters and brothers. Insinuations of ruination and/ or redemption prevail, and evidence of a topical mythology can be found, but there is indeed a paucity of systematic information about outcomes. Assuredly a stress model exists here, yet one more in the long inventory of conditioning human experiences. "Am I my brother's keeper?" is the plaintive inquiry, countered by the platitudinous assurance, "He ain't heavy, Father, he's my brother. . . ."

ELEMENTS OF STRESS

When a child in a family has a troubling degree of handicap, including but not limited to problems of mental retardation, there are many particular emotions which can be generated in a sister or brother. These have been thoughtfully considered by Featherstone[1] and include concern, curiosity, protectiveness, frustration, sorrow, grief, anxiety, longing, unhappiness, jealousy, and resentment. There are other early types of family response, possibly in some degree universal, including

ALLEN C. CROCKER • Developmental Evaluation Clinic, Department of Medicine, The Children's Hospital Medical Center, and Harvard Medical School, Boston, Massachusetts 02115.

guilt, fear, loneliness, and anger. Of special relevance to siblings are matters of potential confusion, identification, and embarrassment.

Disturbing dynamic components that may be operative are as follows:

- Alteration of the normalcy of family rhythms, conformity, and image
- Competition for parental resources and attention
- Misconceptions that the normal child may carry about the causation or outcome of the sibling's handicap, including possible genetic insinuations
- Need for the normal young person to act in some regard as a surrogate parent at an early age, in addition to concerns about implications of future responsibility
- Obligation to meet enhanced parental expectations about personal accomplishments
- Bewilderment in response to parents' conflicts (with the parents, in turn, coming to realize that the children's adjustments mimic their own)

It can be seen that stress on normal sisters and brothers is, in fact, sustained by the presence of a system of contradictions or discordant concepts which they must bridge. These factors potentially place the vulnerable young person in a situation of pressured ambivalence, in which assistance will be needed for the resolution of cultural conflicts.

1. It is customary that parents expect the siblings to accept the special child and to come to realize the intrinsic benefits from his presence. In some contrast, the sisters or brothers may actually have a host of troubled feelings, including anxiety and jealousy. Children's books in this field invariably treat negative feelings of siblings as temporary.
2. As a force for normal mental health, the family (and society) may promulgate the philosophy that virtually any accomplishment is possible for the handicapped child, that she or he is eminently like all the others, and that near-normal interactions and activities should be expected. Lacking some of the special insights which the parents are employing, the siblings are confronted with the reality of the situation, its limitations, and its differences.
3. Families understandably establish a double system for child compliance and performance—with the normal young people required to adhere to rules, while the child with special needs has

a less rigid set of expectations. Furthermore, significant personal accomplishments of the nonhandicapped child may appear to be less valued than seemingly small gains achieved by the handicapped member.

4. It may be assumed by the parents that sisters and brothers have an inherent preoccupation with the needs and progress of their sibling, while, in fact, they have, and need to have, a broad range of interests in other directions as well.

5. It is usually taught that the presence of a common challenge will increase family communication and bonding, and, under ideal conditions, this would be so. For the family with significant turmoil, however, tensions may be rampant and communication blocked.

Elements of Strength

The listing above of confounding factors in the area of sibling adjustment could lead to inappropriate conclusions. It is, in fact, the common observation of clinical workers that siblings of children with handicaps have a most reassuring record of salutary accommodation. In Grossman's investigation on college students who had retarded siblings,[2] the classic conclusion is reported: "in our study, about as many students seemed to have benefited as were harmed." Cleveland and Miller[3] comment, "the majority of the siblings, in all of the variables explored, reported a positive adaptation to . . . the experiences surrounding having a retarded sibling." Lonsdale[4] notes that in the Plymouth study the percentage of siblings with behavioral or adjustment reactions is about the same as that occurring in the general population. Grossman, and others, have found that favorable adaptation is particularly well achieved when the sister or brother under discussion is older than the handicapped child, is of the opposite sex, and/or is part of a larger family.

Benefits accrued from a sibling relationship of this sort speak to broadened perspectives and enhanced humanism. In Featherstone's valuable summary,[1] she reports many instances of important final effects. These include (1) increased acceptance of the spectrum of human differences and new perceptions about the meaning of accomplishment, (2) a less casual acceptance of good health, (3) positive feelings about having assisted in the growth and progress of the handicapped child, and (4) security obtained from sharing in the strength of the parents' adjustment. Grossman[2] also speaks of the presence in the college stu-

dents with whom she worked of greater tolerance, compassion, awareness of prejudice and its consequences, and a more focused personality.
There are no good statistics available about the influence of family experience on final occupational or professional choice, but it appears that
a feeling of identification with the plight of a handicapped sibling and
a desire to have more accurate understanding has often led young people
to elect work in education and human services.

INTERPRETATION

It is clear that the circumstance of being a sister or brother of a child
who is handicapped represents a provocative encounter. The elements
of stress assuredly exist and are troubling to consider. Yet it is a docu-
mentation of the quality of the human spirit that, in the balance, the
elements of strength generally allow a positive resolution. As with other
intrinsic or extrinsic stress situations, the outcome will be a measure of
the vitality of the family, plus the supports that have been received.
Professional attention toward assistance for siblings has traditionally
been moderate, at best. In designing team services for families with
handicapped children, more investment should be planned for guidance
to parents about sibling adjustment and in direct work with these sisters
and brothers. Featherstone[1] urges that parents consider special support
for siblings, understanding for their feelings, and a freedom to work out
ways of personal integration. A report by Murphy et al.[5] underscores
the value of group meetings for siblings of children with Down syndrome, with the content of discussion guided by age level.

The hopeful comments of the Cannings[6] capture well the rewards
that can be felt in a family when a strong positive orientation is achieved
with a special child (in this instance their daughter, Martha, who has
Down syndrome):

> When a son, old enough to grow a beard, builds her a tiny chair just her
> size; when our oldest daughter comes home from college just because she
> is lonesome for Martha; . . . when our youngest son passes up play time so
> he can be the one to assemble the baby's Christmas sled, then I just know
> it is a good world as we grow together in love and share an experience which
> will make us better people. . . .

And, again:

> Because of Martha, I feel all members of our family can better cope with life.
> The children are learning valuable lessons that should help them each day
> of their lives. But, most of all, Martha has taught us how to love, and this
> is truly the most enriching and fulfilling experience.

ACKNOWLEDGMENT. This work was supported in part by Project 928, Maternal and Child Health Service, Department of Health, Education, and Welfare.

REFERENCES

1. Featherstone, H.: *A Difference in the Family: Life with a Disabled Child*. New York: Basic Books, 1980, pp. 137–176.
2. Grossman, F. K.: *Brothers and Sisters of Retarded Children*. Syracuse: Syracuse University Press, 1972.
3. Cleveland, D. W., and Miller, N.: Attitudes of life commitments of older siblings of mentally retarded adults. *Ment. Retard.* **15**:38, 1977.
4. Lonsdale, G.: Family life with a handicapped child: The parents speak. *Child Care, Health, Dev.* **4**:99, 1978.
5. Murphy, A., Pueschel, S., Duffy, T., and Brady, E.: Meeting with brothers and sisters of children with Down's syndrome. *Children Today* **March–April**:20, 1976.
6. Canning, C. D., and Canning, J. P., Jr.: *The Gift of Martha*. Boston: Children's Hospital Medical Center, 1975.

DISCUSSION

MODERATOR: MARY S. CHALLELA

M. HEHIR (Associate Director, Department of Programs and Division of Practice, Massachusetts Nurses Association): Dr. Mary Challela has well documented the role of the nurse in handling families and the handicapped child. She has just touched on the important work she is doing out at the Fernald State School. I wish she had time to document the importance of nursing within this role. But at a period in time last year, it was brought to the attention of the Division of Maternal–Child Health Nursing Practice at the Massachusetts Nurses Association that we needed some sort of a position statement to justify the role of the nurse on these early intervention programs. It took a long time to write this position statement which we now have. I wanted to share that with you. If anybody is interested in a position statement of this type justifying the role of the nurse on early intervention programs, please write to our association. You will be interested in knowing that the American Nurses Association has picked up this position statement in this, the International Year of the Child, and is going to publish it nationally.

C. AGUILAR: I think that the degree of the handicap has a lot to do with our reaction as parents, as health professionals, or just as people to that child who is affected. We have a little Mexican-American patient who is a deaf-mute and who has a malformed face. People often make the automatic though often erroneous assumption that when you have a physical handicap, there must be mental retardation associated with it. It seems to be a common reaction. We started following Brenda when she was 4 years old (she's 9 now). She came from a very supportive, very simple, very poor family, but they utilized and pooled all their resources to help this child. They had to go back to a little village in Mexico because the mother's aunt had died. We didn't hear from them. Six months later we found out that Brenda had been retained in Mexico. She had a strange-looking face, couldn't talk, and was considered to be retarded. They would not allow her mother to bring a retarded child into the United States. Part of our function as helping and caring professionals is to get involved with family units. It took about 6 months to convince the authorities that she was not retarded and should be allowed back into the United States.

P. ROSBOROUGH (Friends of the Sensorially Deprived, on the Health Planning Council of Greater Boston as a consumer): I am here because of the need for preventive medicine in relationship to specific learning disabilities. I have appreciated being here immensely. I would also like to point to the people who are not physically handicapped. To date the medical profession, generally, does not seem to realize the role that they should be playing in preventing learning disabilities. Some feel that the alternative health service needs of these youngsters does not belong in the school, and rightly, it does not, because these things should be dealt with prior to a child being three years of age. The only way we are going to have any cost containment is through prevention, and some of these means are available. If they were not inexpensive I would have never been able to rehabilitate myself. I spent 14 years in steel braces and dog-collars and 7 years unable to ride in an automobile. I have suffered the discrimination of parent, of town, and of situations, even today.

S. CROSSMAN (Coordinator, local Respite Care Agency for the Greater Marlboro Program): I would like to voice the great need for continuing education for the field of respite and to let the parents know, especially people in the professional field, and the student nurses, that it's a very needed area. Respite care serves to provide a short-term break for the parent, and this includes foster parents, or special guardians of a child with special needs, whether it be emotional, physical or mental needs. This can range from an hour or two up to 30 days unless more is sanctioned. There is such a great need for some of these parents. What we are trying to do is to keep the family nucleus together by saying to the parents, "we understand that your job is one of large magnitude and you are very special to be able to do it day after day," and to try and prevent "burn out." We are saying that we are here to help you. We have trained providers that will be glad to take your children for the evening in your home or in our homes, as I do, so that you can resume your super care.

M. CHALLELA: We find that to be a very essential service and we are constantly looking for such places to help families. They don't really need an excuse, they just need a rest.

B. FOSTER (Neonatal Intensive Care Unit, Birmingham, Alabama): I work as a family liaison. If you have a child with a suspected genetic handicap or possible brain damage, do you tell the parents immediately or do you wait until the chromosome test has been done or an exact diagnosis made? Is there a right way? Does it vary from family to family?

A. CROCKER: Dr. S. Pueschell at the Developmental Evaluation Clinic at the Boston Children's Hospital Medical Center did a survey about 5 years ago, of approximately 100 families who had a child with Down syndrome. This questionnaire survey looked into their final resolution and their attitudes and their feelings about how they were given professional services through the years. There were several dozen questions with much variation in the replies. There was only one question that was answered absolutely identically by all families. That question was, "Were you satisfied by the way in which the diagnosis of the child's problem was given to you?" One hundred percent said, "No." Now that could lead to the conclusion that the answer to your question, "Is there a right way?" is "No." However, the general feeling of most people is that the lack of a convincing and credible shared compassion and the mincing of words were probably the two factors that were most disturbing to those families. If the counselors or care-providers had been warm and straight there would have been a greater chance for success. Transmission communication of the early diagnostic information is very challenging and it is very commonly handled ineptly.

A. MILUNSKY: Published data relating to this question, "Were you satisfied with the timing of the communication when you were told your child had Down syndrome," all parents stated that the earlier they were told, the better. I think it would probably be standard practice in most centers where communication rapport is good that even before a diagnosis is established the physician would be communicating at least the concern that some tests will be done and that these are the specific concerns. Trying to protect the parents because they might not be able to handle such a differential diagnosis or the possibility of retardation is inadvisable. Not sharing concerns simply invokes the anger of the patient. Early, clear, warm, concerned communication is the key.

M. CHALLELA: Along with that are the support systems and assistance. There are things you can help them with.

J. HARTMAN: Do families that have a child with profound retardation ever reach a stage in their growth when the parents can cope with the situation in its entirety? I was told by someone that usually by the age of two years the family begins to deal with the situation and can make some future plans. Is this true?

M. CHALLELA: What we have found is that if the families have support from the very beginning, then many times before the age of two they can begin to deal with the situation and make realistic plans and go on with their own lives but incorporate the child into their plans. However I don't think there is any specific time.

D. COOPER (Pediatric Nurse Clinician, Massachusetts Rehabilitation Hospital): We have a family problem with a 14-year-old sibling of a child who is more or less in an unresponsive state from brain damage due to a drowning incident. He's been with us for 8 months. His brother, due to family pressure, has been forced to spend 7 days a week, 8 hours a day at his brother's bedside. To convince the family to let this child go back to lead somewhat of a normal life has been a problem. They refused to let him celebrate a birthday. The mother doesn't let him go out at all to play with friends or have any outside activities. It's really difficult to let go of this child due to their own insecurity about the younger child.

A. CROCKER: The moral of the story in so many of the topics that we discussed this morning is that there are circles of effects and concerns that spread out across parents, siblings, friends, and others whenever there is a striking situation of human exceptionality. The natural solutions may actually be quite troubling and pathologic unless supports can be brought in. This young man is clearly in a pressured circumstance that is going to have very undesirable final ramifications. As you describe it, it is so bizarre as to almost stagger the mind. Here again is a family in crisis requiring a multifactoral solution. The family is preoccupied with the handicapped child, while the normal child has a variety of interests that must be catered to and acknowledged.

T. VARNET (Social Worker, Crystal Springs School, Assonet, Massachusetts): One of my concerns is the lack of interdisciplinary contact and sharing of information among professionals even while they are still in graduate or undergraduate schools. I've recently completed a graduate course in social work, and one of the things I was most dismayed about was the lack of information on handicaps, specifically mental retardation. The old attitudes are still present in schools of social work.

M. CHALLELA: In the Boston area, there are universities that have affiliations with both the Children's Hospital Evaluation Clinic and the Shriver Center, where as part of your graduate studies, you learn to work in an interdisciplinary center with handicapped children.

T. VARNET: It needs to be expanded.

M. CHALLELA: It needs to get into the curriculum is what you are saying. To do that, you have to have somebody on the faculty who is also interested in the field. And that is the problem.

T. VARNET: Perhaps a speakers bureau of parents of handicapped children would be willing to go in. The various self-help groups might be interested in starting one.

C. KING (Nursing faculty, University of Massachusetts at Amherst): What factors are involved when a family has to decide to institutionalize a child who is retarded or handicapped? As professionals, how might we best help those families make that decision? It frequently happens that a family has to finally deal with the decision to institutionalize or not.

M. CHALLELA: There are really many factors. When you talk about institutionalization today there are very few possibilities. It really depends on the severity of the handicap. The availability of support systems such as your respite care facilities for extended family members is important. Another whole area is finances. It can be costly to maintain a child at home. What plans do the parents have in terms of their own development? The sex of the child and the age is important too. Parents should know in the beginning that they are not expected to keep this child at home for the whole life cycle. When children grow up they don't usually stay at home. For retarded or handicapped people, there are such places as respite care for temporary assistance, community residences, boarding homes, and so forth.

A. CROCKER: I think society is in a dilemma regarding this whole issue of the appropriate use of chronic residential care facilities. Generally speaking, the word institutionalization, as you used it, represents so often a nonsolution. We are still in the backwaters of the earlier enthusiasm, and we are still paying some of the price for the inclination toward what was thought to be earnest professional grounds in earlier times. The ultimate way is to go to respite care. Occasionally the nursing home, or the special school that has a residential component, or the community residence, in the long run seems to be a vastly sounder approach than what the word *institutionalization* has conventionally represented.

M. CHALLELA: Then of course there is foster care that can often be used for these children.

M. RINN (Information Center for Individuals with Disabilities): In addition to the lack of schools of social work linking up with the health science colleges nearby is the fact that they are not accepting very many qualified people with disabilities as social workers, or inviting many people as speakers. People in training should be exposed to not only patients or people they were doing something for, but also people they were doing something with. In my work I deal with disabled individuals or their parents who call for assistance. There is a long delay in making connections or services that can be useful. How can available information be more efficiently communicated to the people when they need it? Who starts that process? Is it the doctor, the teacher? Whose responsibility is it first? How do you turn around the undesirable consequences that accumulate when you see a person who has been eligible for rehabilitation services for 20 years and then finally comes for help?

M. CHALLELA: Information should be communicated when the problem is first identified. If it's an infant, information dissemination should start in the hospital. It should start with the nurse and doctor in the newborn nursery. But I agree with you. Getting this up-to-date accurate information out is important. We really have to help each other.

A. CROCKER: I would say that such conferences with press coverage are important. We as a society are just getting some of these topics out of the closet. Information about special circumstances of human handicaps is now receiving better coverage in popular media, but we all have a long way to go. If everyone present in this room serves as a focal point in his or her own personal life, both professionally and in social ways, for information dissemination, the ripple effect should be substantial.

15

WALTER P. CHRISTIAN is Director of the May Institute for Autistic Children, Inc., in Chatham, Massachusetts, and Adjunct Assistant Professor, Department of Human Development, University of Kansas, Lawrence. He received his Ph.D. in Clinical Psychology from Auburn University in 1974 and was a Postdoctoral Fellow at the National Asthma Center in Denver, Colorado. Prior to assuming his present position, Dr. Christian was Chief Psychologist at Children's Behavioral Services, Las Vegas, Nevada. Dr. Christian is the author of the books *Chronically Ill and Handicapped Children: Their Management and Rehabilitation* (with T. L. Creer), *Schedule-Induced Behavior: Research and Theory* (with R. W. Schaeffer and G. D. King), and *Preservation of Clients' Rights: A Handbook for Practitioners Providing Therapeutic, Educational, and Rehabilitative Services* (with J. T. Hannah and H. B. Clark). In addition, Dr. Christian has written many articles for professional journals.

Reaching Autistic Children
Strategies for Parents and Helping Professionals

WALTER P. CHRISTIAN

Reaching autistic children is a problem that has challenged countless parents and helping professionals since Kanner's description of "early infantile autism" in 1944.[1] This challenge has been made more difficult with the proliferation of theories concerning the etiology, prognosis, and preferred treatment of the condition. With no known "cure" for autism and a preponderance of questions still awaiting research, the ground is fertile for misconception.

The individual living and/or working with the autistic child must look at the growing body of literature on autism and determine the most effective philosophy and strategy to employ in reaching the child. A recommended course of action in this endeavor involves: (1) understanding autism and the characteristic handicaps and special needs of autistic children; (2) understanding the basic philosophy of treatment and education that is considered the most effective with autistic children; (3) learning how to apply strategies of documented effectiveness in the behavior management and education of autistic children; and (4) learning to work with others in providing a comprehensive program of progressive services for the autistic child, adolescent, and adult.

WALTER P. CHRISTIAN • May Institute for Autistic Children, Inc., Chatham, Massachusetts 02633.

AUTISM

DEFINITION

In his original diagnosis of autism, Kanner identified four behavioral characteristics: (1) social withdrawal, which he described as "an extreme autistic aloneness" evident from the early months of life; (2) severe communication deficits, such as mutism or the inability to "convey meaning" through whatever speech may exist; (3) desire for "maintenance of sameness" in the environment; and (4) severe limitation in the occurrence and variety of "spontaneous activity," as evidenced by a preoccupation with manipulating objects and stereotyped play habits.[2]

In the years since Kanner's pioneering work, much has been learned about autism, although the basic questions of cause and cure remain unanswered. Perhaps the most comprehensive definition of the condition was recently published under the auspices of the National Society for Autistic Children (Ritvo and Freeman, 1977)[3,4] The abstract of this definition is as follows:

Autism is a severely incapacitating life-long developmental disability which typically appears during the first three years of life. It occurs in approximately five out of every 10,000 births and is four times more common in boys than girls. It has been found throughout the world in families of all racial, ethnic, and social backgrounds. No known factors in the psychological environment of a child have been shown to cause autism. The symptoms are caused by physical disorders of the brain. They must be documented by history or present on examination. They include: (1) Disturbances in the rate of appearance of physical, social, and language skills. (2) Abnormal responses to sensations. Any one or a combination of sight, hearing, touch, pain, balance, smell, taste, and the way a child holds his or her body are affected. (3) Speech and language are absent or delayed, while specific thinking capabilities may be present. Immature rhythms of speech, limited understanding of ideas and the use of words without attaching the usual meaning to them is common. (4) Abnormal ways of relating to people, objects, and events. Typically, they do not respond appropriately to adults and other children. Objects and toys are not used as normally intended. Autism occurs by itself or in association with other disorders which affect the function of the brain such as viral infections, metabolic disturbances, and epilepsy. On IQ testing, approximately 60 percent have scores below 50, 20 per cent between 50 and 70, and only 20 percent greater than 70. Most show wide variations of performance on different tests and at different times. Autistic people live a normal life span. Since symptoms change, and some may disappear with age, periodic reevaluations are necessary to respond to changing needs. The severe form of the syndrome may include the most extreme forms of self-injurious, repetitive, highly unusual, and aggressive behaviors. Such behaviors may be persistent and highly resistant to change, often requiring unique management, treatment, or teaching strategies. Special educational programs using behavioral methods and designed for specific individuals have proven most helpful. Supportive counseling may be helpful for families with autistic members, as it is for families who have members with other severe life-long disabilities. Medication to decrease specific symptoms may help certain autistic people live more satisfactory lives." (p. 146)

DIFFERENTIAL DIAGNOSIS

Despite the apparent clarity of this definition, differential diagnosis of autism for a particular child may be the subject of much dispute. Many of the behaviors identified as characteristic of autistic children are exhibited by nonautistic children as well. For example, self-stimulation and mild levels of social withdrawal and language delay are not unusual in otherwise normal children. Moderate to severe social withdrawal, a high rate of self-stimulation, and/or pronounced communication and academic deficits are seen in children with a number of other diagnoses, including: (1) mental retardation; (2) specific sensory deficit; (3) childhood schizophrenia; (4) physical or psychological trauma; (5) degenerative organic brain syndrome; and/or (6) congenital, developmental, and acquired disorders in the mechanisms associated with language processes.

This confusion regarding differential diagnosis makes *accurate* and *timely* (as early as possible) diagnosis of autism all the more important: Faulty diagnosis may encourage the utilization of a philosophy of treatment and/or program of services effective with many of these other problems while not appropriate for the effective treatment of the autistic child. Parents and helping professionals should familiarize themselves with the specific behaviors characteristic of childhood autism and evaluate the particular child's problem behavior against these criteria. This procedure has been facilitated by the publication of several rating scales and surveys that might be referred to as "autistic behavior checklists."[5,6]

ETIOLOGY

While this discussion is not primarily concerned with hypothesized causes of autism, a brief examination of theories of causation is important since, historically, one's theory of causation determined one's philosophy of treatment. Focusing on the search for some underlying problem or internal pathological state as the cause for an apparent problem in human behavior is characteristic of the "medical model." The medical approach assumes, therefore, that unless the pathogen is removed, symptomatic problems can be expected to increase in severity. Two examples of the medical model are the "psychogenic" and "biological" theories for the causation of autism.

The psychogenic hypothesis considers autism to be the result of an internal disturbance produced by environmental factors, with particular attention given to the mental health of the parents. Kanner offered one such hypothesis, referring to parents of autistic children as "refrigerator-type."[1] Bettleheim, a major proponent of the psychogenic theory, has described autism as due to "the parents' wish that the child should not exist."[7] Psychotherapy directed toward parents, therefore, becomes the treatment of choice.

However, numerous studies have refuted the psychogenic hypothesis. Among the many findings cited by Oppenheim[8] is that parents of autistic children are typically well-adjusted and most have normal children as well.[9] But perhaps the most serious shortcoming of the psychogenic approach has been that countless autistic children in desperate need of services have been left in waiting rooms while their parents have been the focus of expensive, time-consuming, and often unnecessary psychotherapy.

In contrast, the biological or "biogenic" theory of causation considers autism the result of physical disorders to the brain, biochemical deficiencies, and/or metabolic disturbances. The biological view is supported by evidence that autism is seen in the first months of life and is associated with signs of neurological dysfunction (e.g., disturbances of developmental rates, responses to stimuli, and speech/language capacities). Remedial education for the child and chemotherapy to improve biochemical/metabolic deficiencies become the treatments of choice for proponents of this view.

While the biological approach has generated a wealth of promising research, as yet limited insight has been gained into new technologies of treatment. As Kozloff observed, results of numerous studies in the areas of biochemistry, chemotherapy, and cognitive dysfunction have remained ambiguous.[10] In summarizing the present status of this research, Ritvo has concluded[11]:

> Follow-up studies are just beginning to appear in the medical literature. They will eventually answer the question as to the natural history of the disease. Since we do not know the specific cause, nor neuroanatomical or neurobiochemical pathology involved, specific etiologically based therapy is unavailable. (p. 5)

The Behavioral Approach to the Education and Rehabilitation of Autistic Children

The remainder of this discussion will be directed toward increasing the reader's understanding of what has been demonstrated to be the treatment modality of choice for autistic children. As noted in the National Society for Autistic Children's[12] definition of autism, "special education programs using behavioral methods and designed for specific individuals have proven most helpful" in work with autistic children.

However, the use of behavioral methods requires one's commitment to an intrusive, structured approach. Knowing that the child is severely withdrawn and concerned with maintaining "sameness" in his or her environment, one must decide that such a situation must not be allowed to continue. The child must *attend, comply,* and *respond* before learning can occur.

This decision—to thoughtfully but deliberately *intrude* into the child's "autistic aloneness"—is perhaps the most difficult for parents and helping professionals to make. Determined to maintain "sameness," the autistic child typically resists intrusion as if locked in a life and death struggle (e.g., tantrums of high intensity and long duration, aggression, destruction of property, and even self-mutilation). Consider Oppenheim's experience with her nonverbal, autistic son, Ethan[8]:

> Ethan, then four and one-half years old, did not *look,* when I tried to show him something; he gave no indication that he was *listening* to what I was telling him; and he made not the slightest effort to *imitate* the procedures I was trying to teach him. . . . I realized, simply on a common sense basis, that there was one fundamental pre-requisite which had to be accomplished before teaching of any kind could begin. . . . I had to require Ethan to *attend.* The ensuing battle of wits between Ethan and myself was a shattering experience for me. . . . But the resultant change in Ethan taught me a lesson I have never forgotten: the establishment of control is the crucial step in any educational program for autistic children. (p. 34)

Rationale for Behavioral Intervention

As we have seen, the medical approach to the treatment of autism is primarily concerned with the cause of the condition, which is considered to be within the child, i.e., an emotional disturbance, a perceptual disorder, a disturbance of biochemistry or metabolism, or some inherited predisposition. The autistic child's behavior is important, therefore, as a symptom or manifestation of some greater, underlying problem.

However, while the medical model may ultimately result in a greater understanding of the etiology of autism, it is not regarded as the most effective approach to the remediation of the severe behavior problems characteristic of autistic children. In addition, many of the basic tenets of the medical approach to the treatment of autism have been disputed in recent research. As Ritvo has noted, no known factors in the psychological environment of the child have been shown to cause autism.[11] In addition, as previously discussed, no etiologically based treatment is available that can alter the course of the condition.

In direct contrast to the medical model, the *behavioral approach* is primarily concerned with the child's observable behavior rather than some inferred internal disturbance. The search for cause is focused on the environment—on the environmental conditions that act as antecedents and consequences for the child's "autistic" behavior. The approach to treatment is educational, i.e., structuring the environment in such a way as to increase the probability of adaptive behavior while decreasing the probability of maladaptive behavior.

Specifically, a desirable consequence ("reinforcement") for a behavior increases the likelihood that the behavior will occur again; an undesirable consequence ("punishment") decreases the likelihood that the behavior will recur. In addition, antecedent conditions associated with desirable consequences for a behavior tend to occasion the occurrence of the behavior in the future; antecedent conditions associated with undesirable consequences for the behavior tend to discourage the future occurrence of the behavior. For example, if a child receives more reinforcement (praise, eye contact, interest) from social interaction with mother than father, mother's presence will occasion more social interaction from the child than will father's.

It is also important to note that these operations are powerful in their effects on behavior regardless of whether the behavior being effected is appropriate or inappropriate. Therefore, if attention is reinforcing to the child and tantrum behavior brings attention from mother, there is an increased probability that tantrums will occur again in similar circumstances, i.e., when child seeks attention and mother is present.

The fundamental components of the behavioral approach include (1) assessing the special needs of the child, (2) developing plans and goals for training based on the child's needs, (3) analyzing the behavior as a function of its environmental context, (4) applying procedures of proven effectiveness when used with children displaying severe behavior problems, and (5) evaluating the effectiveness of each training effort.[13]

Determining the Child's Special Needs

It would be a mistake to expect that all autistic children will display the same behaviors and thus present the same needs. Those with experience working with autistic children are constantly amazed at the diversity of skills and deficits they exhibit. Some children display "special abilities" such as the "autistic savants" described by Rimland.[14] For example, Rimland describes how one 8-year-old autistic child is capable of rapid, accurate multiplication of four-digit numbers. Others may display severe deficits with no apparent special skills. Some autistic children may be self-injurious to the extent that they must be protected from themselves. In some cases, children labeled "autistic" actually display only a few of the characteristic behaviors typical of autism.

"Baseline" Observation. Therefore, the needs of the individual child must determine what problems are targeted for intervention, what goals are set, and what strategies are implemented. For example, a high rate of stereotyped, self-stimulatory behavior is considered by many to be a "classic" autistic behavior (e.g., flapping hands or fingers, rocking, facial grimacing). In our work with Johnny, we note that he engages in self-stimulation and decide to "change" the behavior. However, if we first take a preliminary or "baseline" record of how frequently Johnny engages in the behavior, we find that he actually spends little time in self-stimulation. Such observation often reveals other needs that are more deserving of intervention. One also learns *when* and *in what context* problem behaviors are displayed—information important in determining what strategy to employ.

This preliminary "baseline" assessment also provides an important index for use in evaluating the effectiveness of one's intervention. For example, if Johnny makes no spontaneous sounds at baseline and is capable of five spontaneous sounds after two weeks of intervention, one can see that progress is being made. This reinforcing feedback for the parent or professional is critical since the child may be resisting even the most effective intervention, thus providing little encouragement and, indeed, increasing the difficulty of one's task.

Standardized Assessment. The use of a standardized assessment instrument can provide information concerning a child's adaptive behaviors, e.g., skills necessary for independent day-to-day living. For example, the Adaptive Behavior Scale[15] assesses skill levels in a large number of categories, including those of particular relevance to the autistic child, i.e., *adaptive behaviors* (such as self-care, socialization, and communication) and *maladaptive behaviors* (such as aggression and a high rate of self-stimulation). Such an instrument provides one with a large number of specific baseline records against which the effectiveness of behavior change and education efforts can be determined.

Measures of adaptive behavior are much preferred over instruments yielding an intellectual quotient (IQ scores) when working with autistic children. Investigators have noted that IQ scores, while of value in assessing the potential of academic performance for average, middle-class children, often fail to provide the most useful (i.e., prescriptive) information in the case of the handicapped child.[15] Gardner and Giampa have observed that handicapped children are typically unable to perform tasks required in standardized psychological assessment.[16]

In summary, as the important first step in designing strategies for reaching autistic children, *baseline assessment* is possible only when one uses objective terms to describe the target behavior. *Objective description of the problem behavior* includes a determination of such parameters as the frequency, duration, and context of the behavior. For example, the description "Johnny engaged in 18 tantrums (defined as crying, screaming, and/or head slaps); each tantrum had an average duration of 70 seconds; 6 occurred in the living room, 12 during meal time" employs wording that is quite different from the more subjective terms one typically uses to describe a problem behavior, e.g., "Johnny tantrums all the time" or "Johnny was a bad boy today."

STRATEGIES FOR BEHAVIOR CHANGE

Functional analysis of the child's behavior thus becomes a critical component of effective behavior change. This analysis enables one to understand the *operation* of the environment in occasioning and maintaining the child's behavior. This understanding makes it possible for one to learn how to use the reinforcing and punishing aspects of the environment in establishing structure and control, motivating the child, and giving the child feedback concerning his or her behavior.[13]

These environmental operations include the following:

1. Operations that increase a behavior
   a. *Providing* desirable consequences for the behavior (e.g., a food item or praise for good work)
   b. *Removing* an undesirable consequence for the behavior (e.g., letting the child come out of his/her room when screaming ceases)
2. Operations that decrease a behavior

    a. *Presenting* an undesirable consequence for the behavior (e.g., a scolding, sending the child to his/her room for a brief "time-out" period)

    b. *Withholding* a desirable consequence that the child is accustomed to receiving (e.g., ignoring rather than attending to temper tantrum)

    c. *Removing* a desirable consequence (e.g., temporary, ten-second removal of a child's plate until screaming at the table ceases)

    d. *Requiring some effort* or change in the child's behavior as an undesirable consequence (e.g., requiring that the child engage in a brief period of calisthenic exercise contingent upon the inappropriate behavior)

However, knowing the operations that bring about behavior change is only one aspect of effective intervention with autistic children. Applied researchers (e.g., ref. 17) have found that education and rehabilitation of autistic children is possible only when a number of critical issues and procedural guidelines are adequately addressed.

Early Intervention. Identification of handicaps and intervention should take place as early as possible with autistic children, preferably before 3 years of age and always before 5 years of age. The most important reasons to begin work with the child as soon as possible are that language plays a critical role in the development of other skills and should be taught as soon as possible, and the longer the history of deviant behavior the more difficult it is to change. Research indicates that the prognosis for near normal development is very poor if progress in language development has not been made by about 5 years of age.

Proper Training Environment. As previously discussed, instructional control must first be established, i.e., the child must comply with such simple requests as "sit down," and "look at me." Training sessions should be kept brief (5 to 10 minutes) at first and gradually extended. An attempt should be made to minimize extraneous and potentially distracting stimuli in the environment where training is to take place. Autistic children tend to "overselect" certain stimuli in their environment for attention.[18] Indeed, many investigators have reported that when the autistic child is trained to respond to a complex of multiple stimuli, only one element of the complex acquires any degree of control over the child's responding. For example, given the task to choose one of two full-face photographs, the child may have difficulty because he attends only to the left eyebrow on one of the pictures.

Proper Use of Instructions. Instructions must be brief, clear, consistent, and to the point. One must be sure that the child is attending to the instruction; requiring *eye contact* from the child as the first step in giving instructions is a way to ensure that he or she is attending and thus more likely to follow instructions. The child must be required to attend and to perform if training is to be effective. Requiring that the child "practice" the same response or behavior a number of times can help to ensure that the child is attending to and complying with instructions.

Effective Use of Reinforcers. A first step in the effective use of reinforcers is to determine what is reinforcing for a particular child by "sampling," e.g., presenting several food items to the child and seeing which one he prefers. Although food reinforcers are typically very effective, one must choose a type and amount of food that does not lead to rapid satiation (e.g., sunflower seeds, raisins, dry cereals).

When teaching a new skill, *immediately* reinforce *each occurrence* of the desired behavior. Later in the teaching process, one may go to an intermittent schedule of reinforcement, i.e., reinforcing every other behavior, or perhaps reinforcing the child for every 5 minutes of the desired behavior.

Using Attention as a Reinforcer. When attempting to decrease the frequency or duration of maladaptive behaviors (e.g., tantrums, a high rate of self-stimulation such as posturing or screaming, and mild self-abusive behavior such as head slaps and pinches), one must realize the power of one's attention as a reinforcer. If close observation of the behavior indicates that it is consistently followed by attention, then systematically withholding attention contingent upon the behavior becomes the treatment of choice. It is also important to concurrently make a special effort to attend to the child when he or she is not engaging in the undesirable behavior, i.e., use of attention to reinforce nonoccurrence of the behavior.

One must be prepared for te child to actively resist this procedure, i.e., the frequent report by parents that "ignoring it just made it get worse." Many parents and helping professionals may abandon the procedure in response to heavy resistance from the child, out of fear that the problem will continue to escalate. However, abandoning the procedure will no doubt be perceived as a desirable consequence by the child, thus reinforcing his high-rate tantrum behavior. Resistance by the child is a predictable result of withholding reinforcement and the target behavior will decrease if one is patient and willing to stick to the procedure. However, this may not be possible if the child resorts to unacceptable levels of self-injury, aggression, and/or destruction of property.

"Social reinforcers" such as attention (e.g., a look, a nod, an answer to a question), affection (e.g., a smile, a touch, a hug), and verbal praise (e.g., "good work," "O.K.") are also powerful consequences for maintaining appropriate behavior. Attention and affection can also increase the effectiveness of another reinforcer (e.g., food) when the two are used together. When using praise in combination with food reinforcement, attention often becomes effective in maintaining the behavior when food reinforcement is faded out. All of the points discussed above concerning the effective use of reinforcers are important to consider in the use of attention as a reinforcer, i.e., reinforcer sampling and immediacy.

Successively Approximating New Behaviors. A particularly important strategy in working with autistic children is "shaping," or the reinforcement of successive approximations to a desired behavior. One begins with a response already in the child's repertoire. A number of behavioral steps or approximations are then identified between this initial behavior and the terminal behavior (or what one wants the child to be able to do). Each step becomes a target behavior and only after stable performance is achieved at each step should the child be taken to the next step. If the child fails to master a step along the way, return to the previous step, regain stable performance, and then proceed in approximating the terminal behavior. The shaping procedure is complete when stable performance of the terminal behavior is achieved.

The shaping procedure is useful in remedying all of the major skill deficits of autistic children—deficits in areas of language, social, self-help, motor, and academic/pre-academic skills. For example, shoe tying is easily broken down to a number of distinct steps, each approximating the terminal behavior of being able to tie one's shoe laces without assistance. Teaching social interaction may begin with reinforcing fleeting face-to-face contact, then moving to eye-to-eye contact, and so on. Similarly, teaching the word *mama* begins with reinforcing the child for emitting any sound and proceeding with reinforcing approximations to *mm*, then *ma*, then *mama*.

Despite the apparent logic and simplicity of the techniques involved, autistic children often require the patience and perseverance of many trials. Even the simplest of skills may require the patient training of a large number of approximations. Keeping the goal in mind and learning not to spend too much or too little time at any one step are important elements of successful shaping.

When a particular behavior has been learned via the shaping procedure, it can be linked with other behaviors to form a behavioral "chain." Establishing chains or routines ("chaining") that are under the control of a single instruction is important when working with autistic children. "Get dressed" thus occasions a number of specific behaviors (finding clothes, putting on shirt, pulling on pants, etc.). "Time to get up" initiates a long morning routine of behaviors that are essential to independent functioning.

Use of Prompts and Cues. Prompts and cues are stimuli that may occasion a desired response prior to training or with little training. Prompts also ensure that the learning situation is successful, i.e., increasing appropriate responding thereby maximizing the reinforcement that the child receives.

In presenting a prompt, one is attempting to ensure that the child engages in the desired response in the presence of the relevant stimulus or antecedent event. When prompts are no longer needed they are "faded" out so that only the training stimulus remains. Knowing when to employ prompts and when to fade them out is of critical importance in working with the autistic child since prompting is an essential ingredient of every training effort.

Prompts may include physically assisting the child in the performance of the desired response (e.g., holding a child's lips in the appropriate position to produce an *oo* sound), by giving the child the correct answer (e.g., in a discrimination task), or by modeling the desired response for the child (e.g., producing speech sounds).

The most effective training programs for autistic children employ a combination of shaping, chaining, prompting, and fading procedures. For example, Lovaas and his colleagues have made effective use of these procedures in their development of a language acquisition training protocol for autistic children.[19] Lovaas's approach consists of a number of programmatic steps, which include: (1) building verbal responses, (2) labeling discrete events, (3) relationship between events, (4) abstract terms, (5) conversation, (6) giving and seeking information, (7) grammatical skills, (8) recalls, and (9) spontaneity. This highly effective program represents both the critical elements of the behavioral approach and the success possible when these elements are correctly utilized.[20]

Consistency of Application. Structure, control, and intervention procedures must be consistent across both home and school environments to be maximally effective. Consistency of target behaviors, goals, and procedures will ensure that skills learned in one environment will "generalize," or be maintained, in another environment. Consistency is particularly important with autistic children, since the child's difficulty in accepting environmental change and other characteristic problems, such as "overselectivity" to certain environmental stimuli, often interact to hinder generalization of skills or control.

Parent training, close communication between parents and helping professionals, and continuing education for teachers and other helping professionals are essential in promoting consistency and generalization. Communication can be facilitated by developing a system of written correspondence between teacher and parent, between other helping professionals and teacher, and so on. For example, Creer and Christian have described the use of a daily school report, hospital behavior checklists, and home behavior checklists in improving communication, fostering a "team approach," and ensuring consistency of effort.[21]

OVERVIEW: WORKING TOWARD A CONTINUUM OF SERVICES

In summary, autistic children require a behaviorally oriented educational approach to treatment if they are to make adequate gains in communication, social, self-help, and academic skills. This approach to treatment requires that one develop an understanding of the child's needs, the relationship between the child's behavior and his or her environment, and the way in which aspects of the environment can be used in educating and rehabilitating the child.

However, solving the child's immediate needs is not the only challenge facing those working with autistic children. Parents and helping professionals must continue to meet the needs of the child as he or she enters adolescence and adulthood. Parents of children who require residential or day treatment and vocational/educational services are finding it difficult to obtain these services. For example, Sullivan has observed that "there is a sudden drop in the number of available programs both day and residential, once the child reaches puberty or late adolescence" and that "suitable programs for adult autistic persons are so rare as to be practically nonexistent" (p. 296).[22]

However, as Sullivan[22] and others have noted, parents of autistic persons are becoming more militant, and there is the prospect of parents and helping professionals working together to bring about an increase in the quantity and quality of services available to autistic persons. In addition, the increased attention being given to the rights of clients and the rights of students[23] will no doubt benefit the autistic individual.[24] For example, with the requirements of new legislation (P.L. 94-142), local education agencies are more likely to join parents and professionals in the development of a continuum of services for autistic children—particularly in terms of special education programs through adolescence.[25] It is encouraging that the unifying goal of these efforts appears to be one of "mainstreaming"—teaching autistic individuals of all ages to better cope with the demands of their environment and, thus, be capable of leading more productive lives.

REFERENCES AND NOTES

1. Kanner, L.: Early infantile autism. *J. Pediatr.* **25(3):** 211, 1944.
2. Kanner, L.: "Autistic disturbances of affective contact." *Nerv. Child.* **2(3):** 217, 1944.
3. Ritvo, E. R., and Freeman, B. J.: National Society for Autistic Children definition of the syndrome of autism. *J. Pediatr. Psychol.* **2(4):**146, 1977.
4. For additional information concerning autism and the characteristic behaviors and handicaps of autistic children, the reader should consult the following:

   Rimland, B.: *Infantile Autism. The Syndrome and Its Implications for a Neural Theory of Behavior.* New York: Appleton-Century-Crofts, 1964.
   Ritvo, E. R., Freeman, B. J., Ornitz, E. M., and Tanguay, P. E.: *Autism: Diagnosis, Current Research, and Management.* New York: Spectrum Publications, 1976.
   Rutter, M., and Schopler, E. (Eds.): *Autism: A Reappraisal of Concepts and Treatment.* New York: Plenum, 1978.
   Wing, L. (Ed.): *Early Childhood Autism* (second edition). Oxford, England: Pergamon Press, 1976.

5. Wing, L.: Diagnosis, clinical description, and prognosis, in L. Wing (Ed.): *Early Childhood Autism* (second edition). Oxford, England: Pergamon Press, p. 15, 1976, pp. 49–52.
6. Rimland, B.: *Infantile Autism. The Syndrome and Its Implications for a Neural Theory of Behavior.* New York: Appleton-Century-Crofts, 1964, pp. 221–236.
7. Bettleheim, B.: *The Empty Fortress.* New York: The Free Press, 1967.
8. Oppenheim, R. C.: *Effective Teaching Methods for Autistic Children.* Springfield, Illinois: Charles C Thomas, 1974.
9. Wing, L.: Counseling and principles of management, in J. K. Wing (Ed.): *Early Childhood Autism: Clinical Educational and Social Aspects.* Oxford, England: Pergamon Press, p. 257, 1966.
10. Kozloff, M. A.: *Reaching the Autistic Child: A Parent Training Program.* Champaign, Illinois: Research Press, 1973.
11. Ritvo, E. R.: Autism: From adjective to noun, in E. R. Ritvo, B. J. Freeman, E. M. Ornitz, and P. M. Tanguay (Eds.): *Autism: Diagnosis, Current Research, and Management.* New York: Spectrum Publications, p. 3, 1976.
12. National Society for Autistic Children, 1234 Massachusetts Avenue, Washington, D. C. 20005.
13. For additional information concerning behavior modification, parents are encouraged to consult the following:

    Patterson, G. R.: *Families: Applications of Social Learning to Everyday Life.* Champaign, Illinois: Research Press, 1975.

Patterson, G. R., and Guillion, M. E.: *Living With Children: New Methods for Parents and Teachers.* Champaign, Illinois: Research Press, 1971.

Helping professionals are encouraged to consult the following:

Leitenberg, H.: *Handbook of Behavior Modification and Behavior Therapy.* Englewood Cliffs, New Jersey: Prentice-Hall, 1976.

Sulzer-Azaroff, B., and Mayer, G. R.: *Applying Behavior Analysis Procedures with Children and Youth.* New York: Holt, Rinehart and Winston, 1977.

14. Rimland, B.: Inside the mind of the autistic savant. *Psychology Today* **August:** 69, 1978.
15. Nihira, K., Foster, R., Shellhaas, M., and Leland, H.: *Adaptive Behavior Scale Manual.* Washington, D.C.: American Association of Mental Deficiency, 1969.
16. Gardner, J. M., and Giampa, F. I.: Utility of three behavioral indices for studying severely and profoundly retarded children. *Am. J. Ment. Defic.* **76:**352, 1971.
17. Rincover, A., Koegel, R. L., and Russo, D. C.: Some recent behavioral research on the education of autistic children. *Educ. Treat. Child.* 1(4):31, 1978.
18. Wilhelm, H., and Lovaas, O. I.: Stimulus overselectivity: A common feature in autism and mental retardation. *Am. J. Ment. Defic.* 1976, **81**(1):26031.
19. Lovaas, O. I.: *The Autistic Child: Language Development through Behavior Modification.* New York: Irvington Publishers, 1977.
20. For additional information concerning the use of behavioral treatment methods with autistic children, the reader is referred to the following:

    Kozloff, M. A.: *Reaching the Autistic Child: A Parent Training Program.* Champaign, Illinois: Research Press, 1973.

    Lovaas, O. I.: *The Autistic Child: Language Development through Behavior Modification.* New York: Irvington Publishers, 1977.

    Lovaas, O. I., Schriebman, L., and Koegel, R. L.: A behavior modification approach to the treatment of autistic children. *J. Autism Child. Schizophr.* 4(2):111, 1974.

    Oppenheim, R. C.: *Effective Teaching Methods for Autistic Children.* Springfield, Illinois: Charles C Thomas, 1974.

21. Creer, T. L., and Christian, W. P.: *Chronically Ill and Handicapped Children: Their Management and Rehabilitation.* Champaign, Illinois: Research Press, 1976.
22. Sullivan, R. C.: Autism: Current trends in services, in E. R. Ritvo, B. J. Freeman, E. M. Ornitz, and P. M. Tanguay (Eds.): *Autism: Diagnosis, Current Research, and Management.* New York: Spectrum Publications, p. 291, 1976.
23. Hannah, J. T., Christian, W. P., and Clark, H. B. (Eds.): *Preservation of Clients' Rights: A Handbook for Practitioners Providing Therapeutic, Educational, and Rehabilitative Services.* New York: Macmillan/Free Press, 1981.
24. McClannahan, L. E., and Krantz, P. J.: Program accountability systems as a protection of the rights of autistic children and youth, in J. T. Hannah, W. P. Christian, and H. B. Clark (Eds.): *Preservation of Clients' Rights: A Handbook for Practitioners Providing Therapeutic, Educational, and Rehabilitative Services.* New York: Macmillan/Free Press, 1981.
25. The Education for All Handicapped Children Act (P.L. 94-142) was signed into law in 1975 and is designed to assure that all handicapped children have available to them a free, appropriate education, emphasizing special education and related services designed to meet their unique needs. In addition, it assures that the rights of children and parents are safeguarded. These safeguards include (1) due process; (2) nondiscriminatory testing; (3) least restrictive environment (the requirement that handi-

capped children be educated with nonhandicapped children to the maximum extent possible); (4) education using the child's native language; (5) confidentiality (the assurance that any information contained in school records will not be released without the permission of the parent); and (6) the right to representation (the assurance that the child is to be represented by his/her parents, guardian, or surrogate parent). The law further requires that Individual Education Plans (IEPs) be prepared for each handicapped child, with parents participating on the team that draws up the plan. IEPs are to be developed or revised at least once every six months.

Discussion

MODERATOR: MARY S. CHALLELA

J. THOMAS (Pediatrics, Tufts New England Medical Center): I'm curious to know what happens to these children when they become adults. Do they get married? Can they lead a normal life? Do they always need treatment?

W.P. CHRISTIAN: It depends on what type of intervention they had in childhood. Service programs for young adults are rare. The best trend in the development of services is toward community-based group homes serving autistic individuals. There is an excellent group home program in Princeton, New Jersey. The Princeton Child Development Institute is just phenomenal. They are working toward getting affected adults to live at home. If you can do away with these seriously aggressive and self-injurious behaviors, these children and adolescents can live at home quite well. Supported day treatment services and community based programming are really critical.

M. O'CONNOR (Hospital Teacher, Shriners Burns Hospital): You quoted a figure that a large percent of these children test under 50.

W.P. CHRISTIAN: 60%.

M. O'CONNOR: Do you think that's fair? Valid?

W.P. CHRISTIAN: It's as fair as the controls used when they were tested.

M. O'CONNOR: Do you mean testing like on a WISC?

W.P. CHRISTIAN: Traditional IQ tests, the type I'm not advocating. Many autistic children do not respond at all, even after 2 hours of testing. Tremendous patience over sometimes an entire day may sometimes yield a response.

M. O'CONNOR: I'm confused. Were these children pretreatment or posttreatment, after behavior had been brought under control?

W.P. CHRISTIAN: This was prior to treatment and is based on published data. It is difficult to know how many autistic children can reach normal IQ levels, because it all has to do with how inadequate present instruments are when working with these IQs. I wish I could give you a better answer.

SPEAKER: I cared for a 5½-year-old autistic child. My children were wrestling and playing with him on the floor and they were tickling him. He had no speech for several years according to his parents. Suddenly he laughed and he said, "Stop, stop, don't do that." Now was that as if he let down his guard or was that a memory response?

W.P. CHRISTIAN: It's very tempting to offer interpretations. You may see this with normal children who haven't yet spoken either. They provide a few verbal interactions you might not see again for a year. It's cruel because some parents are encouraged by such responses and do not bring the child in for treatment. The earlier the intervention the better, unlike *Son-Rise*, which you may have seen on TV. They failed to say that that was a dated broadcast—a dated episode. They gave the impression that people weren't into early intervention. We try to initiate treatment as early as age three.

16

LUDWIK S. SZYMANSKI is a graduate of Hebrew University Medical School, Jerusalem, Israel. After completing his internship and military service in the Israeli Army Medical Corps, he received further training in pediatrics in Israel and at Bellevue Hospital, New York; in adult psychiatry at Mount Sinai Hospital, New York; and in child psychiatry at the Children's Hospital Medical Center, Boston. He is currently Director of Psychiatry at the Developmental Evaluation Clinic and an Associate in Psychiatry at Children's Hospital Medical Center, Boston, and Assistant Professor of Psychiatry at the Harvard Medical School. Most of his professional time is devoted to work with developmentally disabled persons of all ages and with a variety of handicaps, including mental retardation. As the Developmental Evaluation Clinic is a teaching facility, much of his time is devoted to training professionals in a variety of disciplines in the psychiatric aspects of developmental disabilities.

Dr. Szymanski's research interests include infantile autism, mental illness in retarded adults, the sexuality of developmentally disabled persons, and psychotherapies with retarded persons. He is the author of publications on such topics as the psychiatric diagnosis of retarded persons, the sexuality of retarded persons, and mental health issues in the care of the multiply handicapped child. He has also coedited a book on mental illness in retarded persons. Dr. Szymanski chairs the Committee on Mental Retardation of the American Academy of Child Psychiatry and represents psychiatry on the Interdisciplinary Council of the Association of University-Affiliated Facilities.

Coping with Sexuality and Sexual Vulnerability in Developmentally Disabled Individuals

LUDWIK S. SZYMANSKI

INTRODUCTION

This presentation will focus on the sexuality of retarded persons, including their vulnerability to sexual abuse and exploitation. The material presented here is based on the author's experience in providing interdisciplinary diagnostic and treatment services to retarded individuals and their families in a large urban developmental disabilities clinic (Developmental Evaluation Clinic at the Children's Hospital Medical Center, Boston, Massachusetts).

It may be asked why sexuality and sexual abuse of retarded individuals should be treated as a separate subject. There are several reasons for this approach. The first reason is related to the relative uniqueness of at least certain aspects of the environments in which these individuals are brought up, and of their life experiences which may lead to sexual abuse. Included here are: prolonged, often lifelong dependency on their caretakers; cultural myths and misunderstandings about mental retardation; parents' emotional reactions to their retarded child; isolation from the mainstream of society's life, to the point of virtual incarceration

LUDWIK S. SZYMANSKI • Developmental Evaluation Clinic and Department of Psychiatry, The Children's Hospital Medical Center, Boston, Massachusetts 02115.

in special institutions. All these present special opportunities for sexual abuse to arise.

The second reason for separate discussion is the fact that mentally retarded persons are subjected to a unique form of sexual abuse, which is denial of their sexual rights. This may range from punitive suppression of masturbation to enforced segregation by sex and forced sterilization.

The third reason is the fact that most professionals in health services disciplines have had no training or clinical experience in working with mentally retarded patients. They usually have no skills in interviewing and assessment techniques necessary here and thus may be of little help when called upon to consult in cases of the sexual abuse of a retarded child or an adult.

Retarded adults are not children and should not be treated as such. There are, however, certain conceptual and legal similarities concerning the sexuality of these two groups, the most important of which relates to their ability (or inability) to give meaningful consent to participate in sexual activity.

From a legal point of view a child is a person under a certain arbitrarily chosen age. Sexual activity with such a person, even if the child willingly participates, is forbidden by law. For instance, Massachusetts statute 265, section 23 states: "Whoever unlawfully has sexual intercourse or unnatural sexual intercourse, and abuses a child under sixteen years of age shall, for the first offense, be punished . . ."

It is commonly believed that the person who attains this arbitrarily chosen age has sufficient psychological maturity to give free and meaningful consent to participate in the sexual act. Yet, in the light of modern knowledge of child development, such belief is a fallacy, since it does not take into consideration individual differences in psychosocial development, which are not correlated with age.

When one now turns to the mentally retarded individual, it becomes even more obvious that there is no correlation between chronological age and attainment of various skills, in particular skills necessary for social adaptation. This would include judgment necessary for giving meaningful consent for sexual activity. In this respect a mentally retarded adult (by chronological age) is considered by some laws, to be in the same category as a child below the age of majority. For instance, Massachusetts statute 272, section 5 states: " . . . whoever has unlawful sexual intercourse with a female who is feebleminded, an idiot, an imbecile, or insane, under circumstances which do not constitute rape, shall, if he had reasonable cause to believe that she was feebleminded, or idiot, or imbecile, or insane, be punished . . . ." This effectively forbids all unmarried retarded females to have intercourse, although similar laws are not enforced with nonretarded persons.

PSYCHOSOCIAL DEVELOPMENT AND SEXUAL ABUSE OF RETARDED PERSONS

The causes of sexual abuse of the retarded persons, their vulnerability and the long-term effects of such abuse can be best understood in the context of their development. Therefore, psychosexual development of retarded persons will be reviewed briefly. More in-depth discussion of this subject can be found in the works of Bernstein, Webster, and Morgenstern.[1-3]

INFANCY AND PARENTAL ADAPTATION TO A CHILD'S RETARDATION

Environmental influences assume a major importance in the development of a retarded person due to his innate deficiencies and need to rely on external sources of support. This starts with birth (if not already during pregnancy), when the child–mother relationship is established. The dynamics of psychological developments in this crucial period have been discussed by Solnit and Stark.[4] The mother's reaction to the birth of the defective child is seen as partially similar to reaction to the death of a child, in that it entails a loss (of an image of a perfect, expected baby), but different, in that the defective child who stays alive continues to require mother's care and love. They state, "when the mother wards off her feelings of grief by establishing a guilty, depressed attachment to the retarded child, she may fail to relate adequately to other members of the family because she feels she must give her life to the care of the damaged child. Conversely, the mother may identify with her defective child. In identifying with her defective offspring, the mother feels narcissistically wounded. This narcissistic injury is often intolerable, because the mother feels painfully defective as she is caring for her retarded child. The mother's withdrawal thus becomes a denial of the needs of this child. . . . " Thus, an ambivalent child–mother relationship, as pointed out by Tisza et al.[5] is established. In early infancy this may lead to child–mother attachment with elements of symbiosis and rejection. It may be further influenced by the parents' perplexity about the child whom they do not understand and do not know what to expect from; lack of appropriate support from professionals; and slow development of the child with lesser than expected gratifying feedback from him. These disorders of maternal–child bonding have been recently discussed by Emde and Brown,[6] and Stone and Chesney.[7]

EARLY AND LATER PRESCHOOL YEARS

In the toddler stage or, using Eriksonian terms, the stage of acquiring the sense of autonomy while combating a sense of doubt and shame, the retarded child's failures in achieving developmental milestones are prominent. This may further deepen the ambivalent nature of the child–parent attachment and lead to desperate "rescue" efforts through excessive stimulation, or to minimal expectations and understimulation. Most importantly, the retarded child develops and solidifies his low self-esteem, which may be seen as an almost universal characteristic of this population.[1] This is primarily a result of negative feedback which the child perceives from parental and other important figures. The separation–individuation process is interfered with, as well, or even not accomplished at all. On the other hand, if the child attempts to develop some independence, the parents may be concerned that he may become uncontrollable. Even at this early developmental stage a belief in the myth that retardation is associated with a lack of sexual inhibition may be manifested. In some cases the parents may even seek professional advice because of behaviors of their retarded child such as "immodesty," friendliness to strangers, and masturbation.

THE SCHOOL YEARS

This period is marked by the child's confrontation with his learning and social deficits. Mastering tasks such as separation, academic learning, self-control, frustration tolerance, age-appropriate independence, and peer relationship may be difficult for an individual who has not acquired, as yet, earlier developmental skills. Parental ambivalence and pressure to achieve may deepen, and the child becomes even more aware of their dissatisfaction with him. Depression, anxiety, hyperactivity, and school phobia may appear. The defensive and compensatory strategies employed by the retarded child may include aggressive behavior, regression, intensified dependency, and withdrawal. These children may become sexually misused in a variety of ways. Nonretarded peers may use them as objects for sexual exploration and experimentation. The retarded child's wish to be accepted and his lack of awareness of what is socially acceptable may make him a vulnerable, if not a willing, subject.

Interesting dynamics may sometimes be observed in cases of retarded girls. As mentioned before, Solnit and Stark[4] and Tisza et al.[5] have pointed out the ambivalent nature of the relationship between the mother and her defective child. The attachment to the child may be a guilty and depressed one. The mother may, however, develop also an identification with the child, but with negative connotations, since the existence of the retardation is a narcissistic injury to her. As a result the mother withdraws her emotional investment from the child, although she may continue to overprotect her in a guilty and angry way. Thus the mother will not be available to her daughter as an object for feminine identification and as a protective figure, which of course will hamper the little girl in her successful resolution of the oedipal conflict. A ripe opportunity for sexual misuse arises, if the father is a basically passive person with poor masculine identification, unable to establish a mature relationship with his wife. For such a man the sexualized relationship with a compliant, passive, affection-seeking daughter may serve an important role in maintaining his self-esteem. Under these circumstances the child may willingly cooperate, finding in such relationship a source of affection and a reality fulfillment of her unresolved oedipal wishes.

Such cases are some of the most difficult to detect, since families, in general, tend to keep them as a "family secret." Only more enlightened ones will seek professional help (providing it is readily available). The perpetrator may be a family member other than the father. Since in these cases the sexual relationship is usually not forced violently upon the child and often answers her needs for affection, the child may not see it as something bad. However, the reaction of other members of the family and, not infrequently, of professionals may convey the message that she has done something terrible, although the nature of this behavior may be unclear to a girl, who as a rule has not received any sex education.

> Fifteen-year-old Joan was referred for evaluation of her educational status and future plans. At the age of 20 months she had had severe lead poisoning, following which she developed seizures, and her previously normal development became compromised. Since the age of five she had attended ungraded classes for mentally retarded children. Her intelligence was within the mildly retarded range.
>
> She was a constricted and isolated girl, without friends. Her parents, warm and concerned people, had overprotected her. Joan started menstruat-

ing at 11 years of age, but the subject of sexuality had never been discussed with her. At 14 years of age her 17-year-old brother, an isolated and withdrawn boy, impregnated her. The family sought help from their physician, priest, lawyer, and court. As the result of their advice the parents took Joan to a hospital where an abortion and tubal ligation were performed. Joan was told that she had an "infection" and "a cyst." After surgery her mother talked to her about sex and Joan admitted to having done something like that with the brother. In a psychiatric interview during the evaluation she appeared anxious and moderately depressed. She still referred to "an infection" but was reluctant to discuss it partly because she felt embarrassed and partly because her mother had told her not to talk about this subject. There was no evidence that she knew she had been pregnant and that her tubes were tied, but clearly she believed she had done something bad and shameful.

Joan was referred to a group therapy program, which consisted of groups for retarded youngsters and included family meetings. Both she and her parents did well, verbalizing their concerns about the future, discussing the problems of socialization and work, and learning much-needed communication skills. In the process, good communication between Joan and her parents was developed. However, the issue of her pregnancy and the tubal ligation were never brought up.

ADOLESCENCE AND ADULTHOOD

Adolescence is probably the most trying developmental crisis for the retarded individual and his family. First of all, they have to face a confrontation with the permanence of the intellectual defect. It is hard to deny it after years of special education or to maintain a fantasy that, though different teaching, the youngster will "catch up" with his normal peers. The retarded youngster has not usually mastered earlier developmental tasks; thus meeting the challenges of adolescence may be most difficult.

The predominant concern at this stage is about social competence. The educational–rehabilitational facilities for retarded youngsters provide them with vocational training and jobs, but not age-appropriate friends. In multiple family therapy groups for retarded adolescents conducted at the Developmental Evaluation Clinic, the parents would invariably bring up, with underlying anger, their child's isolation, inability to have friends, and resulting dependence on them. Friendship with other retarded people is often difficult because they may live far away. Also, the youngsters themselves, as well as their parents, are often still unwilling to identify with other retarded individuals. Contrary to the usual image of the retarded person as innocent and happily not aware of his problems, retarded adolescents (except profoundly retarded) are greatly preoccupied with their deficiencies. They see their level of functioning as a status symbol and as a question of being good or bad. This apprehension and preoccupation with their retardation and social status has been pointed out by Bernstein.[1] In institutions this has led to development of a sort of caste system, with high-functioning older individuals intimidating and exploiting lower-functioning ones. Young adults in the rehabilitation program at the Developmental Evaluation Clinic have refused to request half-price public transportation passes for the handicapped, since on the application form they had to identify their handicap.

In most mildly and moderately retarded adolescents the physical and physiological changes of puberty occur normally. Delayed cognitive and emotional growth leaves these youngsters puzzled and anxious about their bodily changes and the sexual excitement they may experience, especially if they have not received sex education. They are under considerable pressure, both internal and external, to emulate their nonretarded peers. Their families often ambivalently wish that their retarded child would have friends of the opposite sex and eventually marry, yet fear him or her becoming sexually active. Some parents resort to close supervision of their children, to the point of isolating them at home. Sex education becomes a taboo subject. At the Developmental Evaluation Clinic it has not been unusual that parents would ask for help in obtaining sterilization for a retarded adolescent, or even latency-age child.

Retarded adolescents thus have little opportunity to satisfy their sexual curiosity and drive. They may become victims of sexual exploitation and misuse by unscrupulous "normal" persons. Often they may actively or passively cooperate, in order to gain attention or peer acceptance. A retarded woman may see pregnancy as a status symbol and delivering a normal child as an evidence of her own normalcy as a female. In one of the large state institutions, visited by this author, some young females would regularly escape and after a period of time would return pregnant, becoming "heroines," since they had proven that they were desired (and impregnated) by normal men in the outside world, rather than by retarded residents of the institution.

This behavior of retarded persons should not be seen as an evidence of their increased sexual drive or of decreased moral standards, but as result of the indirect sexual abuse by society, through not providing them with education, vocational and social opportunities, sex education, or protection in a normalized environment.

DENIAL OF RIGHT TO SEXUAL EXPRESSION

In a strict sense the term *sexual abuse* usually denotes that certain sexual activity has been perpetrated upon a person, without a valid consent. A more proper and general term would be *abuse of sexual rights*, which would include both active abuse, as described above, and denial of sexual rights (such as the right to engage in sexual relations, to marry and have children, to have access to sex education, birth control, and voluntary sterilization, and the right to refuse involuntary sterilization). Szymanski and Jensen[8] have pointed out that this denial of sexual rights is actually the most common form of sexual abuse of retarded individuals (who share this dubious distinction with prisoners and hospitalized chronic mental patients). They also point out that this form of sexual abuse has to be understood in the context of society's attitudes toward retardation in general, at the given period. (Concise and excellent reviews of the evolution of these attitudes can be found in papers by Potter,[9] Menolascino,[10] and Donaldson and Menolascino.[11])

From early times retarded individuals were looked upon as unable to inhibit sexual drives and apt to procreate indiscriminately and thus perpetuate their retardation (considered at those times a hereditary condition). These views led to the eugenic movement, involuntary mass sterilizations, and sex-segregated "warehousing" of retarded individuals in large custodial institutions. In the past decade the societal attitudes have been progressively influenced by the principle of normalization, suggested first by Nirje[12] and further elaborated by Wolfensberger,[13] which recognizes retarded persons' right to live in conditions as close to normal as possible. However, old, negative views are slow to change. Morgenstern,[3] who studied community attitudes toward retarded persons as sexual beings, divides them into: (1) the view of the retarded person as a subhuman, animallike, sexual being; (2) the child–innocent view, essentially of an asexual being; and (3) the view of a developing person, with potential for growth, who has a right to normalized functioning in all areas, including the sexual.

Szymanski and Jansen[8] have pointed out that psychodynamically the view that retarded persons are eternal children may lead to suppressing their sexual expression, as sex is seen as the privilege of parents (adults), and the notion of a child engaging in sex may be threatening, as it evokes one's own suppressed oedipal fantasies.

Through the years, a frequently heard rationalization for suppressing the sexuality of retarded persons has been the theory that if given even as little as sex education, because of their lack of normal judgment and of inhibitions, they will progress rapidly to procreation, and to propagation of their (hereditary) retardation. However, there is no evidence to support this view.

Edgerton[14] suggested that retarded persons might exhibit rigid Victorian moral standards. In our own clinical practice we invariably find that (1) sexual acting out by retarded adolescents and young adults is relatively rare; (2) if it happens it is usually the result of a combination of low self-esteem, lack of resources and environmental opportunities to succeed, depression, sexual ignorance, or exploration by "normals," and not of innate "moral retardation"; (3) sexual acting out can be prevented by early counteracting of the above factors; and (4) good response of retarded persons to appropriate sex education is the rule, rather than the exception. Sexual (and delinquent) acting out among "low-functioning" and school-failing teenagers should be seen in a different class, as a function of poverty and other social factors, rather than of retardation, regardless of how they score on intelligence tests. Cushna[15] pointed out this relationship between these social factors and the spurious diagnosis of retardation, especially if it is based on IQ tests geared for middle-class persons, which require high verbal skills.

The study of Reed and Anderson[16] points out that, although there is an increased risk of retardation in offspring of retarded parents, if all the retarded persons procreated at the same rate as their normal siblings (which is unrealistic, since many are sterile), the total percentage of the retarded persons in the population would increase only from 2.0% to 2.2%. Besides, only a minor proportion of mental retardation can be proven to be due to genetic etiology. In our clinic this has been about 4%.

Vitello[17] discusses two other arguments against procreation by retarded persons. The sociological argument claims that they are unfit parents, unable to provide a proper environment for the child. However, there is no strong supportive data to that effect, no clear definition of "proper environment," nor any requirement of persons not labelled retarded to prove, before they procreate, that they will be fit parents.[18] However, as the President's Committee on Mental Retardation[19] points out, every child has the right to be well born. Professionals in this field are aware that providing comprehensive care, and particularly guidance, to the child may be very difficult for a retarded parent, and with each additional child there is an increase in stress, threatening the stability of the marriage. However, there are no good research data on this subject. Mattinson,[20] in follow-up study of 32 marriages of retarded people, reported that they had a total of 40 children, 34 of whom still lived with their families. Neglect was found or suspected in 7, and 3 were "retarded in their development," but no evaluation of the development of all the children in this sample had been done. Floor *et al.*[21] surveyed 54 couples with at least one partner previously institutionalized who had a total of 32 children. One was removed from the home by court order and two were placed voluntarily by the mother. Again, the children's development was not assessed.

In summary one can well agree with Vitello[17] that retardation *per se* does not necessarily mean that the individual in question will be an unfit parent. However, it may also be stated that as a group retarded parents have considerably more difficulty in providing appropriate parenting, and the stress of parenting may be the critical factor contributing to deterioration of a previously adequate adjustment.

Clearly, in light of the above, sex education and planned parenthood for retarded persons are of major importance. For years, their procreation was prevented (besides by institutionalization in unisex environments) by sterilization, usually involuntary. The current developments in this respect are discussed by Bass[22] and Vitello.[17] Owing to continued administrative and legal developments, the status of sterilization is currently in flux. Some courts hold that no retarded persons can give valid informed consent for sterilization, and that no one can give it on their behalf. However, many mildly and even moderately retarded persons, while incapable of total independence, can well understand the nature of the sterilization procedure to give valid agreement (or refusal). One could see this issue as one of denying retarded persons the right to obtain sterilization; indeed, a 1978 California regulation forbids such automatic denial. However, recent banning of public funds payments for sterilization (and abortion) may have a similar denying effect, since many retarded persons depend on public support for their medical care. Some states will still permit sterilization of persons judged incompetent, if proper due legal process is observed, safeguarding their rights. Therefore the professional in the field will be well advised to check current legal status of this issue in his state.

Thus, in summary, one may agree with Bass[22] that the right to procreate should also consider a child's right to be born to parents who are able to give adequate care; of grandparents' right not to be forced to rear a grandchild; of society's right not to be forced to assume financial responsibility for the child. One can also agree with Friedman[18] that any statutes limiting procreation are unconstitutional unless equally applied to everyone, and not to a group of persons on the basis of the label of retardation per se. One may also state that no retarded person should be refused sterilization on the grounds of the retardation per se. Voluntary sterilization should not be denied if the person can give valid informed consent, regardless of his/her competency in other respects, but rigid procedural safeguards which may include court permission should be observed to protect them from abuse or exploitation. No sterilization should be done against a person's will.

Considering retarded persons' vulnerability to exploitation and the general lack of resources and services awarded to them, the importance of providing them with sex education, and, whenever necessary, contraceptive advice and materials, is essential. Until recently sex education for retarded persons was an exception rather than the rule. In the study of Backer,[23] only slightly more than half of retarded adolescents had concrete knowledge pertaining to sexuality, but even fewer knew the facts of conception, contraception, and venereal disease. Closely related is the need for access to contraceptive advice and materials for sexually active individuals. Bass,[22] who discusses this subject and reviews the available literature, points out that for retarded individuals unable to use the temporary methods consistently, birth control may be difficult since the recent Food and Drug Administration ban on long-acting hormonal preparations (Depo-Provera). Intrauterine devices and voluntary sterilization may be the remaining solutions; however, the legal obstacles to the latter may be insurmountable.

Marriage of retarded persons has often evoked opposition, even among professionals in the field, on the grounds that it leads to procreation and that retarded persons cannot understand the responsibilities inherent in the marital contract. Yet no such requirement is made of "normal" persons, even if one is sure that they are irresponsible and will procreate indiscriminately. Bass[22] points out that marriage of retarded persons is restricted in 40 states. However, according to Friedman[18] the right to marry is fundamental to all citizens. Obviously, most retarded persons denied such rights will not resort to legal battle. Yet even more than "normals" they need friendship, companionship, and a long-lasting relationship, which marriage can provide. With sex education, contraception, and support services, many can function well as marriage partners.[20,21]

MANAGEMENT OF THE SEXUALITY OF RETARDED PERSONS

Any professional called upon to counsel on the sexuality of retarded individuals should be aware of, and able to work through, his/her misconceptions and conscious and unconscious biases about mental retardation.[24]

Such professionals should be trained and experienced in the field of mental retardation and sexuality, regardless of their own basic disciplinary background. Training in human development and in biological and behavioral sciences (such as the training of child psychiatrists) may be the most helpful. The best approach is one of warmth and empathy, eclectic, and employing different perspectives, such as sociological, behavioral, psychodynamic, and educational.

DIAGNOSTIC ASSESSMENT

The sexuality of a retarded person can be understood only within the context of the total development, adjustment, and abilities of that individual, as well as his/her family situation. The latter includes the family's attitude toward mental retardation in general and this retarded person in particular and the parents' ability and motivation to foster their child's independence and to see him as a sexual being.

The retarded person's knowledge of sexuality should be assessed thoroughly in the context of the general developmental level, life experiences, and education. It should not be inferred from the intelligence level. As pointed out by Johnson,[25] there is little correlation between sex IQ and general IQ. If necessary, comprehensive developmental evaluation may be requested first. The interview with the retarded individual (evaluating his/her sexuality) is essentially like mental health evaluation. (For a review of the principles of the latter the reader is referred to Szymanski.[24])

SEX EDUCATION AND COUNSELING

These two topics cannot be separated. A competent professional in the mental retardation field, regardless of discipline, should broach the subject of sexual development with the parents of a retarded child when the latter is quite young, e.g., in the preteen period or even earlier. In other words sexual development should be considered an inseparable part of the long-term management of mental retardation.

Counseling the family and/or other caretakers or educators is essential. Its principles have been recently reviewed by Hagamen.[26] In general, these include, among others, (1) working through one's own ambivalence, guilt, and other feelings about having a retarded child; (2) modifying such inappropriate compensatory reactions as overprotection, overstimulation, or rejection; and (3) guidance in bringing up the child and in securing necessary services. Only within this context should the parents be helped to work through their misconceptions and biases about the sexuality of their retarded child. They should receive concrete advice and training to provide basic sex education to their child. This should be closely coordinated with similar educational programs that the child may receive at school.

The subject of sex education for retarded persons has been reviewed in recent years by a number of authors. Very good guides for professionals and parents are available.[27-29] In particular, the National Association for Retarded Citizens provides excellent pamphlets on this subject.

An effective sex education program has to be conducted in a directive fashion, like any psychotherapeutic intevention with retarded persons, yet with warmth, support, and empathy. One has to anticipate the retarded individual's questions, verbalize them for him if necessary, and lead him to find out the answers in a clear, concrete, and unequivocal way. Gordon[29] points out that often a relatively small amount of information is sufficient, if it is given repeatedly. Aids such as concrete models, films, and simple drawings are helpful. Most importantly, one should remember that sex education includes more than just facts of anatomy and physiology of reproduction. The retarded person has to be taught relevant social skills, such as in which situations a particular behavior is socially appropriate, principles of contraception, dating, and how to verbalize his newly acquired knowledge. He has to be led to recognize the responsibilities connected with sexual behavior, marriage, and parenthood.

One has to realize that many retarded young adults may have strong wishes to parent children, even if they are not capable of fulfilling such responsibility. Procreation has for them an importance similar to that which it has for a nonretarded person; it may also mean that in this respect they may be like "normal" persons and thus deny the fact of being retarded. In order to insure their cooperation in a contraception program, they may have to be taught that certain responsibilities, e.g., supporting a family or caring for a child, may be too difficult for them.

Most importantly, retarded individuals should be provided with proper education, habilitation, a normalized living environment, social opportunities, and health (particularly mental health) services. All of these should focus not only on their deficiencies but on their strengths. This should ultimately result in the development of a positive self-image and constructive acceptance of their own limitations. Only then will the individual not need sexual acting out in order to prove his/her own self-worth. For instance, a retarded adult longing to care for a child of her/

his own may recognize that the responsiblities of parenthood may be too difficult, and may be able, without loss of self-esteem, to relinquish the self-image as a parent and substitute for it one of a person successful in another area. Sublimation, e.g., training to become a child-care aide, may also be helpful.

In summary, retarded persons are first of all individuals who have normal human attributes, which include sexual needs. These have to be recognized, supported, and helped to develop in a way commensurate with the person's best interests, which do not need to be mutually exclusive of the needs of society. As in any other aspects of human development, health, and illness, prevention of maladaptation is most important, and providing preventive intervention is a foremost responsiblity of professionals in this field.

REFERENCES

1. Bernstein, N. R.: Intellectual defect and personality development, in *Diminished People*, N. R. Bernstein (Ed.). Boston: Little, Brown, 1970.
2. Webster, T. G.: Unique aspects of emotional development in mentally retarded children, in *Psychiatric Approaches to Mental Retardation*, F. J. Menolascino (Ed.). New York: Basic Books, 1970.
3. Morgenstern, M.: Community attitudes toward sexuality of the retarded, in *Human Sexuality and the Mentally Retarded*, F. F. de la Cruz and G. D. LaVeck (Eds.). New York: Brunner/Mazel, 1973.
4. Solnit, A. J., and Stark, M. H.: Mourning and the birth of a defective child. *Psychoanal. Study Child* **16**:523, 1961.
5. Tisza, V. B., Irwin, E., and Scheide, E.: Children with oral facial clefts. *J. Am. Acad. Child Psychiatry* **12**:292, 1973.
6. Emde, R. N., and Brown, C.: Adaptation to the birth of a Down's Syndrome Infant: Grieving and maternal attachment. *J. Am. Acad. Child Psychiatry* **17**:299, 1978.
7. Stone, N. W., and Chesney, B.H.: Attachment behaviors in handicapped infants. *Ment. Retard.* **16**(1):8, 1978.
8. Szymanski, L. S., and Jansen, P.: Assessing sexuality and sexual vulnerability of retarded persons, in *Emotional Disorders of Mentally Retarded Persons*, L. S. Szymanski and P. E. Tanguay (Eds.). Baltimore: University Park Press, 1980.
9. Potter, H. W.: The needs of mentally retarded children for child psychiatry services. *J. Am. Acad. Child Psychiatry* **3**:352, 1964.
10. Menolascino, F. J.: Psychiatry's past, current and future role in mental retardation, in *Psychiatric Approaches to Mental Retardation*, F. J. Menolascino (Ed.). New York: Basic Books, 1970.
11. Donaldson, J. Y., and Menolascino, F. J.: Past, current and future roles of child psychiatry in mental retardation. *J. Am. Acad. Child Psychiatry* **16**:38, 1977.
12. Nirje, B.: The normalization principle and its human management implications, in *Changing Patterns in Residential Services for the Mentally Retarded*, R. Kugel and W.

Wolfensberger (Eds.). Washington, D.C.: President's Committee on Mental Retardation, 1969.
13. Wolfensberger, W.: *The Principle of Normalization in Human Services.*Toronto: National Institute on Mental Retardation, 1972.
14. Edgerton, R. B.: Sexual Behavior and the Mentally Retarded: Some Sociocultural Research Considerations. Paper presented at the NICHD Symposium on Human Sexuality and the Mentally Retarded, Hot Springs, Arkansas, 1971.
15. Cushna, B.: Psychological definition of mental retardation, in *Emotional Disorders of Mentally Retarded Persons,* L. S. Szymanski and P. E. Tanguay (Eds.). Baltimore: University Park Press, 1980.
16. Reed, S. C., and Anderson, V. E.: Effects of changing sexuality on the gene pool, in *Human Sexuality and the Mentally Retarded,* F. F. de la Cruz and G. D. LaVeck (Eds.). New York: Brunner/Mazel, 1973.
17. Vitello, S. J.: Involuntary sterilization: Recent developments. *Ment. Retard.* 16(6):405, 1978.
18. Friedman, P. R.: *The Rights of Mentally Retarded Persons.* New York: Avon Books, 1976.
19. President's Committee on Mental Retardation: *Mental Retardation: Century of Decision.* Washington, D.C.: U.S. Government Printing Office, 1976.
20. Mattinson, J.: Marriage and mental handicap, in *Human Sexuality and the Mentally Retarded,* F. F. de la Cruz and G. D. LaVeck (Eds.). New York: Brunner/Mazel, 1973.
21. Floor, L., Baxter, D., Rosen, M., Zisfein, L.: A survey of marriages among previously institutionalized retardates. *Ment. Retard.* 13(2):33, 1975.
22. Bass, M. S.: Surgical contraception: A key to normalization and prevention. *Ment. Retard.* 16(6):399, 1978.
23. Szymanski, L. S.: Psychiatric diagnostic evaluation of mentally retarded individuals. *J. Am. Acad. Child Psychiatry* 16:67, 1977.
24. Johnson, W. R.: Sex education of the mentally retarded, in *Human Sexuality and the Mentally Retarded,* F. F. de la Cruz and G. D. LaVeck (Eds.). New York: Brunner/Mazel, 1973.
25. Hagamen, M. B.: Family adaptation to the diagnosis of mental retardation in a child and strategies of intervention, in *Emotional Disorders of Mentally Retarded Persons,* L. S. Szymanski and P. E. Tanguay (Eds.). Baltimore: University Park Press, 1980.
26. Bass, M. *Developing Community Acceptance of Sex Education for the Mentally Retarded.* New York: SIECUS, 1972.
27. Fischer, H. L., Krajicek, M. J., and Borthick, W. A.: *Teaching Concepts of Sexual Development to the Developmentally Disabled.* Denver: Development Unlimited, 1973.
28. Kempton, W.: *Guidelines for Planning a Training Course on Human Sexuality and the Retarded.* Philadelphia: Planned Parenthood Association of Southeastern Pennsylvania, 1973.
29. Gordon, S.: A response to Warren Johnson in *Human Sexuality and the Mentally Retarded,* F. F. de la Cruz and G. D. LaVeck (Eds.). New York: Brunner/Mazel, 1973.

DISCUSSION

MODERATOR: MARY S. CHALLELA

N. CURRAN (Federation for Children with Special Needs): I have a 12-year-old daughter with Down syndrome. I think it's also important to keep in mind that there are many alternative lifestyles today, one of which could be nonmarriage, which could apply equally well for mentally retarded people.

P. WOODWARD (Association for Retarded Citizens, Worcester, Massachusetts): I do sex education with retarded adults. I know many parents who have their retarded children living at home with them. Their ages range from 17 to 35. These parents absolutely refuse to discuss the issue of sex with their children. They think that if they bring up the subject, their child will go out and do everything inappropriately in public. I just would like to know if you have any suggestions on how to make these parents receptive to talking about sex education for these children?

L.S. SZYMANSKI: There are no magic answers. What you describe is a quite classic situation. The first rule is to consider this issue as only part of the total counseling provided. This means one has to develop a long-lasting good rapport with them. They have to trust you. Secondly, one has to talk with them about this issue in a general sense. If they are talking about a child's inability to behave well socially, then we work on that within the general framework, and this can also raise the question of the inappropriate verbalization of sexual knowledge. I've found that the best way is to encourage the parents to express more and more of their concerns and then face them—confront them with the reality. You are not going to keep this child at home forever. What shall we do later? Only when they are led to verbalize this concern, will they come to you asking for help. What they really wait for is confrontation. We tell them as professionals to take an action and they refuse. One has to understand these people and offer them assistance on their own terms rather than demand that they accept our values. There is no reason why they should.

K. NEMETH (Mental Retardation Coordinator, Department of Mental Health): I found the last presentation interesting mainly because it included issues that were directed toward the mentally retarded adult and can be brought to the mentally retarded or handicapped person. Many of the other presentations focused on the care and issues of children, and I think, perhaps I'm prejudiced because I do work with adults, but they are a group that are often neglected. Perhaps they don't receive as many services and research as children but serve as a reminder that these children do grow up. They become adults and their problems don't magically end at the age of 22 when they are dropped by many of the services provided to children. I just think they are an important group to remember.

M. CHALLELA: We couldn't agree with you more. We are very concerned also with what is happening to the adults in the community. There is a movement afoot to look into the needs of the adults in the community and how we can be effective using the same principles that apply to adults as well as children.

A. MILUNSKY: I have a feeling that many people in the audience have not really had the opportunity of caring for children with retardation over long periods. I would like to respond to the question put to Dr. Szymanski about guiding parents in their handling of sexuality. Perhaps one of the most useful approaches is to start early with the care of children with mental retardation. In that way at puberty the parents of a child with Down syndrome have, for example, anticipated menstruation. I think the way to go about this is to introduce in early childhood discussions about the anticipatory developments that are perfectly normal. The question that I have for Dr. Szymanski is about a family that I care for where the boy with Down syndrome entered puberty and exhibited tremendous sexuality. He is constantly going at his mother and his sisters and it is an enormous problem. The matter has come to the point where the mother in particular is demanding castration. How would you handle this?

L.S. SZYMANSKI: I'm glad you brought this matter up. In raising a retarded person, consciously or not we see that person as a child, and it's typical that when the person grows up and is not really a child, he/she is still considered one. It is usually very clearly expressed in the way the person receives affection or attention. Go to any institution. They come and hug you. They sit on your lap or kiss you. They do it because they are asked to do it. They are tolerated and rewarded for doing that. Quite often the family or caretakers cannot even accept that particular youngster who at 18 or 20 sits on their laps and rubs him/herself against them, not simply looking for affection but getting sexually excited. When they suddenly realize that, all hell breaks loose. First thing, obviously, would be to see what this behavior means within their family—how they approach it, how excited they really get about it, and what they can offer this particular youngster as a substitute in terms of more appropriate affections and attention. The second aspect would be to arrange more behavioral techniques for this particular person. If a person learned that a certain pattern of behavior could give the most reward, whatever that is, he is going to do it.

M. CHALLELA: I think another whole issue is that sex is not what you do, it's what you are. Sexuality is the whole issue in terms of general development in talking about what is appropriate feminine behavior, what's masculine behavior? It starts in the beginning. What's it like to be a girl? What do girls do? What kinds of clothes do girls wear? How do they take care of themselves? Taking pride in the care of their bodies will help them gradually learn about themselves. This also applies for boys.

17

ALAN J. BRIGHTMAN, after receiving his Ph.D. in Education from Harvard University in 1975, created and directed The Workshop on Children's Awareness, a nonprofit organization dedicated to increasing children's understanding and acceptance of disabled peers. In this capacity, he was responsible for the development of numerous television, film, and print materials, including FEELING FREE—a six-part children's television series that aired nationally over PBS to more than 4 million viewers weekly—and IN CELEBRATION!—a videodocumentary of the First National Very Special Arts Festival, involving more than 750 disabled children from across the country.

Dr. Brightman is the author or co-author of many books for children and adults, all designed to increase the quality of life for disabled children and their families. These include *Like Me* (Little, Brown), photographs and verse to explain retardation to young readers; *Feeling Free* (Addison-Wesley), a compendium of stories, games, activities, and personal introductions based on the characters and themes of the television series; *Hollis, Ginny,* and *Laurie* (Scholastic Magazines), photo-essays and first-person accounts of children's experiences of disability; and the *Steps To Independence* series (Research Press), nine manuals designed to train parents of retarded children in techniques of home teaching and behavior management.

Dr. Brightman is presently Executive Director of Educational Projects Incorporated, the nonprofit organization of which the Workshop on Children's Awareness is a division.

A Little Bit of Awkward
Children and Their Disabled Peers

ALAN J. BRIGHTMAN

INTRODUCTION: THE MAINSTREAMED TRIANGLE

Fall, 1978. It's the first Friday of the new school year, and the bell will be signaling "weekend" in twenty minutes. Mrs. Simpson quiets her fourth-grade class. "I want to remind you all that next Monday we'll be welcoming someone new into our room . . . Her name is Sarah and she's very, very special . . . Sarah's in a wheelchair . . . but she's just like you . . . and just like me . . . I'm sure we'll all get along just fine."

The silence is soon broken with the simultaneous thudding of books being lumped together, an embarrassed giggle, an "I don't know" and, quickly, a "Don't say that!" Finally the bell. Mrs. Simpson thumbs through Sarah's file on her desk, barely acknowledging the excited exit.

To those of us who have followed the progress of disabled children, particularly during the past ten years, this scene is becoming increasingly familiar. To Mrs. Simpson and her class, the scene is totally new. This mainstreaming, this implementation of federal legislation which provides free and appropriate education to all children, is even happening to them. In their classroom. And what to do?

Mrs. Simpson is most teachers. She has hardly ever known a Sarah in her life. And she certainly was never trained to teach one. In her seven years of teaching (it could as easily be twenty), she has never quite felt so unprepared.

She had a friend who had a retarded boy in his class once. It did not last too long though; it did not work out. The other kids did not

---

ALAN J. BRIGHTMAN ● Educational Projects, Inc., Cambridge, Massachusetts 02139.

know what to make of him; neither did the friend. Now it is her turn. "Just like you . . . and just like me. . . . " But Sarah is in a wheelchair. How much "like" really?

Mrs. Simpson never asked to have Sarah in her class. She did not have the special training. Nor, however, did she have a choice. Hers was the classroom chosen, for some administrative reason, as most appropriate to comply with the legislation.

She never thought too much of the legislation to begin with—except in spirit. Now, in substance, it scares her a little. Mainstreaming. It should be quite a Monday, quite a year.

As for Mrs. Simpson's students, only one of the class of 20 has ever before spoken with someone in a wheelchair. And that someone was an adult.

All of them have questions: "Just like you . . . and just like me. . . . " What's the teacher talking about? Can Sarah toss a ball? How does she go to the bathroom? How come in our class? It must really be sad to live in a wheelchair. What does she do? How does she dress? Reach things? Pledge allegiance? (She'll probably get out of a lot of things, you'll see.) How did it happen? What's she got? How do you ask her? (You don't, for crying out loud; it'll embarrass her.)

But each kid thinks that his or her particular question is too dumb to ask (even though it is not) or that s/he is probably the only one who is thinking it (even though they all are).

Mrs. Simpson's kids are most kids.

Then, of course, there is Sarah; there are many Sarahs. Some are in wheelchairs; others talk by using their hands or they use a gray typewriter that puts dots, not letters, on stiff paper; or they talk funny, kind of slow, and sometimes act weird. Sarah has cerebral palsy, and, like Mrs. Simpson (but not really "like" her at all), she too is a little afraid, a little unprepared for Monday.

Sarah never quite understood why she had to wait an extra week to start school, but she did not question it too much. In her eleven years there was lots not to understand. Like why she was sometimes never in school at all. Or why she was almost never left alone with other kids. Or why things sometimes really hurt.

She knew one thing, though, for sure. She knew that those other kids would have a lot of questions, a lot of stares, and that no one would say too much to her at all.

She also knew that her new teacher would probably treat her just a little too special, a little too uncomfortably special, like she would break or something. It would be embarrassing, again. The teacher would mean well, probably, but, especially in the beginning, things would not just be ordinary; they would be awkward.

There would be a little bit of awkward for everybody.

LITTLE "PROFESSIONALS"

Mrs. Simpson's classroom unquestionably represents a timely and an intriguing problem area for any of a wide variety of professionals to sink their credentialed teeth into. There might, for example, be the medical doctor deciding which combinations of chemicals would most effectively fortify Sarah's resistance to the pressures she is likely to confront. Then of course there are opportunities aplenty for one or another school of psychologists and educators to determine which label best fits her and which battery of tests will most effectively justify that determination. There are, too, the program planners, the evaluators, the parent counselors, maybe even the dentist. And there are the fellow teachers—Mrs. Simpson's close friends—prepared to offer support, advice, and comforting mythology, on demand, and anxious to provide her with intervention strategies for most graciously putting up with Sarah for as long as she needs to.

Last year, no one paid too much professional attention at all to Mrs. Simpson's classroom. It was just an ordinary place, like the ordinary place it had been the year before that. Now, along comes some new legislation and all kinds of professional interest quickly follows.

Notice, though, the particular, almost exclusive, focus of that interest. It is not on the classroom really; it is on Sarah. She is the oddity. She is what is "in." She is the new variable to be worked on, shored up, and fit into the otherwise understood environment. For so many of the professions, it seems, this one point in the triangle is what mainstreaming is all about.

At the Workshop on Children's Awareness, our attention to the mainstreamed classroom remains divided. Sarah and the many children like her necessarily occupy some of our concerns, as do the many Mrs. Simpsons. But, increasingly, in our efforts to facilitate understanding and constructive interactions in mainstreamed settings we have come to rely on a different group of "professionals" for gaining insight and perspective. This is a group of individuals who are, in their own way, no less prone to classifying others and only too eager to intervene. This is a group that has, itself, developed an impressive litany of labels and a concomitant wealth of understanding. It is a group that intervenes automatically in situations like Mrs. Simpson's classroom because its members have learned no other tactic. It is a group that does not ponder or tarry too long before taking action. It is a group that carries around in its collective head one template into which the Sarahs of the world are made to fit. It is a group that has the power to welcome, to reject, to become involved with, or (perhaps the most extreme form of intervention) to ignore.

I refer of course to the non-disabled peer—Mrs. Simpson's regular students—and I propose that these children have more to teach us about the complex system of the classroom than any of the professional groups mentioned above. This is a group which, for us, has acquired nothing but increasing interest and respect from the moment we began developing our series of materials called FEELING FREE. It is a group that taught us, that confused us, that provided us with a rich and useful way to look at and to understand what is going on in mainstreamed classrooms.

My purpose here is to share some of the lessons that these children continue to teach us: lessons about labels, about platitudes, about understanding, and about intervention. I also would like to illustrate how some of what this group has taught us—about themselves and about their roles in Mrs. Simpson's classroom—has so far guided us in the development of a unique and we believe responsive set of materials.

But let me start at the beginning by providing an overview of FEELING FREE, a federally funded project that we began about eighteen months ago.

FEELING FREE: UNDERSTANDING THE TARGET AUDIENCE

FEELING FREE is the name of a package of video, film, and print materials developed by the Workshop on Children's Awareness (a division of the Educational Projects, Inc.) to help introduce third and fourth grade children (primarily) to their disabled peers. The heart of the series consists of six half-hour television programs, which have just completed their premiere run nationally over PBS.

FEELING FREE also includes:

- A series of six 15-minute films developed from the television programs for independent distribution to classrooms
- Seven children's books based on some of the individuals and some of the themes from the television series
- Four different resource books for teachers and educational administrators

- Five posters presenting slogans like: "If you thought the wheel was a good invention, you'll love the ramp" and "Sticks and stones may break my bones but words will really hurt me"

FEELING FREE also included a strong formative evaluation component, which, from preproduction to the premiere of the series, involved screenings and interviews of various kinds with more than 2000 Greater Boston children.

Our approach to understanding who our target audience was led us first to the literature that professed to shed some light on children's attitudes toward disabled peers and on attempts to modify these attitudes in the *right* direction. This was a short stop for us, not a very useful visit. For it seems that in this area, as in so many others, no one can agree on what an attitude is in the first place, never mind how to measure one. Also the right kind of change seemed usually to be indicated by a child saying, very soon after being exposed to one or another kind of intervention, that he or she now *liked* a disabled child better than had been the case only 20 minutes earlier, or that he or she would sure like to sit next to one more often in the future. (There was almost an indiscriminate kind of "Have you kissed a disabled child today?" mentality behind much of what we examined, as if *not liking* were by definition wrong, inappropriate or unacceptable.)

This led us, then, to the kids themselves, our potential third and fourth grade audiences. Our concern was not so much what these kids had to say about their feelings toward disabled peers, or about what they reported their so-called attitudes to be. We were curious, instead, to learn what their direct, personal experiences with disabled peers had been, and what kinds of understandings they had developed as a result. We questioned about 150 of them. We asked them first to give us their definitions for certain disability labels and then to describe people whom they knew who fit these definitions.

In retrospect, what we found out should not have surprised us. Most kids, it seems, have very limited, if any, direct personal experiences with disabled peers. Some will make reference to "that class down the hall" or to "a wheelchair" that they saw in a supermarket one day. One girl even told us that she used to have a guinea pig who she thought was blind (but she did not think that this really counted, and she did not like the guinea pig that much anyway).

Of the children, 26% reported knowing no disabled people at all; 51% reported knowing one or two at most.

No doubt, what the children meant by "knowing" varied considerably, but it appears safe to conclude that whatever *information* these children had about disabilities, it was not based on being with disabled peers very much. More likely, it derived from television and storybook imagery and from the same kinds of backyard sources that deliver up "facts" about sex and other taboo topics of sensational interest.

I should point out that, with only one glaring exception, these children were able to put forward more or less acceptable definitions of disability labels. They knew that blind meant you could not see, deaf meant you could not hear, and so on. What virtually none of them could represent, however, was any kind of appropriate understanding for the term *retarded*. This is where the mythology really came out, and where words like *stupid, weird,* and *broken* got their full play. In fact, in no other case did the children not at least venture a guess as to what a particular label might mean. In the case of *retarded*, almost 20% left the space blank.

One other finding that emerged from this first round of interviews was a beginning sense for the kinds of misconceptions that our young target audience would be bringing to the screen. We asked them to respond with either *yes, no,* or *not sure* to a number of statements concerning the experience and capabilities of disabled people, statements like: "Deaf people cannot get married," "A blind person can teach school," and so on.

As one might imagine, there were a fair number of *not sures* in response to these kinds of items, but in one instance in particular there was hardly any equivocation at all. In response to the statement: "Disabled people are generally sad most of the time," only 17% of the children were unsure. Almost 60% said *yes*. The influence of Jerry Lewis and his telethon colleagues is indeed stunning.

Our next interactions with the target audience revolved around the television set, specifically around a selection of four 6-minute portraits of disabled peers shown in more or less typical daily situations. We were interested to learn, first of all, whether nondramatized glimpses into the life of a disabled peer would even hold our audience's attention, and then what kinds of questions and reactions were elicited by what was shown.

The results are easy to summarize. No matter which portrait we screened (I might add that the four films we used were ones originally produced for the popular ZOOM program and were, as a result, fairly similar in format and production value), the children paid rapt attention; they were riveted to the screen. In addition, the young viewers were quite consistent in their reactions. Almost always they would report wanting to know more about the kids they saw, and, in particular, how these children's days, moment by moment, compared and contrasted with their own.

Less easy to document, but no less vivid to those of us who observed these sessions, was the flood of animated curiosity that was expressed by these children once they learned it was not wrong to talk about subjects like this, that it really was all right to ask questions and to sound "not smart." It was as if the short films began to establish a climate of openness, as if they worked to stimulate rather than satisfy curiosity. All we had to do was to introduce the film's effects in the post-viewing discussion, and it seemed like they would have opened up for hours. It really felt like no one had ever talked with them about this subject before, and they had a lot they wanted to say.

At the same time this research activity was going on, we had to begin wrestling with the overall structure and format of what was to become FEELING FREE. By the third month of the project, we had determined several givens:

1. The show would be assembled like a magazine, with sections of varying length and content tied together throughout each half-hour.
2. There would be a small group of regular cast members who would appear in every show, and with whom the audience could identify over time. This group would also link the elements of the magazine format together and give each program some sense of continuity.
3. The show would not be didactic; it would not feel like school. We were less concerned, for example, that our audience discovered all the different things that can go wrong with an eye or an ear than that they felt they were meeting individuals whom they would like to get to know better. If FEELING FREE was going to be an educational show, then it would be educational because it was social. (Anyway, there was no evidence to suggest that learning about the eye or the ear accomplished anything in the way of increasing kids' understanding about each other.)
4. With the exception of the opening song, nothing in the show would be scripted. We would give the kids of FEELING FREE our structure, but we would not impose upon them our words.

Children have also shown us that while they are confused about the extent to which they can relate successfully to their disabled peers, they are more than willing to wrestle with their confusion. The case for relating does not have to be laid out neatly before them. The opportunities for relating do. Children will not be forced into liking a disabled peer, particularly if the mandate comes from above. But, after all, they are not typically *forced* into liking any peer. Liking, accepting, tolerating, disliking—these ways of relating will naturally differ among children and will need time to be developed. Children need occasions to err and to discover for themselves around themselves.

It has also become clear to us that our young audience will not accept, certainly they will not act upon, platitudes. They can speak the appropriate words (the Golden Rule seems to have been universally memorized among children), and they know how to reap the nodding praises of teachers, parents, and other adults. But they are clearly dissatisfied with generalities. They are asking for information, for answers, for feedback. They want very much to understand, and, as a result, they will inevitably recognize a pleasant-sounding put-off for what it is.

Perhaps most importantly, we have come to respect as never before the tremendous sophistication of children, the significance of their individual and collective reflections, their articulate concerns. Like many other adults, we may well have underestimated how "savvy" these little professionals really are—how much they should be listened to, learned from, and consulted.

Young Teachings

The remainder of this paper will review the teachings of more than 2000 children who participated with us almost daily in the making of FEELING FREE. These young colleagues, in groups ranging from 2 to 25, screened, critiqued, questioned, and discussed every one of the 45 different segments (and, eventually, the sequence of the segments) that comprise the six-program series. Their ongoing direct influence on the making of FEELING FREE was unquestioned and invaluable. So, too, was their influence on our understanding of mainstreaming, of the many Mrs. Simpsons, and of peers like Sarah.

To begin with, our young colleagues reminded us repeatedly that they are not junior adults. Particularly with respect to their disabled peers, their concern is not with issues or semantics; their interest, often passionately expressed, is with the disabled peer as potential friend: Who is the new *person?* Certainly they are confused, uncertain about how to find out what they want to know, but they show no reluctance to barge in where their older counterparts have learned to tread softly or not at all.

We learned, too, that children are fascinated by what it must be like to experience disability—to live in a wheelchair, to need frequent assistance, to not see or hear. But, at the same time, it appears that children have learned to control their curiosity in the presence of adults. They have learned that "we don't talk about things like that."

There are many Mrs. Simpsons in the real world, many who are anxious, reluctant, afraid. Their intentions are almost always good; it is their comfort level which is bad. In the face of new federal legislation, they are feeling put-upon.

What we have discovered throughout the making of FEELING FREE is that Mrs. Simpson has constantly available to her an unacknowledged source of assistance, right there in her classroom, and at no cost to anyone. The help is not in the form of professional intervention, but rather in the children, themselves. If given the opportunity, if consulted, if encouraged, they will enlighten in very important—and, we would argue, necessary—ways.

AFTERWORD

Only a child asks, "Why is the sky blue?" and expects an answer. The child has not yet learned the subtle social graces involved in not generating awkward silences or discomforting glances among adults. The child asks naturally what is on his or her mind—easily, candidly, directly. Only gradually does he or she learn to distinguish the acceptable from the taboo, the nice from the not-so-nice, the easy to answer from the we'll-talk-about-it-later. The child learns that growing up means keeping certain things to oneself, just like everybody else.

FEELING FREE has tried to present a picture of people, primarily young people, whom most of us grew up never knowing. These are the people we were not supposed to talk about, so we stared at and pointed at them instead. These are the people we were not supposed to ask about, so we made up stories and created our own understandings. These are the people who were always kept separate, away from the rest of us, included not in our classrooms, our playgrounds, our storybooks, or our television programs. These are the people we grew up only hearing about or noticing occasionally. We did not know them; we did not have to.

ACKNOWLEDGMENTS.   Support for FEELING FREE was provided by the Bureau of Education for the Handicapped and the Office of Career Education (Department of Health, Education, and Welfare, Office of Education). The author is grateful to Kim Susan Storey, whose efforts as Director of Evaluation contributed significantly to the FEELING FREE program.

18

MARGARET A. O'CONNOR is a home and hospital instructor for the Boston Public Schools. She has been specially assigned to the Shriners Burns Institute in Boston since 1974. She received her B.A. at Boston College and hopes to receive her masters degree in Education from Boston University.

She is a member of the Council for Exceptional Children and is active in the Association for the Care of Children in Hospitals (ACCH) and the American Burn Association. She has served on the Special Needs Advisory Council of the Cleveland–Marshall Community School in Dorchester, Massachusetts.

She has had extensive experience in teaching severely burned and disfigured children as well as aiding them and their families in the returning-to-school process. Her summers have often been spent as Special Education Director of the Easter Seals Project Summertime. This has been a summertime intervention program for multihandicapped children together with nondisabled inner-city children. Ms. O'Connor has had extensive experience in establishing hospital–school liaisons for the burn-injured child as well as using the law to secure appropriate placements and services for burned children.

Teaching Teachers to Cope

MARGARET A. O'CONNOR

On November 29, 1975, then-President Gerald Ford affixed his signature to the Federal Education for All Handicapped Children Act of 1975 (P.L. 94-142). This historic legislation, with its mandate for drastic change in the educational policies and services which directly affect handicapped children, was somewhat anticlimactic in Massachusetts, where similar legislation had already been implemented, beginning in September, 1974. As a result of the landmark legislation noted above, the educational community received a mandate to put into effect almost immediately programs which would provide a comprehensive approach to the evaluation, placement, and servicing of all handicapped children. Furthermore, the new law was quite specific about the nature and kind of approach to be taken. To the parents, guardians, and advocates of handicapped individuals aged 3 to 21 (the ages included under the law) the new policies gave, for the first time in many if not most cases, a decisive role in the evaluation and placement process. For the handicapped individuals serviced, the law provides "free appropriate public education which emphasizes special education and related services designed to meet their unique needs. . . . " (P.L. 94-142). In addition, and for the purpose of this paper, it is most significant to note that the law also required state and local educational agencies to develop procedures to "ensure that, to the maximum extent appropriate, handicapped children in public or private institutions or other care facilities are educated with children who are not handicapped, and that special classes, separate schooling or other removal of handicapped children from the regular educational environment occurs only when the nature or severity is such that education in regular classes with the use of supplementary aids and services cannot be achieved satisfactorily" (P.L. 94-142).

MARGARET A. O'CONNOR • Shriners Burns Institute, Boston, Massachusetts 02114.

While all aspects of the implementation of P.L. 94-142 presented a major challenge to educational professionals everywhere, it is the doctrine of "least restrictive environment," commonly referred to as "mainstreaming," which directly and drastically affected the everyday routines of all teachers and schoolchildren everywhere. This provision appeared in direct contrast to the widespread and widely accepted procedure of classification and subsequent segregation of most children who deviated from the accepted norm intellectually, educationally, physically, and/or functionally. This type of special education service delivery system generally placed such children in smaller classroom units often with caring teachers where, it was widely held, they were better serviced and all parties involved benefited. The law not only challenged or changed but also revolutionized this method of servicing children.

If, as was the case for some veteran special educators, the changes and the threat of the end of a self-contained, supportive, and protective environment for "their kids" seemed difficult, for many regular classroom teachers the idea seemed impossible. The language of the legislation was indeed strong and rumors quickly spread that all special classes would be shut down. This notion, though never the intent of the law, panicked many teachers, administrators, and parents of both handicapped and nonhandicapped children, as well it should. The intent of the reformers was not to shut down, but to open up, that is to open the doors of the "regular" classrooms to handicapped children who could benefit from them.

The debate over the practicality of the new regulations continues, even after five years of implementation. Yet one thing is clear. Many handicapped children are receiving some or most of their instruction in regular classroom settings. For the classroom teachers, who had always been led to believe that "special" children required "special" teaching by "special" teachers, the idea of assuming responsibility for even part of their instruction was often anxiety-provoking. Even teachers who embraced mainstreaming in theory, felt that the new regulations asked for too much too soon. In many school systems where "special" and "regular" class structures operated entirely independent of each other, teachers were receiving special needs children with inadequate support and/or preparation from special education professionals. In many cases integration of special needs students into the mainstream was occurring in advance of the administrative reorganization needed to insure the success of the program. In those beginning years, between the radical new procedures, paperwork, and sheer magnitude of change, it is no wonder that classroom teachers were having difficulty coping.

The educators of Massachusetts have come a long way since the fall of 1974, and many children have been successfully returned to their rightful place alongside their peers, while still receiving the special educational services they require. In addition, special educators and their regular classroom counterparts have, in many communities, learned to work together to achieve the objectives of a child's Individualized Educational Plan. Resistance to integration has begun to be replaced by acceptance of integration as standard educational practice. With appropriate administrative and interdisciplinary support, children can be successfully serviced in the mainstream of educational services.

Certainly then the kinds of coping difficulties teachers are experiencing have changed in recent years. From a generalized sense of helplessness and resistance to servicing handicapped children, teachers are currently experiencing frustrations and difficulties in dealing with specific children in specific situations. For those of us who deal with children with special needs outside a school setting, it is not a question of *teaching* teachers to cope; rather we can provide support and information to enable them to cope and better service children with abilities, limitations, and/or special needs with which they may not be familiar. In addition, professionals with expertise related to the handicapped population can and should be available to schools and individual teachers as they struggle with their personal feelings and emotions about children's disabilities. This kind of support is especially necessary in the case of severe handicaps and/or observable physical deformities. The following is a brief description of one support model, as practiced by the Rehabilitation Team at Shriners Burns Institute in Boston, on behalf of children returning to school after a disfiguring burn injury:

> When planning for a child's discharge from Shriners, school considerations are second only to plans aimed at returning a child home to his family. Actually, the hospital staff has in mind, almost from the time of admission, the necessity for a strong hospital–school liaison effort if the child is to recover from his injury and return to his former social circle. For the school-aged child society is virtually synonymous with home and school. Thus, initial contact with the school is made, often within days of admission. Depending on the nature and severity of the injury, this contact is made by the social worker, the hospital teacher, or both. The effect of this initial contact can open lines of communication, provide the school with accurate information regarding the child's condition, obtain pertinent background information which may be helpful to the hospital staff as they develop a care plan for the child, or set up an educational program for the child in the hospital. Teachers are encouraged to visit the hospital whenever possible. In any event, mail and telephone contact with teachers and schoolmates can serve as the child's link to the "normal" world. In some cases, conference calls between a child and his class have proven valuable.
> As discharge time approaches, school contact intensifies. Often, what

was initially a teacher-to-teacher relationship revolving around a child's hospital instructional program expands to include social workers, nurses, and therapists at the hospital, as well as administrators, nurses, guidance counselors, and others at the school. The ease or difficulty with which a child makes the initial readjustment to his community depends in large measure on the actions and reactions of the adults in his home and school, as well as his friends, siblings, and classmates. It has been our experience that it is the trusted adults in a child's life who, if accepting and supportive, will be able to transmit those same emotions to siblings and schoolmates. Likewise, if adults are overcome with fear, guilt, anger, or pity, children, especially the younger ones, will pick up on it, and respond accordingly, with the potential for disastrous consequences. It is often the hospital social worker who helps both parents and teachers deal with their feelings in this regard.

Once the school staff has come to terms with their own feelings, they are better able to accept and assimilate practical suggestions and pertinent information with regard to the child's individual needs in a school setting. In addition, hospital staff can offer materials and suggestions for answering the questions of classmates, coping with the occasional taunt or awkward moment, as well as providing schoolmates with a broader understanding of burn injuries and their consequences.

The extra effort expended in helping teachers deal with the unique needs of a disabled child results in a greater ability to cope on the part of all involved.

BIBLIOGRAPHY

Beery, K.: *Models for Mainstreaming.* San Rafael, California: Dimension Publishing, 1972.

Cahners, S.S.: A strong hospital–school liaison; A necessity for good rehabilitation planning for disfigured children. *Scand. J. Plas. Reconstr. Surg.* 13:167, 1979.

MacGregor, F.C., Abel, T., Bryt, A., Lauer, E., and Weissman, S.: *Facial Deformities and Plastic Surgery: A Psychological Study.* Springfield, Illinois: Thomas Publishing, 1953.

Weintraub, F., Abeson, A., Ballard, J., and La Vor, M. (Eds.): *Public Policy and the Education of Exceptional Children.* Reston, Virginia: The Council for Exceptional Children, 1976.

Wolfensberger, W.: *Normalization.* Toronto: National Institute on Mental Retardation, 1975.

19

SUE S. CAHNERS is Director of the Social Service Department at the Shriners Burns Institute in Boston. She obtained her B.A. at Smith College and her masters degree in social work at the Simmons College School of Social Work. She is a member of the Academy of Certified Social Workers. Her internship was spent in the Psychiatry Department at the Massachusetts General Hospital and at the Faulkner Hospital in Boston.

She has had long experience in working with and caring for burned children. Her work and publications have been especially oriented toward the rehabilitation of these children and working with their parents and families.

Coping Strategies of Children and Their Families

SUE S. CAHNERS

If we recognize that the family is the most pervasive influence in a child's development, a critical factor in the successful or unsuccessful adjustment of a badly disfigured child is his family's reaction to this problem and its ability to support him, to help him pursue the long course of treatment and its many associated problems, and also to help him navigate the social world into which he must go. The treatment plan for a disfigured child must therefore include careful and sensitive work with the family from the very day of admission, on through the years of reconstruction and rehabilitation. Adults with disfiguring scars see themselves in the mirror and know they have changed and may ask for help in adjusting. Children on the other hand do not depend on glass mirrors; they see themselves as they are reflected in other people's eyes and they feel the shocks of social rejection bit by bit as they grow up and encounter new people and new experiences. Along the way, they seek to be nurtured, protected and influenced by a family, be it their natural family, foster family or relatives. If this family is emotionally healthy, the child may reap enormous benefits as he matures to adulthood, though bearing a handicap.

Therefore, in addition to paying very close attention to a disfigured child's emotional reactions to hospitalization, pain, school, playground, and family, it is most important to recognize the continuing burden carried by parents whose children have endured a physical trauma.

In burn accidents specifically, a high incidence of emotional disturbance in the families prior to the burn has been reported, as noted

SUE S. CAHNERS • Shriners Burns Institute, Boston, Massachusetts 02114.

in 1961 by Cope and Long,[1] while Jackson and Woodward's studies[2] revealed over 80% of children showed some emotional disturbance following a severe burn injury. We continue to find a high level of disorganization and emotional disturbance which often has contributed directly to the accident. Most often, this disorganization or emotional disturbance which contributes to neglect in child care or acting out behavior has gone unnoted by helping agencies, so the admission to hospital may occur as the culmination of a series of incidents and is the first time that action toward prevention and redirection in care of the child is initiated. But even when the injury is a true accident, careful assessment of family psychodynamics is essential to assure successful reentry and normal growth.

As the social worker, I work openly with the staff and intensively with all parents from the time of admission until discharge. An atmosphere of trust and confidentiality must be established before voluntary participation in working toward change can be expected. This is best facilitated through mental health workers who are trained to communicate in a nonthreatening, nonjudgmental way. The nurse and the doctor are sincerely concerned and are dedicated in their well-meaning intentions to help, but they are often unable to ally themselves with a parent when they are daily caring for the child, nor do they have the time. However, they must show support and interest in carrying through the team effort in understanding what the social worker is trying to achieve.

In the first days after admission, while attention is focused on survival, the social worker gets to know the family and the circumstances of the injury, building clear pathways of communication. This enables us to clarify treatment plans and give much needed support during painful procedures and life and death issues. After the long weeks in the hospital when it becomes necessary to prepare for the return of the injured child to society, and to plan for the protracted period of reconstructive surgery, we have a clear understanding of the family's psychodynamics and can therefore be more effective.

During this reentry period into the home and community we recognize different patterns of adaptation, some related to psychiatric disorders and some to economic class. But no matter what the source of their difficulty in coping and following through with appropriate care, *grief* and *guilt*, as well as *hopelessness* and *helplessness*, are experienced at some time by all. It is these emotions whose negative forces must be eased in order to clear the path for progress in rehabilitation. The helping professionals need to be alert to this and to help support people without forcing the expression of feelings too quickly.

Guilt is expressed by fathers who wonder if this would have happened if they had not been absent. Bystanders feel they should have done more, peers wonder why they escaped, but mostly mothers wonder if they were remiss in supervision of the child who was injured.

To ease their guilt parents usually begin by looking for some logical explanation for the catastrophe, and those working with them help suppress their feelings of culpability by discussing how it was a true random event, an incident about which nothing could have been done, that could happen to anyone. One must also be wary of some parents' need to blame each other either covertly or consciously, in an effort to minimize their own guilt.

Actively involving parents in the care of their children is a very useful means of helping them deal with their guilt and feelings of helplessness. They can help feed, aid in physical therapy, and change dressings. This investment of time and energy in the hard work of care and rehabilitation restores positive feelings and enhances their self-esteem. A free flow of information from medical staff concerning the patient's condition also helps restore some much needed control over the destiny of one's child and need not be a threat to the staff. We recently had a patient, critically ill, whose mother obsessively kept track of the figures for her child's gases—it gave her a sense of sharing with the nurses in the constant monitoring of his condition. On the other hand, some parents are unable to face the daily medical routine and prefer to keep a distance from the treatment plan. This should be respected as well. However, both staff and family members must be wary of the tendency to become overinvolved with the damaged child, which results in his being overprotected and pampered—made to feel set apart from his normal peers—and also deprives other siblings of attention to their needs. A very fine balance must be maintained as we too often hear of acting out behavior developing in the sibling who was unharmed and suffers from deprivation of attention.

The role of denial in maintaining adjustment to serious disfigurement and loss is an important one and serves many useful functions for all members of the family. Their use of denial is often discomforting to staff who fear that families are not facing the reality of disfigurement, but the patient and his relatives need the opportunity to ignore the problem from time to time, giving them a rest to renew strength to face situations where they cannot ignore the disfigurement and its problems.

During the long period of acute hospitalization the social worker can see the parents regularly on an individual basis in an effort to help them deal with the new demands on their emotions and energy. So it is understandable that for many the return home from the supportive hospital staff leaves many families feeling frightened and isolated.

A challenge then is how to most effectively reach out to help these parents help their children and how to keep up with the patients' own feelings when they are out in the community.

Brodland and Andreason[3] wrote that group support meetings for beleaguered relatives during hospitalization served to educate as well as provide emotional support through the most traumatic periods. We have chosen to conduct weekly parent meetings for those who have been discharged and are dealing with the disfigurement on their own in the community. It is often in these parent-to-parent discussions that we learn the most about the hurt to the family. In the classic practice of group therapy, the leader would try to determine most carefully who would be a suitable member by personality, background, and presenting problem. After careful selection of group members, the next requisite would be continuity of attendance. These preliminary requisites are not possible for us; our families come to clinic from great distances and with varying frequency. But, as with Brodland and Andreason, we have found that the existence of one target symptom, that of being the parent of a burned child, gives the group immediate cohesiveness and at times even allows rather intensive work to be done.

Most often one parent will begin by saying "I'm so glad to be back with people who really understand," which blatantly reveals the continuing feelings of isolation. Discussion about a child's refusal to participate in social activities may ensue, or difficulties in handling the name-calling of playmates and the offensive curiosity of neighbors. Or one eager participant may enter the group stating, "I'm exhausted," and is instantly recognized as the parent of a recently discharged patient who is coping with dressing changes, exercises, and strained family relationships. The others recognize the symptoms because they've been there—they reach out and offer help—mutual sharing, which is a great source of comfort and strength.

At times group meetings can help with the mourning process, which may endure many months, even years after the traumatic loss, be it loss of function, loss of self-image or loss of a dream. They gain strength from the opportunity to retell the tragic experience. One mother tearfully shared her story, and added, "It's been two years and this is the first time I've been able to talk about it!" They bring to the group their guilt, their depression, fears, rage, resentment, helplessness, and feelings of isolation. They reach out to each other supported by the belief that only another parent of a burned child really understands.

Group meetings are also held on the Reconstructive Ward—one for teenagers, and one for preteens—and again we learn that mutual sharing of coping strategies among the young people is also very beneficial. Generally we learn that children tend to forget their long painful experience in intensive care and focus on the present. The parents never forget and some anticipate a difficult future. Those children who are natural "fighters"—the feisty ones—seem to do best in dealing with the public-at-large, who shoot rejecting looks and remarks.

A return to the hospital for yet another reconstructive procedure is often a means to have a brief respite as well—a chance for some time of total acceptance. Acceptance as a normal person is the key to successful reentry and normal growth and may lead to acceptance of the person as well as acceptance of the incurability of the disfigurement. The need to accept reality is often overwhelming; every practical effort should be made to involve whole families in the active coping of the patient. There is a balance that must be established between misleading a patient to keep him cheerful and crushing all his hopes and aspirations. The myths that total health will be restored are often a functional part of the coping system and can be used to help in adjustment to life, though it may be an altered life.

So many factors go into a person's ability to cope and to adjust to change. We have learned from observation and from direct communication that it is the adult who has the most difficulty in adapting and therefore needs the most help. Yet children are dependent on these adults for direction and for strength through their acceptance and love. Erikson[4] followed Anna Freud in citing that children who come to feel loved become more beautiful. Disfigured children can be helped to feel more beautiful from within, if their caretakers are able to provide that love.

REFERENCES

1. Cope, O., and Long, R.T.: Emotional problems of burned children. *N. Engl. J. Med.* **264:**22, 1961.
2. Jackson, D., and Woodward, J.: Emotional reactions in burned children and their mothers. *Br. Med. J.* **i:**1009, 1959.
3. Brodland, G.A., and Andreason, N.J.C.: Adjustment problems of the family of the burn patient. *Soc. Casework* **January:**165–172, 1974.
4. Erikson, E.H.: *Insight and Responsibility. Lectures on the Ethical Implications of Psychoanalytic Insight.* New York: Norton, 1962.
5. Bernstein, N.R.: *Emotional Problems of the Facially Burned and Disfigured*, Boston: Little, Brown, 1976.
6. Bernstein, N.R., and Cahners, S.S.: Rehabilitating families with burned children. *Scand. J. Plas. Reconstr. Surg.* **13:**173, 1979.
7. Cahners, S.S.: Group meetings benefit families of burned children. *Health Soc. Work* **3**(3), 1978.

DISCUSSION

MODERATOR: GUNNAR DYBWAD

S. WALSH (Pediatric Nurse Practitioner, Medford, Massachusetts): One of our patients was a participant in the FEELING FREE Program. She was an 11-year-old asthmatic. After filming the program the children were taken to New Hampshire on a camping trip. From that camping trip she was admitted to the hospital. We first assumed that it was due to the camping conditions, her allergies, and the molds, etc., but after she was discharged from the hospital, I sat down and talked with her because she was very upset. She told me how uncomfortable she felt with the other children and this took both her parents and myself totally by surprise. She could not get beyond the differences. What came out about this was her guilt that she was not as different as they were. She was with them and had all these feelings but couldn't express them. Would you please comment?

A. BRIGHTMAN: I remember the weekend well. We had about 40 children at Camp Freedom, 20 who were disabled and 20 who were not disabled. We didn't really know what to expect during that weekend. It was a weekend that we filmed. It was not at the end of making the series. I can't explain her particular problem but our experience with most children has shown that that's going to be the case on occasion. One of the regulars in the show named John is dyslexic. He has one of the "invisible" disabilities, and every time we pulled a regular group of five kids together he increasingly felt a little bit different. His disability is qualitatively different. We took the kids to a fair one afternoon. Five of the regulars went. Four of them have visible handicaps; John doesn't. At every stand we went to throw a ring toss, etc., the barkers were falling over themselves to give away things to the four visibly disabled children, but not to John. I think those things exist in the real world and I don't know what to do about it. It is, however, unequivocally the case that kids, more often than adults, after having an experience with disabled peers come away with a sense of differences generally, rather than a sense of disability. Nondisabled children can look at themes that came up for example, in the FEELING FREE show such as feeling left out or having a tough time making friends. Those are things that are not specific to disability, but rather what you think about when you are a young adolescent. The role that the disabled child has is to put that in italics. To make that really something up front that we can talk about. That's a comment, not really an answer. I didn't know that situation, in fact, existed.

A. McKAY (Worcester, Massachusetts): I have a friend whose child was burned over 80% of her body. She was terribly disfigured and somewhat disabled, having lost fingers on one hand. When she was to return to her school her parents and her grandmother met with great resistance to her returning to the public school system. They were fortunate enough in being able to provide private schooling for the child, but when you meet with that kind of resistance in the school system, what do you do?

S. CAHNERS: We do try to recognize that teachers are people and they bring with them their own frustrations, their own experiences in life with illness and disfigurement, and their own fears. When we first contact a school we usually try to determine their feelings about the thought of a disfigured child coming back to school. We offer, in the Greater Boston area, to bring a program to the school which would help the teachers in preparing the classmates and also at the same time to run a little group meeting with the teachers to discuss with them after school hours, what they can expect. We find through education people are much more relaxed if they understand what has gone on. They are better able to deal with it. What we do ideally is to have the teacher visit in the hospital. If the child's teacher comes to the hospital at some point close to discharge or along the way and has shared the experience with the child, then the teacher has a better understanding and can take the information home and help her fellow teachers and principal understand. Basically what they need to know is that a burn is an injury that can happen to anyone. They have to understand what happens to the skin, and therefore why the skin scars. With a little scientific background, they make a science project out of this, quite often for classes. But people do become much more relaxed if they understand everything about the injury and that can take a little sensitive work. Most often it does work with support and help.

M. O'CONNOR: That is exactly what social workers do in this situation. It is a team effort. I try to be very supportive in my statements. When there is resistance like that, what I do is quote paragraph and verse from the law. It's really important to determine where that resistance is coming from. If it is coming from the direct service level, the people that are having to deal with that, then we ask Ms. Cahners to help. To give you a brief example, we had a typical case of a small child who was burned and who was returning to school. (We have had problems with 5- and 6-year-olds who were burned at 2 and 4 who were entering school for the first time. Many parents resist because of concern about their own child's adjustment. We had parents who said their kids were having nightmares. What we found out was that the parents were saying to their children every day, "Good morning, did you have a nightmare about Timmy last night?" After a week or so, the child had nightmares about Timmy.) So what we did in that case is, understanding the law, helped the parents. There are also advocacy agencies that can do this. We help the parents fight through the due process procedures. We went through a hearing and initially lost. Then we went to a hearing officer and we won, but the school system had the right to reject that. Approaching the Commissioner was next. It took a lot of time and a lot of effort but basically we got that child the right to a seat. Now at the same time we used the kind of resources that Ms. Cahners was talking about.

C. HOURIHAN (Wrentham State School): Listening all day yesterday and part of today I felt that something was missing. On death and dying yesterday no one mentioned death and dying as far as mental retardation is concerned. I have had an experience where I've been in a room where a mentally retarded patient has died. This was my first experience concerning death and dying while working in the infirmary. How do parents of a mentally retarded person feel? Do they die two deaths if their mentally retarded child dies, or is it a sense of relief? Could someone comment?

M. CHALLELA: We have been involved with the parents of retarded children for a long time. Recently there was a family with a severely involved little boy who had a debilitating disease along with his mental retardation. When that little boy died we all felt very, very sad. The nurse who was responsible for helping that mother went to the wake at the funeral home. It happened to be a child of an Italian family and they experienced the same kind of grief and mourning that every family goes through. It was really important for the nurse to follow-up afterwards because there was another child in the same family with the same type of disability. We all grieve, whether we are attendants, families or professionals. It's very hard to take, and the important thing is that you are there when they need you and that you go back afterwards to offer them support.

G. DYBWAD (Professor Emeritus of Human Development, Florence Heller Graduate School of Social Welfare, Brandeis University): Some may have seen newspaper reports of a case now active in California. A child with Down syndrome has some cardiac defect and the parents are refusing permission for surgery even though the physicians want to do it. They say they did not want him to survive them because of what might happen to him. Finally the father said the child would be better off dead. It's always easy to say it's the parents' opinion. What I wonder is, how many professional people, how many psychologists, social workers, pediatricians, nurses, etc. have been feeding this parent misinformation about that child, or have distorted the relationship from the very moment the parents were informed. It's very hard to judge retrospectively. I have the feeling that, in any setting or community you find parents who have difficulty with a child regardless of disability. On the whole I think, what we are seeing in the parents today is the result of misinformation and incorrect advice and they finally don't know how to help themselves at all. My sympathy, in general, is with the parents and I just wonder (even in this case out in California), if it would make a difference if someone would take the time to let these parents see what children with Down syndrome can do. It would be useful if physicians could see what children with Down syndrome can accomplish and become later on. So I am very gentle with parents because I have been so impressed that they are the victims of such very poor information.

B. GINHAM: I am the parent of a child with Down syndrome and this conversation is very upsetting to me. How could anybody possibly think that the death of your retarded child would not make any difference to you, or that you would care less or feel relieved. We were just today discussing my child. I no longer see her as a child with Down syndrome. I don't see her facial features as a child with Down syndrome. She is my child, my daughter.

A. MILUNSKY: I think it has to be said that there are parents whose children are profoundly retarded and whose IQs are barely measurable and who, as they grow, are unable to achieve more than a vegetative state. In these situations the parents may well feel relief. And that's not wrong.

SPEAKER: Even in the case of a child who is profoundly retarded, the grief may be the same. I only know of one instance. My best friend is a twin and she comes from a rural Maine family. Her twin died at 23. He was profoundly retarded. This was several years after he had been institutionalized. He was totally blind, nonverbal, was climbing the furniture, and nobody could restrain him. I know that the grief of that family was the same because I comforted the mother several months after he died. No one else in the family would let her grieve, saying it was better this way since he was profoundly retarded. No one had let her cry for him but he was still her child and that was the main thing.

B. GUILBAULT: I have a brother who died 35 years ago at the age of one. He was profoundly retarded. We kept him alive for almost a year by force feeding, because he didn't have a swallowing reflex. I also have an aunt who is very much alive, 62 years old and institutionalized at the Fernald State School. I recall that looking back on my brother's death (I was 7 when he died), there was an immense relief because he had been so hard to take care of. But through the years the grieving for my brother is the same as the grieving for anyone else who has died. You have to deal with your exhaustion and with the guilt. My brother was handicapped because my mother was exposed to German measles while she was pregnant.

M. O'CONNOR: I just want to comment about your anger response that people would even consider that a child would be better off dead. We are not only talking about retardation. When a child suffers a 90% burn, resulting in no face, no hands, no nose we get the same responses. The answer is not for us to recoil in anger. We have to show people, whether they be the teachers or other professionals, why it is that the child would not be better off dead. Why wouldn't Timmy be better off dead? Because Timmy likes birthday parties, he likes tricycles, and so forth. Children handle discussion about death better than adults. Stop and ask children why Timmy would be better off alive, and they will see the positive things. I think the answer is not to get angry, but to work on it and let other people know why it is that these kids have the right to live and enjoy their lives.

SPEAKER: I would like to comment on the way that this mother spoke about her child. She said my child with Down syndrome rather than my Down syndrome child. I think that if our attitudes are to place the child first, we should also do that in our language.

HENRY A. BEYER is Associate Director of the Boston University Center for Law and Health Sciences. His responsibilities there include directing a project providing training and technical assistance to advocates for developmentally disabled people in New England. He also teaches a seminar on handicap law at the Boston University School of Law.

Mr. Beyer currently serves on the Human Rights Committees of the Massachusetts Developmental Disabilities Council and the Walter E. Fernald State School. He is also a Trustee of the Fernald School and a Director of the Massachusetts Association for Mental Health. He has published a number of articles on advocacy and the rights of handicapped people.

Mr. Beyer has degrees from Duquesne University and Harvard Law School. He and his wife, Janet, reside in Concord, Massachusetts, where they are raising five sons and assorted garden vegetables.

Law and the Handicapped

HENRY A. BEYER

Other contributors to this volume have focused primarily on what "can" be done and what "ought to" be done to help people with handicaps cope with those handicaps. I would like to discuss what "must" be done—that is, what are some of the legal rights of persons with handicaps?

Things that "ought to" be done are sometimes referred to as moral rights or human rights. When there exists a sufficient societal consensus concerning a moral right, and when it is the type of right that lends itself to enforcement by the legal system, then, sometimes, a state legislature, or the United States Congress, or a court will transform that moral right into a legal right—a right that is enforceable in a court of law.

In the decade of the 1970s, this transformation has occurred with a number of specific rights of people with mental and physical handicaps. A significant portion of this progress was in the area of public education. For example, in 1971 and 1972, in two important class action law suits in Pennsylvania[1] and the District of Columbia,[2] federal courts recognized that *all* children (including those who are mentally retarded) are capable of benefitting from education and training and that they have a legal right to free public education, appropriate to their capacities to learn. Those courts held further that such children have a right to the most "normalized" school placement possible—that is, that placement in a regular class in a public school is preferable to placement in a special class, but placement in a special class in a public school is preferable to placement in a special school. Some basic moral rights had become legal rights for children in Pennsylvania and Washington, D.C.

Following these two decisions, a number of state legislatures en-

HENRY A. BEYER • Center for Law and Health Sciences, Boston University School of Law, Boston, Massachusetts 02115.

acted special education statutes, which translated these educational principles into legal rights for children in several more states. Massachusetts "Chapter 766"[3] is one example of such statutes. And in 1975, Congress enacted Public Law 94-142,[4] which extended these legal rights to children in all states receiving federal funds for education (which, I understand, is now all states except New Mexico).

People with mental handicaps have also gained significant legal rights during this decade in the area of institutions—in acquiring rights to treatment or habilitation within institutions (and rights to the improvements in physical plant and staffing implied therein).[5] More recently, they have gained recognition of their right to services in the least restrictive setting possible, which generally means the right to live and receive services in the community rather than in an institution. These rights have been recognized in a growing number of court decisions and consent decrees in every section of the country as well as in statutes passed by Congress and by many state legislatures.[6]

But perhaps the law having the potential to effect the most widespread change in the legal rights of handicapped people is one passed by Congress as Section 504 of the Rehabilitation Act of 1973. It reads, in pertinent part:

> No otherwise qualified handicapped individual in the United States . . . shall solely by reason of his handicap, be excluded from the participation in, be denied the benefits of, or be subject to discrimination under any program or activity receiving Federal financial assistance or activity conducted by any Executive agency or by the U.S. Postal Service. . . . [7]

A legal articulation of a basic moral right: Programs and activities receiving federal assistance (and they are ubiquitous) may not discriminate against otherwise qualified people merely because they are handicapped. What are some of the ways Section 504 has been applied?

There have been a number of judicial decisions concerning public schools. Several courts have held that the failure of school districts to provide mentally and emotionally handicapped children with appropriate educational programs constitutes discrimination in violation of Section 504.[8] Another application of 504—one of special interest to those in the health professions—concerned the New York City school system's proposal for educating certain mentally retarded children recently discharged from the Willowbrook Institution. Under the school system's plan, 48 children who had been identified as carriers of hepatitis B were to be segregated in special classes. But a federal court ruled that this would amount to discrimination in violation of Section 504.[9] The court was persuaded by testimony that the risk of infection was low, that simple prophylactic and classroom management measures were available, and that the stigma resulting from segregating the children could cause regression in their development and undermine employment and community placement efforts on behalf of retarded persons generally. The court also noted that, among a million New York public school children, no other groups had been tested for hepatitis B, nor was any action planned against any hepatitis B carriers other than the former Willowbrook residents. The court considered all of these factors in deciding that the segregation plan constituted discrimination on the basis of handicap in violation of Section 504.

Courts have also found that 504 requires universities to provide interpreters for deaf students[10] and that a partially-sighted university student had a right to play basketball.[11] Although there is no doubt that Section 504 applies to the field of public transportation, courts appear to be less willing to recognize broad rights there—particularly with regard to existing transportation systems.[12] This can no doubt be explained by the high costs involved in making existing systems accessible to persons in wheelchairs. The United States Department of Transportation, which has written federal regulations to implement 504 in this area, has backed off somewhat from its original proposal that *all* subway and commuter train stations be made accessible. But key station accessibility and accessible buses are to be phased in over the next two decades.[13]

Section 504 rights have also been recognized in the area of employment.[14] A United States Court of Appeals has held that failure to employ a blind teacher solely because of her handicap is a violation of 504.[15] And a federal District Court found that 504 was also violated when the City of Tampa, Florida, refused to hire a man as a police officer because of his history of epilepsy.[16]

The case which has received widest attention, however, is the one decided by the Supreme Court, *Southeastern Community College v. Davis.*[17] The Court held that the school did *not* violate Section 504 when it refused to admit Ms. Davis to its clinical program because of her serious hearing impairment. The Court's opinion was certainly not a victory for handicapped people's rights, but neither did it sound their death knell, as some sensationalized news stories implied. What did the Court actually say in *Davis*? It upheld the authority of a professional school to require reasonable physical qualifications for applicants to its clinical training programs[18]; it said that 504 does not require "affirmative action" by the recipient of federal funds[19]; and it interpreted the statute's phrase "otherwise qualified" handicapped individual to mean "one who is able to meet all of the program's requirements in spite of his handicap."[20]

The Court did, however, recognize as law the central thrust of

Section 504—that an otherwise qualified handicapped individual cannot be excluded from participation in federally funded programs solely by reason of his or her handicap.[21] "It is possible," said the Court, "to envision situations where an insistence on continuing past requirements and practices might arbitrarily deprive genuinely qualified handicapped persons of the opportunity to participate in a covered program."[22] The gist of the opinion appears to be that, although recipients of funds need not make substantial alterations to programs, they do have a duty to make reasonable accommodations to the needs of handicapped persons.[23] Admittedly, the distinction between "substantial change" and "reasonable accommodation" will not always be clear. It is thus safe to predict that there will be much more 504 litigation in the next few years as courts attempt to draw this line in specific situations.

One more point should be noted in *Davis*. In deciding the case, the Supreme Court was apparently influenced considerably by the belief that Ms. Davis's disability would endanger the safety of patients under her care.[24] It is quite possible that the Court might reach a decidedly different conclusion in a case where such safety considerations were not present.

Well, what legal work remains to be done? In a word, much. One area receiving increasing attention by handicapped people and their advocates is that of state nondiscrimination statutes. Remember that Section 504 applies only to activities receiving federal funds. Only slightly more than half of the fifty states now have statutes banning discrimination on the basis of handicap in employment, housing, or public accommodations where federal funds are *not* involved. And these statutes vary considerably—some applying in only one or two of those contexts, and some not covering persons with mental disabilities.[25]

Other efforts must continue in the implementation and enforcement of legal rights which already exist on paper. Laws, as we know, are not self-enforcing. For example, Section 504 Regulations promulgated by the Department of Health, Education, and Welfare (HEW) required organizations receiving federal funds to have completed by June 1978 a self-evaluation of their physical and programmatic barriers, and they must have modified any policies and practices not meeting the requirements of those regulations. If structural changes are required to make the organization accessible, a transition plan setting forth the steps which will be taken to complete the changes must have been developed by December 1977.[26] Have all of these organizations completed these studies and plans? The HEW Regulations also require that recipient hospitals establish procedures for effective communication with persons who are hearing-impaired for the purpose of providing emergency health care.[27] Do all hospitals have such procedures? Too often, action to comply with such requirements is not taken until after a handicapped person files a complaint or brings a lawsuit.

But perhaps the greatest continuing need is the need to raise society's consciousness and understanding of the problems faced by handicapped people. A mentally retarded person's legal right to live in the least restrictive environment may have little value if public ignorance, fear, and resulting opposition prevent the establishment of community residences.

Our laws, and how they are observed, evaded, or ignored, provide a rather clear reflection of our society, of the sort of people we are and wish to become. When it comes to helping handicapped people cope with their handicaps, what do we wish that reflection to show?

References and Notes

1. *Pennsylvania Association of Retarded Citizens (PARC) v. Pennsylvania*, 343 F.Supp. 279 (E.D. Pa. 1972).
2. *Mills v. Board of Education*, 348 F.Supp. 866 (D.C. Cir., 1972).
3. Mass. G.L.A. ch. 71B.
4. 20 U.S.C. §§1401 *et seq.*
5. E.g., *Wyatt v. Stickney*, 344 F.Supp. 387 (M.D. Ala. 1972), affirmed in pertinent part sub nom. *Wyatt v. Aderholt*, 503 F.2d 1305 (5th Cir. 1974).
6. See., e.g., *Halderman v. Pennhurst*, 446 F.Supp. 1295 (E.D. Pa. 1977); P.L. 95-602, Developmental Disabilities Assistance and Bill of Rights Act, §111.
7. 29 U.S.C. §794. The portion following the words, "Federal financial assistance," was added by the Rehabilitation, Comprehensive Services, and Developmental Disabilities Act of 1978. 92 Stat. 2982.
8. See, e.g., *Howard S. v. Friendswood Independent School District*, 454 F.Supp. 634 (S.D. Tex. 1978); *Lora v. New York Board of Education*, 456 F.Supp. 1211 (E.D. N.Y. 1978); *Mattie T. v. Holladay*, C.A. No. 75-31-S (N.D. Miss., Consent Decree, Feb. 22, 1979). (Previous summary judgment held that lack of educational programs for handicapped children was a violation of Section 504.)
9. *N.Y. State A.R.C. v. Carey*, C.A. No. 72-C-356 (E.D. N.Y., Feb. 28, 1979).
10. *Camenisch v. Univ. of Texas*, C.A. No. A-78-CA-061 (W.D. Tex, May 17, 1978); *Crawford v. Univ. of North Carolina*, 440 F.Supp. 1047 (M.D. N.C., 1977).
11. *Borden v. Rohr*, C.A. No. 75-844 (S.D. Ohio, Dec. 31, 1975).
12. See, e.g., *Snowden v. Birmingham–Jefferson County Transit Authority*, 407 F.Supp. 394 (N.D. Ala. 1975), affirmed 551 F.2d 862 (5th Cir. 1977); *Vanko v. Finley*, 440 F.Supp. 656 (N.D. Ohio 1977); *Atlantis Community v. Adams*, 453 F.Supp. 825 (D. Colo. 1978).
13. 44 Fed. Reg. 31442 (May, 1979).
14. One unusual employment opinion, which runs counter to the general trend, is that of *Trageser v. Libbie Rehabilitation Center* 590 F.2d 87 (4th Cir. 1978), *cert. denied*, 47 U.S.L.W. 3811 (1979), in which the Fourth Circuit Court of Appeals held that a visually handicapped nurse could not maintain a private action under Section 504 to redress her dismissal without showing that the primary purpose of the federal financial as-

sistance was to provide employment. It is uncertain whether this line of reasoning will be followed in the other federal circuits. See *Cannon v. University of Chicago*, 47 U.S.L.W. 4549 (May 14, 1979).

15. *Gurmankin v. Costanzo*, 556 F.2d 184 (3rd Cir. 1977).
16. *Duran v. City of Tampa*, C.A. No. 76-863 Civ. T-K (M.D. Fla. June 15, 1978), previously reported at 430 F.Supp. 75 (M.D. Fla., 1977).
17. 47 U.S.L.W. 4689 (1979).
18. *Id.*, at 4693.
19. *Id.*, at 4692.
20. *Id.*, at 4691.
21. *Id.*, at 4691.
22. *Id.*, at 4693.
23. *Id.*, at 4692.
24. *Id.*, at 2691-92.
25. "Prohibiting Discrimination Against Developmentally Disabled Persons: Statutory Survey; Model Statute," Commission on the Mentally Disabled, American Bar Association (Nov. 1978).
26. 45 C.F.R. §§84.6, 84.22.
27. 45 C.F.R. §84.52.

GUNNAR DYBWAD is Professor Emeritus of Human Development at the Florence Heller Graduate School for Advanced Studies in Social Welfare at Brandeis University, Adjunct Professor of Special Education at Syracuse University, and a Senior Staff member of its Center on Human Policy. He also serves as Visiting Scholar at the National Institute on Mental Retardation, Toronto, Canada.

He received his J.D. from the Faculty of Laws, University of Halle, Germany, and is a graduate of the New York School of Social Work. Following criminology studies in Italy, Germany, England, and the U.S.A., he worked in prisons and in institutions for juvenile delinquents in Indiana, New Jersey, New York, and Michigan. From 1943 to 1951 he directed the child welfare program of the Michigan State Department of Social Welfare and subsequently served as Executive Director of the Child Study Association of America and of the National Association for Retarded Citizens. From 1964 to 1967 he and his wife, Dr. Rosemary Dybwad, were Co-Directors of the Mental Retardation Project of the International Union for Child Welfare in Geneva, working as consultants throughout Europe and in Central and South America. This period included assignments in Spain and in Uruguay for the Rehabilitation Unit of the United Nations and consultations for the World Health Organization.

In this country, Dr. Dybwad has served as consultant to President Kennedy's Special Assistant on Mental Retardation, to the U.S. Public Health Service, the U.S. Office of Education and the Social and Rehabilitation Service, the President's Committee on Mental Retardation, and numerous state governmental agencies. Presently he is Chairman of the Board of the Epilepsy Society of Massachusetts, and President of the International League of Societies for the Mentally Handicapped.

He is a Fellow of the American Orthopsychiatric Association, the American Sociological Association, the American Public Health Association, and the American Association on Mental Deficiency; a member of the Council for Exceptional Children and the National Rehabilitation Association; and Honorary Associate Fellow of the American Academy of Pediatrics, from which he received in 1973 the C. Anderson Aldrich Award for his activities in the field of child development. In 1977 Temple University awarded him the honorary degree of Doctor of Humane Letters.

Societal Perspectives
Where Do We Go from Here?

GUNNAR DYBWAD

Societal perspectives are to a considerable extent related to societal perceptions, and I think it is fair to say that there is a reciprocal relationship. A change in societal perception of persons with handicap will of necessity result in new societal perspectives, but it is equally obvious that once these new perspectives have been implemented, the changed circumstances of societal life will cause us to form changed perceptions.

On the broadest societal scale in the world community, the programmatic implementation of a succession of resolutions by the United Nations will demonstrate this interrelationship. Following one of the most destructive decades in human history, characterized by the Holocaust and a brutal war and culminating in the disasters of Hiroshima and Nagasaki—a decade that resulted in the enslavement, the maiming, and the death of millions of people—the United Nations came into being. One of its first major accomplishments was to create a new societal perspective with the Universal Declaration of Human Rights.

That was in 1948, and it was exactly at that time that in countries around the globe parents of handicapped children began to rise up and demand from society a better life and a greater acceptance for their children. It may seem strange to link together these two developments, but the only clue I have found in trying to determine what caused the sudden uprising of these parents in so many different parts of the world is that, side by side with the incredible human destruction during and after the war, we also observed incredible feats of human reconstruction. The key issue of the emerging rehabilitation and habilitation movement

GUNNAR DYBWAD • Florence Heller Graduate School of Social Welfare, Brandeis University, Waltham, Massachusetts 02154.

was a belief in the value and dignity of all human beings, their right to life, to liberty, and to security of person.

This first very broad Declaration of Human Rights was followed in 1959 by the Declaration of the Rights of the Child, and in that Declaration there was specific mention of the child with a handicap. But what was far more significant was the implied shift in societal perspective as far as the child's role and rights were concerned. For some of our contemporaries the significance of this second Declaration may not be so obvious unless and until they realize that even in our own country in the early years of this century, attempts to curb child labor, through the introduction of legislation that would stop the maiming of children at work in factories, was repeatedly judged unconstitutional. Children were deemed not to have *any* rights.

But what distinguishes, in our context, the Declaration of the Rights of the Child from the Declaration on Human Rights is that it specifically prescribes that the child who is physically, mentally, and socially handicapped shall be given the special treatment, education, and care required by his particular condition.

It was slightly more than a decade later, when, in December 1971, the United Nations General Assembly adopted the Declaration on the Rights of Mentally Retarded Persons, and I emphasize this last word—persons. This Declaration, which spelled out in detail the general and specific rights that must be granted to persons with mental retardation, served as a precursor to a broad Declaration on the Rights of Disabled Persons, adopted by the United Nations in 1975. And not only United Nations personnel, but also national governments throughout the world, including the government of the United States, have planned a program of action and public information and education for the International Year of Disabled Persons, 1981.

I have gone far afield, but I wanted to lay a good foundation for the next point—my perception of the changing societal perspectives for persons with handicaps in our own country. Programs for children and adults with handicaps in the United States were characterized for many years by a "yes, but . . . " attitude. They were exclusionary programs and, to use the language of our vocational rehabilitation bureaucrats, were open only to those deemed "feasible." As a result very often those persons most urgently in need of help would not receive it, and this was particularly true with the individuals who have been the subject of our discussions here today.

The 1950s and 1960s witnessed in our country ever more effective work on the part of organizations interested in the handicapped and increasingly, by persons with handicaps themselves, although at first primarily those with physical or perceptual handicaps. And then in the

1970s we came upon a decade of unprecedented change in societal perspectives. I say unprecedented because it surely was the first time that
a major societal institution, namely the Judicial System, responded in
broad terms to the needs of yet another minority group, those with
severe handicaps who either suffered severe discrimination in school or
even exclusion from it and those who were imprisoned by society in
institutions where they were exposed (and even today are still exposed)
to rampant abuse and neglect, often resulting in serious deterioration
and physical harm.

From Alabama to Michigan, from Maine to Kentucky, in more than
a score of cases, judges in Federal District Courts came to the conclusion,
based on sworn testimony, that societal perspectives in the field of handicap needed a thorough overhaul. It is an interesting commentary on
society in general and on the professional community specifically that
it took courts of law to point up the extreme abuses in testing and
classifications of children with disabilities and, yet more importantly, to
point up that there is no child who is uneducable, that there is no child,
no matter how severe his impairment, who cannot grow and develop.
It was indeed a fascinating development in our society that a federal
judge in California had to explain to the professional community in his
state that if one uses an English-language Stanford–Binet Test with children from Spanish-speaking homes the results are most likely inconclu-
sive.

Added to the impact of this judicial activity came a legislative revolution, the Rehabilitation Act of 1973, the Developmental Disability
Amendments of 1975 and 1978, and, finally, Public Law 94-142, the
Education for All Handicapped Children Act, all stressing the importance of serving persons with severe handicap. Together, these three
laws represent in effect a new congressional perspective on the problems
of handicap and indeed, a far reaching new societal perspective.

Some in this audience may say that courts and legislators are only
a very small part of our societal patterns. Society is made up of people,
and are the people really ready for these revolutionary changes? In
response I would say that, having worked in the fields of mental, physical, and social handicap for more than four decades, I have had the
chance to make observations in 49 states in this country and in 35 countries around the world. What has consistently surprised and delighted
me has been the readiness of the general population, the neighbors
down the street, plain folks, first to tolerate and gradually to accept
persons with disabilities. To be sure, we have occasional troubles, but
so do we have troubles in every phase of life, from the home to the
church, from the marketplace to our place of employment.

Today, not 1000, not 10,000, but hundreds of thousands of substantially handicapped individuals walk on the streets or get about in wheelchairs, shop at supermarkets, go to movies, use buses and street cars, and go to ball games and playgrounds—places where they never used to be seen—and yet the number of untoward incidents is minimal. We have here a clear example of how a new perception, observing one's handicapped neighbors, leads to a readiness to accept new societal perspectives. To use an old adage, "Seeing is believing."

So far, I have presented what some may feel is an almost euphoric picture. Are there no impediments to the implementation of these new perspectives? There are, and very formidable ones at that. And, strange as it seems, I see these major impediments as originating with my professional colleagues in psychology, social work, medicine, and the other so-called helping professions, and I would add the somewhat risky generalization that the higher the academic training my colleagues may have enjoyed, the more they are apt to impede persons with severe handicaps individually and collectively, by restrictive views, by a low level of expectation, and by a refusal to accept their educational or rehabilitation potential. I am well aware that these are grim words, but I think, with noble exceptions which always exist, by and large this has been the picture.

Let me be more specific: Over and over I have observed professional persons with advanced academic training, who become not just uncomfortable, but actively angry and even hostile when they are confronted with substantial achievements of persons whom they consider as incapable of making progress.

But that is only half the story. What is still more disturbing is that at this time we are witnessing efforts directed at actual extermination of human beings with severe and profound disabilities for the reason that these individuals are not considered worthy of being alive. Of course, the actual number of severely handicapped individuals who are presently considered as subjects for premeditated killings[1] is small.

Let me single out one prominent source for this trend, a professor of theology and ethics, Joseph Fletcher. Since 1968, Fletcher has time and again advocated the killing of infants with Down syndrome and other conditions because he considers them unworthy of living. In 1972 in a lead article in the *Hastings Center Report*, the journal of the Institute of Society, Ethics and the Life Sciences, he wrote that we need to recognize indicators of humanhood—that it is time to spell out the "which" and the "what" and the "when" in making dispositions ("normative decisions") as to who may live and who must die.[2] And thus he arrives at the astonishing dictum that "any individual of the species homo sapien who falls below the IQ 40 mark on a standard Stanford–Binet test, amplified if you like by other tests, is questionably a person; below the 20 mark, not a person." He adds that "this has bearing obviously on decision making in gynecology, obstetrics and pediatrics, as well as in general surgery and medicine." But it surely is grotesque to think of IQs in relation to infants in a nursery for newborns, unless of course one indulges, as does Fletcher, in long-disproven assumptions that low IQ levels are the rule in children with specific mental disabilities.

Those who read *Time* or *Newsweek* or such medical journals as *Lancet* and the *New England Journal of Medicine* or the excellent journal *Developmental Medicine and Child Neurology* will know how this question of who shall live and who must die (and die by premeditated action or inaction) comes up with increasing frequency.

I cannot pursue here this very important matter, which cries out for attention. Suffice it to say that a position statement of the Canadian Psychiatric Association entitled "Withholding Treatment," written by a committee chaired by Dr. Malcolm Beck and approved by the Association's Board of Directors, will furnish you with a comprehensive view of this problem area.[3]

In case the reader wonders whether my interpretation of this matter has perhaps led me to make too sweeping generalizations, let me quote from a letter sent me by Dr. Beck, Director of Child Psychiatric Services in Prince Edward Island, after I commented on the aforementioned statement of the Canadian Psychiatric Association:

> I consider that active euthanasia of the retarded is now the single most important issue facing us professionals in the field of retardation as we try to promote the welfare of our retarded friends. It is abundantly clear that we have now reached that point in our "advanced" society where this practice is a reality. For older hands like you and myself it is cold comfort to remember that Germany got to this position in the 1930s—a full thirty years before we North Americans got around to practicing it in the late sixties.
>
> This is going to be an even more difficult issue to tackle than that of good medical, educational, and other services for the retarded, as hard a battle as that one was and still is. This issue thrusts us deeply into our basic assumption about the meaning, value and purpose of human existence and this is a field where the going is very rough indeed in this day and age.[4]

As I have stated, the number of those actually recommended for extermination is as yet small, but, obviously, arriving at such a judgment has far reaching implications as to what one thinks about persons who are severely handicapped. One cannot simply isolate a little group and say that only they are our target for extermination. Rather, by implication all severely handicapped human beings are thus denigrated and devalued as are the services now being developed for them.

It is in such a light that one needs to look at the efforts of a small group of rather well-known academicians and administrators who recently prepared for a federal court a position paper in which they strongly recommended continued incarceration of most of the more severely handicapped persons now in our retardation institutions.[5] They specifically object to efforts to provide training and education, feeling that such a regime may constitute "cruel and inhuman treatment." Custodial care is what they see as appropriate for such individuals: kind custodial care that effectively segregates them as, in effect, subhuman. And it is not a very big step from custodial segregation to extermination.[6]

There is another major impediment to the realization of the new perspectives I have presented. It is a political problem, long known by sociologists as "System Maintenance," a line of least resistance. The mental retardation institutions (which, of course, at least in the past, have housed persons with epilepsy, cerebral palsy, autism, and a variety of impairments that classified them as multiply handicapped) represent not only a past investment of hundreds of millions of dollars but are still today considered by legislators and politicians as ideal receptacles for new construction funds, again to the tune of hundreds of millions of dollars. In addition, they employ large groups of people who have come to feel that their employment in the institution is a vested right. As a result, throughout this country men, women, and children are kept imprisoned in these institutions not in consideration of their own needs but rather so that financial advantage will accrue to others.

But this is only one side of the story. Hundreds upon hundreds of millions of dollars are thus spent in this country to effect alterations and repairs in old out-moded institutional buildings without regard to cost effectiveness or minimal economic prudence in what can only be called an insane building spree, considering the cost per bed. And this insane waste of hundreds of millions of dollars to reconstruct unneeded institutional buildings in inappropriate locations is the major obstacle to the development of desperately needed, cost-effective community services.

Much as the general public can and does help with the adjustment of persons with mental retardation in the community, if only by allowing them to find their own way, the ordinary members of the general public are not likely to perceive the unholy conspiracy which perpetuates the institutional complexes, nor would they likely feel prepared to cope with it. A determined, unceasing campaign of convincing legislators of the intolerable fiscal burden they are asked to maintain is what is needed.

This campaign is making and will continue to make heavy demands on the associations for retarded citizens and related advocacy organizations. However, I predict that they will be supported with increasing effectiveness by a new phenomenon, the self-advocacy of persons with mental retardation and other developmental disabilities.

Only a few years ago such a prediction would have been brushed aside as totally unrealistic. However, already we have seen concrete examples of their successful lobbying for state legislation, active participation on official committees and commissions, and effective presentations in public meetings. Marc Gold[7] opened up new perspectives when he demonstrated how effectively he could teach work processes to individuals once considered totally incapable of any part in a production process. In a similar fashion, new societal perspectives are beginning to open up in this field as we see the entrance into public life of individuals once thought to be incapable of reasoning or making choices, let alone of expressing them effectively.

I join those who feel it is safe to predict that this new societal perspective will result in a new and more favorable perception of persons with mental retardation as fellow citizens, endowed with individual rights and capable of learning to contribute to the common weal—no longer just objects of pity and charity.

REFERENCES AND NOTES

1. Euthanasia is by no means an appropriate term because the individuals I am talking about are not dying, but have a good chance for more than mere survival if they get proper medical care.
2. *Hastings Center Rep.* 2:(5), 1972.
3. Obtainable for $1.00 from the Executive Secretary of the Canadian Psychiatric Association, 103-225 Lisgar Street, Ottawa, Ontario.
4. Written communication, February 13, 1979.
5. *Wyatt v. Hardin,* Civil Action No. 3195 N, M.D. Alabama. Defendants' Motion for Modification, October 20, 1978.
6. To counter this regressive proposal, the Center on Human Policy at Syracuse University (216 Ostram Avenue, Syracuse, New York 13210) has issued a widely reprinted paper entitled "The Community Imperative: A Refutation of All Arguments in Support of Institutionalization of Anybody Because of Mental Retardation," 1979.
7. Gold, M. W.: Stimulus factors in skill training of retarded adolescents on a complex assembly task: Acquisition, transfer, and retention. *Am. J. Ment. Defic.* 76:517, 1972.

DISCUSSION

MODERATOR: GUNNAR DYBWAD

A. MILUNSKY: Advances in technology which enable the prevention of serious mental retardation or serious genetic disease move in parallel with the enormously escalating costs of care. Mr. Beyer, do you envisage taxpayer or class action suits that will ultimately force us, the consumers, to take note of the preventive measures that are available?

H. A. BEYER: I'm a poor prognosticator and not competent in that area. When courts have been faced with questions of preserving or maintaining life, they have invariably said that human life is a value to be maintained without attempting to inquire into the quality of that human life. The only exception that I can think of is in cases like the Saikewicz case here in Massachusetts. In that case the life of a severely retarded man would have been maintained only for a short period of time on a rather painful regimen of treatment. In general the legal answer is that life is of inestimable value and must be maintained.

G. DYBWAD: My comment is that nothing is as expensive as our present system. We just settled a case in Michigan where the state paid $43,000 per year per person in an institution. In Massachusetts we paid $30,000 per person per institution. If you gave somebody in Michigan $90,000, and three people from the Fernald State School, they would surely do better. Then what is happening at such institutions? I stake my professional reputation that the home and community care we are planning will be infinitely more effective and more economical. Nothing is as wasteful. I think the programs of early intervention in this and other states are going to show tremendous results. The vast costs which you still find at any of the state institutions are largely the results of gross neglect over years and years.

A. MILUNSKY: Let me clarify the point I was making about prevention. Neither Mr. Beyer or Professor Dybwad really responded to the preventive technology techniques I'm talking about, such as amniocentesis and prenatal diagnosis. The entry of government into these areas has developed in Alabama and in Oregon. In Oregon there are statutes now which declare that if the patient is unable to afford such costs, the state will pay for an amniocentesis and the prenatal genetic studies for women who are 35 years old or over. Now that, of course, is step one. Step two will be the small print, and that is what I am asking my colleagues. When will the small print read (or will it ever), "if you don't have these tests?"

G. DYBWAD: Elective abortion remains a problem at the moment. But there are, of course, other preventive methods such as secondary prevention, which are very effective and which are part of what is being initiated by a program of early intervention. Medicine is also making progress and is more effectively dealing with premature infants. Not only can we prevent problems but we will have more effective management.

Selected Recent Bibliography

CHILDREN AND DEATH

BOOKS, CHAPTERS, TELEVISION PROGRAMS (1973–1978)

1. Morris, A.: Proposed legislation, in Kohl, M. (Ed.), *Infanticide and the Value of Life.* Buffalo: Prometheus Books, p. 221, 1978.
2. Sell, I.: *Guide to Materials on Death and Dying for Teachers of Nursing.* Ann Arbor: University Microfilms International, 1975.
3. Garfield, C. (Ed.): *Psychosocial Care of the Dying Patient.* New York: McGraw-Hill, 1978.
4. Vaughan, V.: The care of the child with a fatal illness, in Vaughan, V., McKay, R., and Nelson, W. (Eds.), *Nelson Textbook of Pediatrics,* Tenth Edition. Philadelphia: W.B. Saunders, p. 142, 1975.
5. Tripp, J.: Ethical problems in pediatrics, in Vale, J. (Ed.), *Medicine and the Christian Mind.* London: Christian Medical Fellowship Publishers, p. 106, 1975.
6. Vaux, K.: Lazarus revisited: Moral and spiritual aspects of experimental therapeutics with children, in Van Eys, J. (Ed.), *Research on Children: Medical Imperatives, Ethical Quandaries, and Legal Constraints.* Baltimore: University Park Press, p. 77, 1978.
7. Holt, J.: The right of children to informed consent, in Van Eys, J. (Ed.), *Research on Children: Medical Imperatives, Ethical Quandaries, and Legal Constraints.* Baltimore: University Park Press, p. 5, 1978.
8. Shaw, A.: The ethics of proxy consent, in Swinyard, C. (Ed.), *Decision Making and the Defective Newborn. Proceedings of a Conference on Spina Bifida and Ethics.* Springfield, Illinois: Charles C Thomas, p. 384, 1978.
9. Beauchamp, T., and Walters, L. (Eds.): *Contemporary Issues in Bioethics.* Encino, California: Dickenson Publishing, 1978.
10. Simmons, R., Klein, S., and Simmons, R.: *Gift of Life: The Social and Psychological Impact of Organ Transplantation.* New York: John Wiley, 1977.
11. Pattison, E. (Ed.): *The Experience of Dying.* Englewood Cliffs, New Jersey: Prentice-Hall, 1977.
12. Spicker, S., and Engelhardt, H. (Eds.): *Philosophical Medical Ethics: Its Nature and Significance.* Boston: D. Reidel, 1977.
13. Veatch, R.: Death and dying, in Veatch, R. (Ed.), *Case Studies in Medical Ethics.* Cambridge, Massachusetts: Harvard University Press, p. 317, 1977.
14. U.S. National Commission for the Protection of Human Subjects of Biomedical and Behavioral Research: *Transcript of the Meeting Proceedings (21st Meeting), 6–7 August, 1976.* Springfield, Virginia: National Technical Information Service, 1976.
15. Veatch, R.: *Death, Dying, and the Biological Revolution: Our Last Quest for Responsibility.* New Haven: Yale University Press, 1976.

16. Holder, A.: The child and death, in Holder, A. (Ed.), *Legal Issues in Pediatrics and Adolescent Medicine*. New York: John Wiley, p. 107, 1977.

17. Poteet, G.: *Death and Dying: A Bibliography (1950–1974)*. Troy, New York: Whitston, 1976.

18. Cooper, I.: Serendipity, science, and morality, in *Social Responsibility: Journalism, Law, Medicine, Volume II*. Lexington, Virginia: Washington and Lee University, p. 59, 1976.

19. Falik, M.: *Ideology and Abortion Policy Politics*. Ann Arbor: University Microfilms, 1975.

20. Kutscher, M. (Ed.): *A Comprehensive Bibliography of the Thanatology Literature*. New York: MSS Information, 1975.

21. Dempsey, D.: *The Way We Die: An Investigation of Death and Dying in America Today*. New York: Macmillan, 1975.

22. Engelhardt, H.: Ethical issues in aiding the death of young children, in Kohl, M. (Ed.), *Beneficent Euthanasia*. Buffalo: Prometheus Books, p. 180, 1975.

23. Kohl, M. (Ed.): *Beneficent Euthanasia*. Buffalo: Prometheus Books, 1975.

24. Annas, G.: *The Rights of Hospital Patients: The Basic ACLU Guide to a Hospital Patient's Rights*. New York: Avon Books, 1975.

25. Kremer, E., and Synan, E. (Eds.): *Death Before Birth: Canada and the Abortion Question*. Toronto: Griffin Press, 1974.

26. Scharper, P.: *On Death and Dying*. New York: National Broadcasting Company, 1974.

27. Sanders, M.: *The Right to Die*. New York: American Broadcasting Company, 1974.

28. Nelson, J.: Human experimentation, in Nelson, J. (Ed.), *Human Medicine: Ethical Perspectives on New Medical Issues*. Minneapolis: Augsburg Publishing House, p. 79, 1973.

29. Goldberg, I., Malitz, S., and Kutscher, A. (Eds.): *Psychopharmacological Agents for the Terminally Ill and Bereaved*. New York: Foundation of Thanatology, 1973.

30. Binger, C.: Jimmy—A clinical case presentation of a child with a fatal illness, in Anthony, E., and Koupernik, C. (Eds.), *The Child in His Family: The Impact of Disease and Death, Volume 2*. New York: Wiley, p. 171, 1973.

31. Wolters, W.: The dying child in hospital, in Anthony, E., and Koupernik, C. (Eds.), *The Child in His Family: The Impact of Disease and Death, Volume 2*. New York: Wiley, p. 159, 1973.

32. Alby, N., and Alby, J.: The doctor and the dying child, in Anthony, E., and Koupernik, C. (Eds.), *The Child in His Family: The Impact of Disease and Death, Volume 2*. New York: Wiley, p. 145, 1973.

33. Vernick, J.: Meaningful communication with the fatally ill child, in Anthony, E., and Koupernik, C. (Eds.), *The Child in His Family: The Impact of Disease and Death, Volume 2*. New York: Wiley, p. 105, 1973.

34. Johnson, R.: Consent in the emergency room, in Wecht, C. (Ed.): *Legal Medicine Annual 1973*. New York: Appleton-Century-Crofts, p. 345, 1973.

35. Hendin, D.: *Death as a Fact of Life*. New York: Norton, 1973.

JOURNAL ARTICLES (1973–1978)

36. Ramsey, P.: The Saikewicz precedent: What's good for an incompetent patient? *Hastings Center Report* **8**(6):36, 1978.

37. I had to play God. *Med. Econom.* **55**(25):53, 1978.

38. Mnookin, R.: Children's rights: Legal and ethical dilemmas. *Pharos* **41**(2):2, 1978.

39. Pauli, R., and Cassell, E.: Nurturing a defective newborn. *Hastings Center Rep.* **8**(1):13, 1978.

40. Raible, J.: The right to refuse treatment and natural death legislation. *Medicolegal News* **5**(4):6, 1977.

41. Brant, J.: The right to die in peace: Substituted consent and the mentally incompetent. *Suffolk Univ. Law Rev.* **11**(4):959, 1977.

42. Stone, B.: Telling the patient he has cancer. *Am. Med. News* **20**(37):1, 1977.

43. Goldstein, J.: Medical care for the child at risk: On state supervention of parental autonomy. *Yale Law J.* **86**(4):645, 1977.

44. Stephen, L., and Billings, M.: The law and death—An overview. *J. of Contemp. Law* **1**(2):224, 1975.

45. Koop, C.: The slide to Auschwitz. *Human Life Rev.* **3**(2):101, 1977.

46. Gaylin, W., Lyons, K., Smith, H., *et al.:* Who should decide? The case of Karen Quinlan. *Christianity and Crisis* **35**(22):322, 1976.

47. Ethical considerations in surgery of the newborn. *Contemp. Surg.* **7**(6):17, 1975.

48. Riga, P.: Compulsory medical treatment of adults. *Catholic Lawyer* **22**(2):105, 1976.

49. Fost, N.: Ethical problems in pediatrics. *Curr. Probl. Pediat.* **6**(12):1, 1976.

50. Nolan-Haley, J.: Defective children, their parents, and the death decision. *J. Legal Med.* **4**(1):9, 1976.

51. Korsch, B.: Children should be consulted concerning their medical treatment. *Am. Med. News* **19**(23):6, 1976.

52. Koop, C.: The right to die (II). *Human Life Rev.* **2**(2):33, 1976.

53. Stone, R.: The research imperative and human rights. *Philadelphia Med.* **70**(2):73, 1974.

54. Lorber, J.: Ethical problems in the management of myelomeningocele and hydrocephalus. *J. R. Coll. Physicians (London)* **10**(1):47, 1975.

55. Hauerwas, S.: The demands and limits of care—Ethical reflections on the moral dilemma of neonatal intensive care. *Am. J. Med. Sci.* **269**(2):222, 1975.

56. Engelhardt, H.: Bioethics and the process of embodiment. *Perspect. Biol. Med.* **18**(4):486, 1975.

57. Duke, P.: Media on death and dying. *Omega* **6**(3):275, 1975.

58. Sharp, T., and Crofts, T.: Death with dignity: The physician's civil liability. *Baylor Law Rev.* **27**(1):86, 1975.

59. Horan, D.: Euthanasia, medical treatment and the mongoloid child: Death as a treatment of choice? *Baylor Law Rev.* **27**(1):76, 1975.

60. Fletcher, J.: Abortion, euthanasia, and care of defective newborns. *New Eng. J. Med.* **292**(2):75, 1975.

61. Levine, M., Camitta, B., and Nathan, D.: The medical ethics of bone marrow transplantation in childhood. *J. Pediatrics* **86**(1):145, 1975.

62. American Heart Association, Committee on Ethics: Ethical implications of investigations in seriously and critically ill patients. *Circulation* **50**(6):1063, 1974.

63. Sidel, V.: The allocation of expensive medical resources: Who should decide? *New Physician* **22**(4):229, 1973.

64. Zachary, R.: Attitudes to the abnormal child. *Doc. Med. Ethics* **3**:4, 1974.

65. Burgert, E.: Psychological management of children with cancer and of their families. *Paediatrician* **1**(4–5):311, 1972–1973.

66. Van Leeuwen, G.: Natural committee for life: Accepting death in our patients. *Clin. Pediatr.* **12**(2):64, 1973.

67. Cantor, N.: A patient's decision to decline lifesaving medical treatment: Bodily integrity versus the preservation of life. *Rutgers Law Rev.* **26**(2):228, 1973.

68. Shaw, A.: Dilemmas of informed consent in children. *New Eng. J. Med.* **289**(17):885, 1973.

HANDICAP AND BIOETHICS

BOOKS (1973–1978)

69. Baker, R.: Social control and medical models in genetics, in Buckley, J. (Ed.), *Genetics Now: Ethical Issues in Genetic Research.* Washington: University Press of America, p. 75,1978.
70. Diamond, E.: *This Curette for Hire.* Chicago: ACTA Foundation, 1977.
71. Reich, W.: Quality of life and defective newborn children: An ethical analysis, in Swinyard, C. (Ed.), *Decision Making and the Defective Newborn. Proceedings of a Conference on Spina Bifida and Ethics.* Springfield, Illinois: Charles C Thomas, p. 489, 1978.
72. Heymann, P., and Holtz, S.: The severely defective newborn: The dilemma and the decision process, in Swinyard, C. (Ed.), *Decision Making and the Defective Newborn. Proceedings of a Conference on Spina Bifida and Ethics.* Springfield, Illinois, Charles C Thomas, p. 396, 1978.
73. Fletcher, J.: Spina bifida with myelomeningocele: A case study in attitudes towards defective newborns, in Swinyard, C. (Ed.), *Decision Making and the Defective Newborn. Proceedings of a Conference on Spina Bifida and Ethics.* Springfield, Illinois: Charles C Thomas, p. 281, 1978.
74. Duff, R.: A physician's role in the decision-making process: A physician's experience, in Swinyard, C. (Ed.), *Decision Making and the Defective Newborn. Proceedings of a Conference on Spina Bifida and Ethics.* Springfield, Illinois: Charles C Thomas, p. 194, 1978.
75. Annas, G., Glantz, L., and Katz, B.: Law of informed consent in human experimentation: Institutionalized mentally infirm, in *U.S. National Commission for Protection of Human Subjects. Research Involving Those Institutionalized As Mentally Infirm: Appendix.* Washington, D.C.: U.S. Government Printing Office, p. 3.1, 1978.
76. U.S. National Commission for the Protection of Human Subjects of Biomedical and Behavioral Research: *Transcript of the Meeting Proceedings (34th Meeting).* Springfield, Virginia: National Technical Information Service, p. 414, 1977.
77. U.S. National Commission for the Protection of Human Subjects of Biomedical and Behavioral Research. *Transcript of the Meeting Proceedings (33rd Meeting).* Springfield, Virginia: National Technical Information Service, p. 335, 1977.
78. Pattison, E. (Ed.): *The Experience of Dying.* Englewood Cliffs, NJ: Prentice-Hall, 1977.
79. Chase, A.: *The Legacy of Malthus: The Social Costs of the New Scientific Racism.* New York: Alfred A. Knopf, 1977.
80. U.S. National Commission for the Protection of Human Subjects of Biomedical and Behavioral Research. *Transcript of the Meeting Proceedings (32d Meeting).* Springfield, Virginia: National Technical Information Service, p. 326, 1977.
81. U.S. Congress, Senate, Committee on the Judiciary, Subcommittee on the Constitution: *Civil Rights of Institutionalized Persons. Hearings.* Washington, D.C.: U.S. Government Printing Office, 1977.
82. Friedman, P.: Rights of mentally retarded persons in institutions, in *Rights of Mentally Retarded Persons: The Basic ACLU Guide for the Mentally Retarded Person's Rights.* New York: Avon Books, p. 57, 1976.
83. Lister, C., Baker, M., and Milhous, R.: Record keeping, access, and confidentiality, in Hobbs, N. (Ed.), *Issues in the Classification of Children, Volume Two.* San Francisco: Jossey-Bass, p. 544, 1975.
84. Reiser, S., Dyck, A., and Curran, W. (Eds.): *Ethics in Medicine: Historical Perspectives and Contemporary Concerns.* Cambridge, Massachusetts: MIT Press, 1977.

85. Veatch, R.: Transplantation, hemodialysis, and the allocation of scarce resources, in *Case Studies in Medical Ethics*. Cambridge, Massachusetts: Harvard University Press, p. 221, 1977.

86. Scheerenberger, R.: Litigation, in *Deinstitutionalization and Institutional Reform*. Springfield, Illinois: Charles C Thomas, p. 97, 1976.

87. Nakao, G.: *Sterilization and the Mentally Retarded*. Ann Arbor: University Microfilms, 1974.

88. Turnbull, H. (Ed.): *American Association on Mental Deficiency. Legislative and Social Issues Committee, 1975–76. Task Force. Consent Handbook*. Washington, D.C.: American Association on Mental Deficiency, 1977.

89. Burns, M.: Wyatt V. Aderholt: Constitutional standards for statutory and consensual sterilization in state mental institutions, in *Law and Psychology Review, Spring 1975*. University of Alabama Law and Psychology Review, p. 79, 1975.

90. Price, M., and Burt, R.: Sterilization, state action, and the concept of consent, in *Law and Psychology Review, Spring 1975*. University of Alabama Law and Psychology Review, p. 57, 1975.

91. Roos, P.: Psychological impact of sterilization on the individual, in *Law and Psychology Review, Spring 1975*. University of Alabama Law and Psychology Review, p. 45, 1975.

92. Jonsen, A., and Garland, M. (Eds.): *Ethics of Newborn Intensive Care*. Berkeley: University of California, Institute of Government Studies, 1976.

93. Holder, A.: Minors, contraception, and abortion, in *Legal Issues in Pediatrics and Adolescent Medicine*. New York: John Wiley, p. 267, 1977.

94. Wing, K.: The state's police powers and involuntary civil commitment, in *The Law and the Public's Health*. St. Louis: C.V. Mosby, p. 30, 1976.

95. Pilpel, H.: Voluntary sterilization: Your human right 1976, in Schima, M., and Lubell, I. (Eds.), *New Advances in Sterilization: The Third International Conference on Voluntary Sterilization*. New York: Association for Voluntary Sterilization, p. 147, 1976.

96. Price, M., and Burt, R.: Nonconsensual medical procedures and the right to privacy, in Kindred, M., et al. (Eds.), *The Mentally Retarded Citizen and the Law*. New York: Free Press, p. 93, 1976.

97. Kramer, J.: The right not to be mentally retarded, in Kindred, M., et al. (Eds.), *The Mentally Retarded Citizen and the Law*. New York: Free Press, p. 31, 1976.

98. Halpern, C.: The right to habilitation: Litigation as a strategy for social change, in Golann, S., and Fremouw, W. (Eds.), *The Right to Treatment for Mental Patients*. New York: Irvington, p. 73, 1976.

99. U.S. National Commission for the Protection of Human Subjects of Biomedical and Behavioral Research. *Transcript of the Meeting Proceedings (17th Meeting)* Springfield, Virginia: National Technical Information Service, 1976.

100. U.S. National Commission for the Protection of Human Subjects of Biomedical and Behavioral Research. *Transcript of the Meeting Proceedings (14th Meeting)*. Springfield, Virginia: National Technical Information Service, 1976.

101. Bajema, C. (Ed.): *Eugenics Then and Now*. Stroudsburg, Pennsylvania: Dowden, Hutchinson and Ross, 1976.

102. Sorenson, J.: From social movement to clinical medicine—The role of law and the medical profession in regulating applied human genetics, in Milunsky, A., and Annas, G. (Eds.), *Genetics and the Law*. New York: Plenum Press, p. 467, 1976.

103. Robertson, J.: Discretionary non-treatment of defective newborns, in Milunsky, A., and Annas, G. (Eds.), *Genetics and the Law*. New York: Plenum Press, p. 451, 1976.

104. Burt, R.: Authorizing death for anomalous newborns, in Milunsky, A., and Annas, G. (Eds.), *Genetics and the Law*. New York: Plenum Press, p. 435, 1976.

105. Baron, C.: Voluntary sterilization of the mentally retarded, in Milunsky, A., and Annas, G. (Eds.), *Genetics and the Law*. New York: Plenum Press, p. 267, 1976.
106. Casey, T., and Robbins, R.: Patients' rights, in *New York University School of Law Annual Survey of American Law 1974/1975*. Dobbs Ferry, New York: Oceana, p. 303, 1976.
107. Weber, L.: *Who Shall Live? The Dilemma of Severely Handicapped Children and Its Meaning for Other Moral Questions*. New York: Paulist Press, p. 138, 1976.
108. Lappé, M.: Reflections on the cost of doing science, in Lappé, M., and Morison, R. (Eds.), *Ethical and Scientific Issues Posed by Human Uses of Molecular Genetics*. New York: New York Academy of Sciences, p. 102, 1976.
109. U.S. National Commission for the Protection of Human Subjects of Biomedical and Behavioral Research. *Transcript of the Meeting Proceedings (11th Meeting)*. Springfield, Virginia: National Technical Information Service, 1975.
110. U.S. National Commission for the Protection of Human Subjects of Biomedical and Behavioral Research. *Transcript of the Meeting Proceedings (10th Meeting)*. Springfield, Virginia: National Technical Information Service, 1975.
111. Gostin, L.: Treatment, in *A Human Condition: The Mental Health Act from 1959 to 1975—Observations, Analysis and Proposals for Reform*. London: MIND, National Association for Mental Health, p. 115, 1975.
112. Birch, C., and Abrecht, P. (Eds.): *Genetics and the Quality of Life*. Elmsford, New York: Pergamon Press, 1975.
113. Ottenberg, P.: Dehumanization and human experimentation, in Schoolar, J., and Gaitz, C. (Eds.), *Research and the Psychiatric Patient*. New York: Brunner/Mazel, p. 87, 1975.
114. Lappé, M.: Can eugenic policy be just? in Milunsky, A. (Ed.), *The Prevention of Genetic Disease and Mental Retardation*. Philadelphia: W.B. Saunders, p. 456, 1975.
115. Great Britain, Department of Health and Social Security: *Sterilization of Children Under 16 Years of Age: Discussion Paper*. London: Her Majesty's Stationery Office, 1975.
116. Goldman, L.: Controversy at Willowbrook, in *When Doctors Disagree: Controversies in Medicine*. London: Hamish Hamilton, p. 63, 1973.
117. Eisenberg, J., and Bourne, P.: *The Right to Live and Die*. Toronto: Ontario Institute for Studies in Education, 1973.
118. New South Wales Humanist Society: *Euthanasia (Compassionate Death)*. Winston Hills, NSW: NSW Humanist Society, 1973.
119. Sorenson, J.: Some social and psychologic issues in genetic screening: Public and professional adaptation to biomedical innovation, in Bersma, D. (Ed.), *Ethical, Social and Legal Dimensions of Screening for Human Genetic Disease*. Miami: Symposia Specialists, p. 165, 1974.
120. Ohio, Legislative Service Commission: *Human Organ Transplantation: Staff Research Report No. 109* Columbus, Ohio: Ohio Legislative Service Commission, 1973.
121. Slovenko, R.: Admission and discharge procedures, in *Psychiatry and Law*. Boston: Little, Brown, p. 201, 1973.
122. Nelson, J.: Human experimentation, in *Human Medicine: Ethical Perspectives on New Medical Issues*. Minneapolis: Augsburg Publishing House, p. 79, 1973.
123. Burt, R.: Legal restrictions on sexual and familial relations of mental retardates—Old laws, new guises, in De La Cruz, F., and LaVeck, G. (Eds.), *Human Sexuality and the Mentally Retarded*. New York: Brunner/Mazel, p. 206, 1973.
124. Ennis, B., and Friedman, P. (Eds.): *Legal Rights of the Mentally Handicapped: Volume Two*. New York: Practicing Law Institute, Mental Health Law Project, 1973.
125. Ennis, B., and Friedman, P. (Eds.): *Legal Rights of the Mentally Handicapped: Volume One*. New York: Practicing Law Institute, Mental Health Law Project, 1973.

126. Bass, M.: Voluntary eugenic sterilization, in Robitscher, J. (Ed.), *Eugenic Sterilization*. Springfield, Illinois: Charles C Thomas, p. 94, 1973.
127. Gullattee, A.: The politics of eugenics, in Robitscher, J. (Ed.), *Eugenic Sterilization*. Springfield, Illinois: Charles C Thomas, p. 82, 1973.
128. Giannella, D.: Eugenic sterilization and the law, in Robitscher, J. (Ed.), *Eugenic Sterilization*. Springfield, Illinois: Charles C Thomas, p. 61, 1973.
129. Tarjan, G.: Some thoughts on eugenic sterilization, in Robitscher, J. (Ed.), *Eugenic Sterilization*. Springfield, Illinois: Charles C Thomas, p. 17, 1973.
130. Robitscher, J. (Ed.): *Eugenic Sterilization*. Springfield, Illinois: Charles C Thomas, 1973.

JOURNAL ARTICLES (1973–1978)

131. U.S. Congress, Senate, President's Commission for the Protection of Human Subjects of Biomedical and Behavioral Research Act of 1978. *Congress. Record (Daily Edition)* **124**(98):S9689, 1978.
132. Humphrey, M., and Anderson, W.: Infant screening for metabolic disorders. *Congress. Record (Daily Edition)* **124**(26):S2527, 1978.
133. Gould, C., and Stiles, T.: The prevention of mental retardation: The physician's changing standard of care and the need for legislative action. *Gonzaga Law Rev.* **13**(3):691, 1978.
134. Dodge, G.: Sterilization, retardation, and parental authority. *Brigham Young Univ. Law Rev.* **1978**(2):380, 1978.
135. Ramsey, P.: The Saikewicz precedent: What's good for an incompetent patient? *Hastings Center Report* **8**(6):36, 1978.
136. Peduzzi, E.: Wrongful conception: A conditional prospective liability to one not yet in being. *Univ. of Dayton Law Rev.* **2**(2):311, 1977.
137. Gressel, M.: Compulsory sterilization: Equal protection and the quality of life. *Univ. of Dayton Law Rev.* **2**(2):327, 1977.
138. Smith, T.: Vaccination/immunisation and compensation. *J. Med. Ethics* **4**(3):152, 1978.
139. Perspectives in spina bifida. *Brit. Med. J.* **2**(6142):909, 1978.
140. Gaylord, C.: The sterilization of Carrie Buck. Case and comment. **83**(5):18, 1978.
141. Jeddeloh, N., and Chatterjee, S.: Legal problems in organ donation. *Surg. Clinics North Amer.* **58**(2):245, 1978.
142. Autonomy, authenticity, ethics, and death. *Emerg. Med.* **10**(2):148, 1978.
143. Bayles, M.: Sterilization of the retarded: In whose interest? The legal precedents. *Hastings Center Report* **8**(3):37, 1978.
144. Neville, R.: Sterilization of the retarded: In whose interest? The philosophical arguments. *Hastings Center Report* **8**(3):33, 1978.
145. Thompson, T.: Sterilization of the retarded: In whose interest? The behavioral perspective. *Hastings Center Report* **8**(3):29, 1978.
146. Vanderpool, H.: B.F. Skinner on ethics and the control of retarded persons. *Linacre Q.* **45**(2):135, 1978.
147. Sherlock, R.: From stigma to sterilization: Eliminating the retarded in American law. *Linacre Q.* **45**(2):116, 1978.
148. Regan, W.: Court blocked sterilization of a retarded minor. *Hospital Progress* **59**(5):96, 1978.
149. Cusine, D.: To sterilize or not to sterilize? *Med., Sci. Law* **18**(2):120, 1978.
150. Donovan, P.: Recent events highlight sterilization's potential as emotional, political issue. *Family Planning/Population Reporter* **6**(6):82, 1977.

151. Moylan, J.: What is ordinary care today? *Postgraduate Med.* **62**(5):21, 1977.
152. Henry, N.: In re Moore: The sound and the fury and the scalpel. *North Carolina Central Law J.* **8**(2):307, 1977.
153. U.S. Dept. of Health, Education, and Welfare: Protection of human subjects: Research involving those institutionalized as mentally infirm; report and recommendations for public comment. *Fed. Register* **43**(53):11328, 1978.
154. Cook, J., Altman, K., and Haavik, S.: Consent for aversive treatment: A model form. *Ment. Retard.* **16**(1):47, 1978.
155. Burgdorf, R., and Burgdorf, M.: The wicked witch is almost dead: Buck v. Bell and the sterilization of handicapped persons. *Temple Law Quarterly* **50**(4):995, 1977.
156. Annas, G.: The incompetent's right to die: The case of Joseph Saikewicz. *Hastings Center Report* **8**(1):21, 1978.
157. Can parents "volunteer" their children into mental institutions: An analysis of Kremens v. Bartley and Parham v. J.L. and J.R. *Ment. Retard. and the Law* **1**:1, 1978.
158. Simms, M.: A sterilization debate. *New Humanist* **92**(4):141, 1976.
159. Turnbull, H.: Individualizing the law for the mentally handicapped. *North Carolina J. of Mental Health* **8**(6):1, 1977.
160. Chernus, R.: Substantive due process—compulsory sterilization of the mentally deficient. *New York Law School Law Rev.* **23**(1):151, 1977.
161. Brant, J.: The right to die in peace: Substituted consent and the mentally incompetent. *Suffolk Univ. Law Rev.* **11**(4):959, 1977.
162. Schultz, S., Swartz, W., and Appelbaum, J.: Deciding right-to-die cases involving incompetent patients: Jones v. Saikewicz. *Suffolk Univ. Law Rev.* **11**(4):936, 1977.
163. Hellegers, A.: 'Incompetence' and consent. *Ob. Gyn. News* **12**(21):20, 1977.
164. Pratt, D., et al.: Autonomy for severely burned patients. *New Eng. J. Med.* **297**(21):1182, 1977.
165. Shaw, M.: Procreation and the population problem. *North Carolina Law Rev.* **55**(6):1165, 1977.
166. Schukoske, J.: Abortion for the severely retarded: A search for authorization. *Mental Disability Law Reporter* **1**(6):485, 1977.
167. Beyer, H.: Changes in the parent–child legal relationship—what they mean to the clinician and researcher. *J. Autism Child. Schizophr.* **7**(1):84, 1977.
168. Roth, L.: Involuntary civil commitment: The right to treatment and the right to refuse treatment. *Psych. Annals* **7**(5):50, 1977.
169. Imbus, S., and Zawacki, B.: Autonomy for burned patients when survival is unprecedented. *New Eng. J. Med.* **297**(6):308, 1977.
170. Darling, R.: Parents, physicians, and spina bifida. *Hastings Center Report* **7**(4):10, 1977.
171. Bice, M.: Is the right to die a solo decision? *Hospital Med. Staff* **6**(7):1, 1977.
172. American Academy of Pediatrics, Committee on Drugs: Guidelines for the ethical conduct of studies to evaluate drugs in pediatric populations. *Pediatrics* **60**(1):91, 1977.
173. Collester, D.: Death, dying and the law: A prosecutorial view of the Quinlan case. *Rutgers Law Rev.* **30**(2):304, 1977.
174. Eugenic sterilization—involuntary sterilization statute upheld as constitutional. *J. Family Law* **15**(2):344, 1976–1977.
175. Fleming, B., and Hunt, J.: Bartley v. Kremens: A step backward? *J. Nat. Assoc. Private Psych. Hospitals* **8**(3):10, 1976.
176. Ennis, B.: Legal rights of the voluntary patient. *J. Nat. Assoc. Private Psych. Hospitals* **8**(2):4, 1976.
177. Gauvey, S., and Shuger, N.: The permissibility of involuntary sterilization under the parens patriae and police power authority of the state: In re Sterilization of Moore. *Univ. of Maryland Law Forum* **6**(3):109, 1976.

178. Mont. eugenic sterilization law applies to retardates. *Pedi. News.* **8**(6):6, 1974.
179. Hillman, H.: After resuscitation—what? *Resuscitation* **4**(2):73, 1975.
180. Wildgen, J.: Rights of institutionalized mental patients: issues, implications, and proposed guidelines. *Univ. of Kansas Law Rev.* **25**(1):63, 1976.
181. Weathered, F.: Constitutional law—due process—North Carolina compulsory sterilization statute held constitutional against challenge that it constituted an unlawful invasion of privacy. *Texas Tech. Law Rev.* **8**(2):436, 1976.
182. Shaw, R.: Constitutional law—legislative naiveté in involuntary sterilization laws. *Wake Forest Law Rev.* **12**(4):1064, 1976.
183. Mason, B., and Menolascino, F.: The right to treatment for mentally retarded citizens: An evolving legal and scientific interface. *Creighton Law Rev.* **10**(1):124, 1976.
184. Corbett, K., and Raciti, R.: Withholding life-prolonging medical treatment from the institutionalized person—who decides? *New Eng. J. Prison Law* **3**(1):47, 1976.
185. Clayton, T.: O'Connor v. Donaldson: Impact in the states. *Hospital and Community Psych.* **27**(4):272, 1976.
186. Parent and child—mother seeks to have mentally deficient minor son sterilized. *J. Family Law* **14**(3):501, 1975–1976.
187. Foianini, R.: Postcommitment: An analysis and reevaluation of the right to treatment. *Notre Dame Lawyer* **51**(2):287, 1975.
188. Bender, H.: A geneticist's viewpoint towards sterilization. *Amicus* **2**(2):45, 1977.
189. Linn, B.: Involuntary sterilization—a constitutional awakening to fundamental human rights. *Amicus* **2**(2):34, 1977.
190. Trenkner, T.: Power of parent to have mentally defective child sterilized. *Amer. Law Reports, 3rd Series.* **74**:1224, 1976.
191. Murray, R.: Psychosocial aspects of genetic counseling. *Social Work in Health Care* **2**(1):13, 1976.
192. Laves, B.: Legal aspects of experimentation with institutionalized mentally disabled subjects. *J. Clin. Pharm.* **16**(10, Part 2):592, 1976.
193. Headings, V.: Ethical dimensions of optimizing human intelligence. *Perspect. Biol. Med.* **20**(1):30, 1976.
194. Wallach, S.: A constitutional right to treatment: Past, present, and future. *Prof. Psych.* **7**(4):453, 1976.
195. Sexual development in mentally handicapped children. *Brit. Med. J.* **2**(6027):71, 1976.
196. Matyas, D.: Anatomical transplants: Legal developments in Wisconsin. *Marquette Law Rev.* **59**(3):605, 1976.
197. Guthrie, D., *et al.*: Data bases and the privacy rights of the mentally retarded: Report of the AAMD Task Force on Data Base Confidentiality. *Ment. Retard.* **14**(5):3, 1976.
198. Bennett, S.: In the shadow of Karen Quinlan. *Trial* **12**(9):36, 1976.
199. Marx, P.: Abortion/euthanasia. Persona y Derecho: Revista de Fundamentacion de las Institutciones Juridicas **2**:383, 1975.
200. Beattie, J.: The right to life. *New Zealand Law J.* **1975**(14):501, 1975.
201. Spriggs, E.: Involuntary sterilization: An unconstitutional menace to minorities and the poor. *New York Univ. Rev. of Law and Social Change* **4**(2):127, 1974.
202. Trenkner, T.: Jurisdiction of court to permit sterilization of mentally defective person in absence of specific statutory authority. *Am. Law Rep., 3d series* **74**:1210, 1976.
203. Walters, L., and Gaylin, W.: Sterilizing the retarded child. *Hastings Center Report* **6**(2):13, 1976.
204. Perrin, J., *et al.*: A considered approach to sterilization of mentally retarded youth. *Am. J. Dis. Child.* **130**(3):288, 1976.
205. Perr, I.: Confidentiality and consent in psychiatric treatment of minors. *J. Legal Med.* **4**(6):9, 1976.

206. Hall, E., and Cameron, P.: Our failing reverence for life. *Psychology Today* 9(11):104, 1976.
207. Hagard, S., Carter, F., and Milne, R.: Screening for spina bifida cystica: A cost-benefit analysis. *Brit. J. Prev. Social Med.* 30(1):40, 1976.
208. Graham, R.: The 'right to kill' in the Third Reich. Prelude to genocide. *Catholic Hist. Rev.* 62(1):56, 1976.
209. Diamond, E.: Coerced sterilization under federally funded family planning programs. *New Eng. Law Rev.* 11(12):589, 1976.
210. Culliton, B.: Genetic screening: States may be writing the wrong kind of laws. *Science* 191(4230):926, 1976.
211. Court, S., and Lloyd, J.: Sterilization of children under 16. *Brit. Med. J.* 1(6013):830, 1976.
212. Bissett-Johnson, A., and Everton, A.: Preserving the status quo: Re D (a minor). *New Law J.* 126(5735):104, 1976.
213. Stone, R.: The research imperative and human rights. *Philadelphia Med.* 70(2):73, 1974.
214. Sterilisation of minors. *New Law J.* 125(5718):925, 1975.
215. Abelson, I.: The right to conceive. *Nursing Times* 71(40):1607, 1975.
216. Stone, A.: Overview: The right to treatment—comments on the law and its impact. *Amer. J. Psych.* 132(11):1125, 1975.
217. Dickens, B.: Eugenic recognition in Canadian law. *Osgoode Hall Law J.* 13(2):547, 1975.
218. Ghent, J.: Validity of statutes authorizing asexualization or sterilization of criminals or mental defectives. *Amer. Law Reports, 3d Series* 53:960, 1973.
219. Varner, J.: Rights of mentally ill—involuntary sterilization—analysis of recent statutes. *West Virginia Law Rev.* 78(1):131, 1975.
220. Lorber, J.: Ethical problems in the management of myelomeningocele and hydrocephalus. *J. R. Coll. Physicians London* 10(1):47, 1975.
221. Bernstein, A.: Approving the sterilization of minors. *Hospitals* 49(13):70, 1975.
222. Gould, D.: Judging is not for doctors. *New Statesman* 90(2323):360, 1975.
223. Gould, D.: The Dr. God syndrome. *New Statesman* 90(2316):166, 1975.
224. Wexler, D.: Behavior modification and other behavior change procedures: The emerging law and the proposed Florida guidelines. *Crim. Law Bull.* 11(5):600, 1975.
225. Porter, G., et al.: Child sterilization. *J. Med. Ethics* 1(4):163, 1975.
226. Hauerwas, S.: The demands and limits of care—ethical reflections on the moral dilemma of neonatal intensive care. *Am. J. Med. Sci.* 269(2):222, 1975.
227. Lister, J.: Private rights and the public good—sterilization of minors—comment or report? *New Eng. J. Med.* 293(22):1135, 1975.
228. Baker, J.: Sexual sterilization—constitutional validity of involuntary sterilization and consent determinative of voluntariness. *Missouri Law Rev.* 40(3):509, 1975.
229. Neuwirth, G., Heisler, P., and Goldrich, K.: Capacity, competence, consent: Voluntary sterilization of the mentally retarded. *Columbia Human Rights Law Rev.* 6(2):447, 1974–1975.
230. LaChat, M.: Utilitarian reasoning in Nazi medical policy: Some preliminary investigations. *Linacre Q.* 42(1):14, 1975.
231. Stetter, R.: Kidney donation from minors and incompetents. *Louisiana Law Rev.* 35(2):551, 1975.
232. Ehrhardt, H.: Abortion and euthanasia: Common problems—the termination of developing and expiring life. *Human Life Rev.* 1(3):12, 1975.
233. Sterilisation of minors. *Brit. Med. J.* 3(5986):775, 1975.
234. Sterilisation of handicapped minors. *Lancet* 2(7930):352, 1975.
235. Noon, C.: The right to reproduce. *Lancet* 2(7936):655, 1975.

236. Childress, J., and Fletcher, J.: Who has first claim on health care resources? *Hastings Center Report* **5**(4):13, 1975.
237. Bokser, B.: Problems in bio-medical ethics: A Jewish perspective. *Judaism* **24**(2):134, 1975.
238. Civil rights—right to treatment–neither due process nor equal protection clause of the Fourteenth Amendment guarantees the 'right to treatment' for mentally retarded children confined in a state. *Fordham Urban Law J.* **2**(2):363, 1974.
239. Perper, J.: Medical experimentation on captive populations in the United States. *J. Forensic Sci.* **19**(3):557, 1974.
240. Flaschner, F.: Constitutional requirements in the commitment of the mentally ill in the U.S.A.: Rights to liberty and therapy. *Int. J. Offender Ther. Comparative Criminology* **18**(3):283, 1974.
241. Wade, M.: Mental health law bibliography. *Univ. of Toledo Law Rev.* **6**(1):314, 1974.
242. De Veber, L.: On withholding treatment. *Canadian Med. Assoc. J.* **111**(11):1183, 1974.
243. Gastonguay, P.: Euthanasia: The next medical dilemma. *Amer.* **130**(8):152, 1974.
244. The right to protection from harm: Legal justice for the mentally retarded? *Albany Law Rev.* **38**(3):540, 1974.
245. Bennett, A.: Eugenics as a vital part of institutionalized racism. *Freedomways* **14**(2):111, 1974.
246. Friedman, P.: Analysis of the opinions in Donaldson v. O'Connor and Wyatt v. Aderholt. *Ment. Retard. and the Law* **1**, 1974.
247. American Association on Mental Deficiency: Sterilization of persons who are mentally retarded: Proposed official policy statement. *Ment. Retard.* **12**(2):59, 1974.
248. Gibson, D.: Involuntary sterilization of the mentally retarded: A western Canadian phenomenon. *Canadian Psych. Assoc. J.* **19**(1):59, 1974.
249. Shaw, A., and Shapo, M.: Legal, moral and ethical dilemmas in the newborn nursery. *Virginia Med. Monthly* **101**(12):1059, 1974.
250. Perlman, J.: Human experimentation. *J. Legal Med.* **2**(1):40, 1974.
251. Barshefsky, C., and Liebenberg, R.: Voluntarily confined mental retardates: The right to treatment vs. the right to protection from harm. *Catholic Univ. Law Rev.* **23**(4):787, 1974.
252. Shapiro, S.: A survey of the legal aspects of organ transplantation. *Chicago-Kent Law Rev.* **50**(3):510, 1973.
253. Murdock, C.: Sterilization of the retarded: A problem or a solution? *California Law Rev.* **62**(3):917, 1974.
254. Fremouw, W.: A new right to treatment. *J. Psych. Law* **2**(1):7, 1974.
255. Jonch'eres, J.: Euthanasia: An account of the CIOMS Round Table. *World Med. J.* **21**(4):63, 1974.
256. Weber, L.: Human death as neocortical death: The ethical context. *Linacre Q.* **41**(2):106, 1974.
257. Riga, P.: Euthanasia. *Linacre Q.* **41**(1):55, 1974.
258. Paul, E.: The sterilization of mentally retarded persons: The issues and conflicts. *Family Planning/Population Reporter* **3**(5):96, 1974.
259. Zaremski, M.: Blood transfusions and elective surgery: A custodial function of an Ohio juvenile court. *Cleveland State Law Rev.* **23**(2):231, 1974.
260. Laurence, K.: Effect of early surgery for spina bifida cystica on survival and quality of life. *Lancet* **1**(7852):301, 1974.
261. Eckstein, H.: Spina bifida: The surgeon's responsibility. *Doc. Med. Ethics*, **3**:4, 1974.
262. Cochran, B.: Conception, coercion, and control: Symposiums on reproductive rights of the mentally retarded. *Hospital and Community Psych.* **25**(5):283, 1974.

263. Richards, R.: The problem of compensation for antenatal injuries. *Nature* **246**(5428):54, 1973.
264. Starkman, B.: The control of life: Unexamined law and the life worth living. *Osgoode Hall Law J.* **11**(1):175, 1973.
265. Levison, C., and Jaffe, F.: Public policy on fertility control. *Scient. Amer.* **229**(5):8, 1973.
266. Walker, O.: Mental health law reform in Massachusetts. *Boston Univ. Law Rev.* **53**(5):986, 1973.
267. Goldman, L.: The Willowbrook debate: Concluded? *World Med.* **9**(2):79, 1973.
268. Furlow, T.: A matter of life and death. *Pharos* **36**(3):84, 1973.
269. Rollins, R., and Wolfe, A.: Eugenic sterilization in North Carolina. *North Carolina Med. J.* **34**(12):944, 1973.
270. McGarry, A., and Kaplan, H.: Overview: Current trends in mental health law. *Am. J. Psych.* **130**(6):621, 1973.
271. U.S. Social and Rehabilitation Service: Medical assistance programs: Sterilization procedures. *Fed. Register* **38**(183):26460, 1973.
272. U.S. Public Health Service: PHS supported programs: Sterilization procedures. *Fed. Register* **38**(183):26459, 1973.
273. Dornette, W.: Consent and living donors of organs. Anesthesia and Analgesia *Current Researches* **52**(3):311, 1973.
274. Ryan, J.: Medical ethics and the civil law. *J. Am. Osteo. Assoc.* **78**(2):153, 1973.
275. Nixon, R., *et al.*: Ethical principles in human experimentation. *New Eng. J. Med.* **288**(23):1247, 1973.
276. Hunt, G.: Implications of the treatment of myelomeningocele for the child and his family. *Lancet* **2**(7841):1308, 1973.
277. Diamond, E.: The Willowbrook experiments. *Linacre Q.* **40**(2):133, 1973.
278. Smith, G., and Smith, E.: Selection for treatment in spina bifida cystica. *Brit. Med. J.* **4**(5886):189, 1973.
279. Smithells, R., and Beard, R.: Research investigations and the fetus. *Brit. Med. J.* **2**(5864):464, 1973.
280. Plotkin, S., and Moore, J.: Ethics of human experimentation (cont.) *New Eng. J. Med.* **289**(11):593, 1973.
281. Cooke, R.: Ethics and law on behalf of the mentally retarded. *Pedi. Clinics North Am.* **20**(1):259, 1973.
282. Avery, G., *et al.*: The right to life: Can we decide? *Clinical Proceedings, Children's Hospital National Medical Center* **29**(11):265, 1973.

DEATH, SUICIDE, AND CHRONIC ILLNESS

JOURNAL ARTICLES (1972–1979)

283. Symposium on child psychiatric nursing. *Nurs. Clin. North Am.* **14**(3):389, 1979.
284. The Guyana tragedy—an international forensic problem. *Forensic. Sci. Int.* **13**(2):167, 1979.
285. Abdurrahman, M.: Parents' understanding of their children's illnesses. *J. Trop. Pediatr.* **24**(5):233, 1978.

286. Ahmed, M.: Suicide and attempted suicide. Etiologic factors and management. *Ohio State Med. J.* **75**(2):73, 1979.
287. Arlow, J.: Pyromania and the primal scene: A psychoanalytic comment on the work of Yukio Mishima. *Psychoanal. Q.* **47**(1):24, 1978.
288. Baker, H., and Wills, U.: School phobia: Classification and treatment. *Br. J. Psychiatry.* **132**:492, 1978.
289. Barba, E.: Attitudes toward the chronically ill and disabled: Implications for the health care systems. *Soc. Work Health Care* **3**(2):199, 1977.
290. Baxter, J.: Recognition and management of emotional problems associated with end-stage kidney disease. *J. Dial.* **2**(2):175, 1978.
291. Beck, L., Lattimer, J., and Braun, E.: Group psychotherapy on a children's urology service. *Soc. Work Health Care* **4**(3):275, 1979.
292. Belle-Isle, J., and Conradt, B.: Report of a discussion group for parents of children with Leukemia. *Matern. Child Nurs. J.* **8**(1):49, 1979.
293. Benjamin, P.: Psychological problems following recovery from acute life-threatening illness. *Am. J. Orthopsychiatry.* **48**(2):284, 1978.
294. Benoliel, J., and McCorkle, R.: A holistic approach to terminal illness. *Cancer Nurs.* **1**(2):143, 1978.
295. Bergmann, A., Lewiston, N., and West, A.: Social work practice and chronic pediatric illness. *Soc. Work Health Care.* **4**(3):265, 1979.
296. Bertwistle, H.: Practical aids for coping with progressive neurological disease. *Nurs. Times* **75**(41):1768, 1979.
297. Bhaduri, R.: Care of the terminal patient. 4. A family's sorrow. *Nurs. Times* **75**(15):638, 1979.
298. Black, D.: The bereaved child. *J. Child Psychol. Psychiatry.* **19**(3):287, 1978.
299. Boll, T., Dimino, E., and Mattsson, A.: Parenting attitudes: The role of personality style and childhood long-term illness. *J. Psychosom. Res.* **22**(3):209, 1978.
300. Bowlby, J.: On knowing what you are not supposed to know and feeling what you are not supposed to feel. *Can. J. Psychiatry.* **24**(5):403, 1979.
310. Burchell, R.: Counseling the formerly married. *Clin. Obstet. Gynecol.* **21**(1):259, 1978.
302. Calvert, P., Northeast, J., and Cunningham, E.: Death in a country area and its effect on the health of relatives. *Med. J. Aust.* **2**(19):635, 1977.
303. Cotter, J., and Schwartz, A.: Malignant diseases of infancy, childhood and adolescence: Psychological and social support of the patient and family. *Major Probl. Clin. Pediatr.* **18**:120, 1978.
304. Craft, M.: Help for the family's neglected 'other' child. *Mon.* **4**(5):297, 1979.
305. Crumley, F.: Adolescent suicide attempts. *JAMA* **241**(22):2404, 1979.
306. Cusine, D.: Artificial insemination with the husband's semen after the husband's death. *J. Med. Ethics* **3**(4):163, 1977.
307. DeVaul, R., Zisook, S., and Faschingbauer, T.: Clinical aspects of grief and bereavement. *Primary Care.* **6**(2):391, 1979.
308. Duberley, J., and King, H.: Cystic fibrosis: The nurse-teacher and carer for mother and child. *Nurs. Mirror* **147**(8):15, 1978.
309. Eisen, P.: Children under stress. *Aust. N. Z. J. Psychiatry* **13**(3):193, 1979.
310. Elliott, B.: Neonatal death: Reflections for parents. *Pediatrics* **62**(1):100, 1978.
311. Essen, J., and Lambert, L.: Living in one-parent families: Relationships and attitudes of 16-year-olds. *Child Care Health Dev.* **3**(5):301, 1977.
312. Farber, M.: Factors determining the incidence of suicide within families. *Suicide Life Threat. Behav.* **7**(1):3, 1977.

313. Fredlund, D.: Children and death from the school setting viewpoint. *J. Sch. Health* **47**(9):533, 1977.
314. Fulton, J.: Death and Dying. *Ohio State Med. J.* **74**(12):769, 1978.
315. Gawchik, S., and Mansmann, H.: Long-term hospitalization in the management of intractable bronchial asthma. *Pediatr. Ann.* **7**(1):112, 1978.
316. Geist, R.: Onset of chronic illness in children and adolescents: Psychotherapeutic and consultative intervention. *Am. J. Orthopsychiatry* **49**(1):4, 1979.
317. Gilder, R., Buschman, P., Sitarz, A., and Wolff, J.: Group therapy with parents of children with leukemia. *Am. J. Psychother.* **32**(2):276, 1978.
318. Glaser, K.: The treatment of depressed and suicidal adolescents. *Am. J. Psychother.* **32**(2):252, 1978.
319. Gogan, J., Koocher, G., Fine, W., Foster, D., and O'Malley, J.: Pediatric cancer survival and marriage: Issues affecting adult adjustment. *Am. J. Orthopsychiatry* **49**(3):423, 1979.
320. Goodwin, J., Goodwin, J., and Kellner, R.: Psychiatric symptoms in disliked medical patients. *JAMA* **241**(11):1117, 1979.
321. Goodykoontz, L.: Touch: Attitudes & practice. *Nurs. Forum* **18**(1):4, 1979.
322. Gottschalk, L., McGuire, F., Heiser, J., Dinovo, E., and Birch, H.: A review of psychoactive drug-involved deaths in nine major United States cities. *Int. J. Addict.* **14**(6):735, 1979.
323. Grant, W.: What parents of a chronically ill or dysfunctioning child always want to know but may be afraid to ask. *Clin. Pediatr. (Phila.)* **17**(12):915, 1978.
324. Gray, C., and Garner, J.: Health and the law conference in Ottawa exposes pitfalls but offers few remedies. *Can. Med. Assoc. J.* **120**(11):1407, 1979.
325. Green, A.: Self-destructive behavior in battered children. *Am. J. Psychiatry* **135**(5):579, 1978.
326. Green, M.: Parent care in the intensive care unit. *Am. J. Dis. Child.* **133**(11):1119, 1979.
327. Grove, S.: Encounters with grief: I am a yellow ship. *Am. J. Nurs.* **78**(3):414, 1978.
328. Harding, R., Heller, J., and Kesler, R.: The chronically ill child in the primary care setting. *Primary Care* **6**(2):311, 1979.
329. Headlam, H., Goldsmith, J., Hanenson, I., and Rauh, J.: Demographic characteristics of adolescents with self-poisoning. A survey of 235 instances in Cincinnati, Ohio. *Clin. Pediatr. (Phila.)* **18**(3):147, 1979.
330. Helmrath, T., and Steinitz, E.: Death of an infant: Parental grieving and the failure of social support. *J. Fam. Pract.* **6**(4):785, 1978.
331. Humphrey, J.: Role interference: An analysis of suicide victims, homocide offenders, and non-violent individuals. *J. Clin. Psychiatry* **39**(8):652, 1978.
332. Jacobs, J.: Social pediatric emergencies. *Paediatrician* **7**(4–5):239, 1978.
333. Jensen, I., and Larsen, J.: Mental aspects of temporal lobe epilepsy. Follow-up of 74 patients after resection of a temporal lobe. *J. Neurol. Neurosurg. Psychiatry* **42**(3):256, 1979.
334. Jurk, I.: Children with terminal disease. *Aust. Nurses J.* **7**(7):26, 1978.
335. Kaffman, M.: Kibbutz civilian population under war stress. *Ment. Health Soc.* **5**(1–2):45, 1978.
336. Kapotes, C.: Emotional factors in chronic asthma. *J. Asthma Res.* **15**(1):5, 1977.
337. Kearney, P.: Ethics, cancer and children. *Med. Hypotheses* **3**(5):174, 1977.
338. Kendra, J.: Predicting suicide using the Rorschach inkblot test. *J. Pers. Assess.* **43**(5):452, 1979.
339. Kenny, T., Rohn, R., Sarles, R., Reynolds, B., and Heald, F.: Visual-motor problems of adolescents who attempt suicide. *Percept. Mot. Skills.* **48**(2):599, 1979.

340. Kerner, J., Harvey, B., and Lewiston, N.: The impact of grief: A retrospective study of family function following loss of a child with cystic fibrosis. *J. Chronic Dis.* **32**(3):221, 1979.
341. Kessler, J.: Parenting the handicapped child. *Pediatr. Ann.* **6**(10):654, 1977.
342. Kittleson, J.: Nursing experience: In spite of everything, Tommy died in peace. *RN.* **41**(2):97, 1978.
343. Knight, J.: Human response to success and failure. *J. Med. Assoc. Ga.* **67**(11):916, 1978.
344. Kodadek, S.: Family-centered care of the chronically ill child. *Aorn. J.* **30**(4):635, 1979.
345. Kraus, R., and Buffler, P.: Sociocultural stress and the American native in Alaska: An analysis of changing patterns of psychiatric illness and alcohol abuse among Alaska natives. *Cult. Med. Psychiatry* **3**(2):111, 1979.
346. Lansky, S., Cairns, N., Hassanein, R., Wehr, J., and Lowman, J.: Childhood cancer: Parental discord and divorce. *Pediatrics* **62**(2):184, 1978.
347. Lascari, A.: The dying child and the family. *J. Fam. Pract.* **6**(6):1279, 1978.
348. Lavigne, J., and Ryan, M.: Psychologic adjustment of siblings of children with chronic illness. *Pediatrics* **63**(4):616, 1979.
349. Leahy, M.: Expectations: Do we always agree? Care of the child with cancer. *Australas. Nurses J.* **8**(4):14, 1979.
350. Lebowitz, M.: The relationship of socio-environmental factors to the prevalence of obstructive lung diseases and other chronic conditions. *J. Chronic. Dis.* **30**(9):599, 1977.
351. Lewis, E.: Inhibition of mourning by pregnancy: Psychopathology and management. *Br. Med. J.* **2**(6181):27, 1979.
352. Lewis, E.: Mourning by the family after a stillbirth or neonatal death. *Arch. Dis. Child.* **54**(4):303, 1979.
353. Lewis, J.: How the nurse copes with grief. Care of the child with cancer. *Australas. Nurses J.* **8**(4):16, 1979.
354. Lewis, S., and Armstrong, S.: Children with terminal illness: A selected review. *Int. J. Psychiatry Med.* **8**(1):73, 1977–1978.
355. Lopez, T., and Kliman, G.: Memory, reconstruction, and mourning in the analysis of a 4-year-old child: Maternal bereavement in the second year of life. *Psychoanal. Study Child.* **31**:235, 1979.
356. Lord, R., Ritvo, S., and Solnit, A.: Patients' reactions to the death of the psychoanalyst. *Int. J. Psychoanal.* **59**(2–3) :189, 1978.
357. Malla, A., and Hoenig, J.: Suicide in Newfoundland and Labrador. *Can. J. Psychiatry* **24**(2):139, 1979.
358. Martinson, I.: Caring for the dying child. *Nurs. Clin. North Am.* **14**(3):467, 1979.
359. Martinson, I., Armstrong, G., Geis, D., Anglim, M., Gronseth, E., MacLnnis, H., Kersey, J., and Nesbit, M.: Home care for children dying of cancer. *Pediatrics* **62**(1):106, 1978.
360. Martinson, I., Geis, D., Anglim, M., Peterson, E., Nesbit, M., and Kersey, J.: When the patient is dying: Home care for the child. *Am. J. Nurs.* **77**(11):1815, 1977.
361. Mathur, G., and Carruthers, I.: Attempted suicide in patients admitted to hospital in Wigan and Leigh Borough, England. *Pahlavi Med. J.* **8**(4):407, 1977.
362. Matus, I., Kinsman, R., and Jones, N.: Pediatric patient attitudes toward chronic asthma and hospitalization. *J. Chronic Dis.* **31**(9–10):611, 1978.
363. McIntier, T.: Hillhaven Hospice: A free-standing, family-centered program. *Hosp. Prog.* **60**(3):68, 1979.
364. McIntire, M., Angle, C., and Schlicht, M.: Suicide and self-poisoning in pediatrics. *Adv. Pediatr.* **24**:291, 1977.

365. Mills, G.: Books to help children understand death. *Am. J. Nurs.* **79**(2):291, 1979.
366. Mills, I.: Self-poisoning—A modern epidemic, in Duncan, W., Leonard, B. (Eds.), *Clinical Toxicology.* Amsterdam: Excerpta Medica, 1977. (W3 EX89 No. 147, 1976.)
367. Nagai, S.: Children in hospitals; two special problems. *Dimens. Health Serv.* **56**(5):30, 1979.
368. Naish, J.: Problems of deception in medical practice. *Lancet* **2**(8134):139, 1979.
369. Neill, K.: Behavioral aspects of chronic physical disease. *Nurs. Clin. North Am.* **14**(3):443, 1979.
370. Neugarten, B.: Time, age, and the life cycle. *Am. J. Psychiatry* **136**(7):887, 1979.
371. Olin, H.: When a child dies. *Am. Fam. Physician* **18**(3):107, 1978.
372. Orbach, I., and Glaubman, H.: Children's perception of death as a defensive process. *J. Abnorm. Psychol.* **88**(6):671, 1979.
373. Orbach, I., and Glaubman, H.: Suicidal, aggressive, and normal children's perception of personal and impersonal death. *J. Clin. Psychol.* **34**(4):850, 1978.
374. Patenaude, A., Szymanski, L., and Rappeport, J.: Psychological costs of bone marrow transplantation in children. *Am. J. Orthopsychiatry* **49**(3):409, 1979.
375. Patrick, D., Coleman, J., Eagle, J., and Nelson, E.: Chronic emotional problem patients and their families in an HMO. *Inquiry* **15**(2):166, 1978.
376. Paulson, M., Stone, D., and Sposto, R.: Suicide potential and behavior in children ages 4 to 12. *Suicide Life Threat. Behav.* **8**(4):225, 1978.
377. Pfeffer, C.: Clinical observations of play of hospitalized suicidal children. *Suicide Life Threat. Behav.* **9**(4):235, 1979.
378. Pfefferbaum, B., and Lucas, R.: Management of acute psychologic problems in pediatric oncology. *Gen. Hosp. Psychiatry* **1**(3):214, 1979.
379. Pond, H.: Parental attitudes toward children with a chronic medical disorder: Special reference to diabetes mellitus. *Diabetes Care* **2**(5):425, 1979.
380. Pope, A.: Children's attitudes toward death. *Health People* **10**(3):22, 1979.
381. Pugh, L.: The child and death. *Issues Ment. Health Nurs.* **2**(2):53, 1979.
382. Ramsey, P.: The Saikewicz precedent: What's good for an incompetent patient. *Inquiry* **8**(6):36, 1978.
383. Ray, R.: The mentally handicapped child's reaction to bereavement. *Health Visit.* **51**(9):333, 1978.
384. Rebele, M.: When we tried to get close to Carrie, she pushed us away. *Nursing (Horsham)* **9**(8):52, 1979.
385. Rodgers, B.: Comprehensive care for the child with a chronic disability. *Am. J. Nurs.* **79**(6):1106, 1979.
386. Rosenheim, E., and Ichilov, Y.: Short-term preventive therapy with children of fatally-ill parents. *Isr. Ann. Psychiatry* **17**(1):67, 1979.
387. Ross, J.: Social work intervention with families of children with cancer: The changing critical phases. *Soc. Work. Health Care.* **3**(3):257, 1978.
388. Sandler, N.: Working with families of chronic asthmatics. *J. Asthma Res.* **15**(1):15, 1977.
389. Satterwhite, B.: Impact of chronic illness on child and family: An overview based on five surveys with implications for management. *Int. J. Rehabil. Res.* **1**(1):7, 1978.
390. Schraeder, B.: A creative approach to caring for the ventilator-dependent child. *McN.* **4**(3):165, 1979.
391. Seeman, M.: Management of the schizophrenic patient. *Can. Med. Assoc. J.* **120**(9):1097, 1979.
392. Sickel, R.: Children who experience parental suicide. *Pediatr. Nurs.* **5**(3):37, 1979.

393. Silverman, S., and Silverman, P.: Parent–child communication in widowed families. *Am. J. Psychother.* **33**(3):428, 1979.
394. Slimmer, L.: Helping parents cope with their child's seizure disorder. *J. Psychiatr. Nurs.* **17**(2):30, 1979.
395. Stanton, M., Todd, T., Heard, D., Kirschner, S., Kleiman, J., Mowatt, D., Riley, P., Scott, S., and Van Deusen, J.: Heroin addiction as a family phenomenon: A new conceptual model. *Am. J. Drug Alcohol Abuse* **5**(2):125, 1978.
396. Stecchi, J.: The death of a child. Looking back from a parent's point of view. *J. Nurs. Care.* **12**(5):13, 1979.
397. Stubblefield, K.: A preventive program for bereaved families. *Soc. Work Health Care* **2**(4):379, 1977.
398. Tonkin, P.: Parent care for the low risk and terminally ill child. *Dimens. Health Serv.* **56**(5):42, 1979.
399. Toolan, J.: Therapy of depressed and suicidal children. *Am. J. Psychother.* **32**(2):243, 1978.
400. Tooley, K.: The remembrance of things past: On the collection and recollection of ingredients useful in the treatment of disorders resulting from unhappiness, rootlessness, and the fear of things to come. *Am. J. Orthopsychiatry* **48**(1):174, 1978.
401. Tudor, M.: The problems of long term care. *Med. Leg. J.* **47**(2):69, 1979.
402. Udelman, D.: Chronic illness: Treatment and the self-concept, Part I. *Ariz. Med.* **36**(9):684, 1979.
403. Veatch, R.: Death and dying: The legislative options. *Hastings Center Report* **7**(5):5, 1977.
404. Vines, D.: Bonding, grief, and working through in relationship to the congenitally anomalous child and his family. *Ana. Publ.* **59**:185, 1979.
405. Walsh, F.: Concurrent grandparent death and birth of schizophrenic offspring: An intriguing finding. *Fam. Process.* **17**(4):457, 1978.
406. Watt, N., and Nicholi, A.: Early death of a parent as an etiological factor in schizophrenia. *Am. J. Orthopsychiatry* **49**(3):465, 1979.
407. Weissman, M., Prusoff, B., and Klerman, G.: Personality and the prediction of long-term outcome of depression. *Am. J. Psychiatry* **135**(7):797, 1978.
408. Wessel, M.: The grieving child. *Clin. Pediatr. (Phila.)* **17**(7):559, 1978.
409. Wilson, J.: Care of children with progressive neurological disease. *Nurs. Times* **75**(41):1766, 1979.
410. No not-non-accidental injury letter. *Lancet* **2**(7991):913, 1976.
411. Ansell, B.: Psyche and Rheuma. *J. Int. Med. Res.* **4**(2 suppl):50, 1976.
412. Apley, J.: Pain in childhood. *J. Psychosom. Res.* **20**(4):383, 1976.
413. Astrachan, M.: Management of a staff death in a children's institution. *Child Welfare* **56**(6):380, 1977.
414. Bedell, J., Giordani, B., Amour, J., Tavormina, J., and Boll, T.: Life stress and the psychological and medical adjustment of chronically ill children. *J. Psychosom. Res.* **21**(3):237, 1977.
415. Begleiter, M., Burry, V., and Harris, D.: Prevalence of divorce among parents of children with cystic fibrosis and other chronic diseases. *Soc. Biol.* **23**(3):260, 1976.
416. Black, D.: What happens to bereaved children. *Proc. R. Soc. Med.* **69**(11):842, 1976.
417. Branson, H.: Grieving and growing. *J. Pract. Nurs.* **26**(12):34, 1976.
418. Burgess, K.: The influence of will on life and death. *Nurs. Forum.* **15**(3):238, 1976.
419. Carey, P.: Grief process: Some implications for nursing practice. *Mich. Nurse* **50**(1):5, 1977.

420. Cashmore, G.: The reduction of soiling behaviour in an 11-year-old boy with the parent as therapist. *N. Z. Med. J.* **84**(572):238, 1976.
421. Codden, P.: The meaning of death for parents and the child. *Matern. Child Nurs. J.* **6**(1):9, 1977.
422. Cornwell, J., Nurcombe, B., and Stevens, L.: Family response to loss of a child by sudden infant death syndrome. *Med. J. Aust.* **1**(18):656, 1977.
423. Daitch, P., and Vianna, N.: The influence of finite observation periods on familial and chronic disease epidemiology letter. *J. Natl. Cancer Inst.* **58**(1):9, 1977.
424. Evans, N.: Mourning as a family secret. *J. Am. Acad. Child Psychiatry* **15**(3):502, 1976.
425. Ghory, J.: The ABCs of educating the patient with chronic bronchial asthma. How we do it. *Clin. Pediatr. (Phila.)* **16**(10):879, 1977.
426. Green, M.: The management of children with chronic disease. *Proc. Inst. Med. Chic.* **31**(3):51, 1976.
427. Grinnell, R., and Kyte, N.: Crisis. *MH* **60**(4):11, 1977.
428. Grollman, E.: Explaining death to children. *J. Sch. Health* **47**(6):336, 1977.
429. Groot, J.: Mourning in a 6-year-old girl. *Psychoanal. Study Child* **31**:273, 1976.
430. Hart, E.: Death education and mental health. *J. Sch. Health* **46**(7):407, 1976.
431. Higgins, G.: Grief reactions. *Practitioner* **218**(1307)):689, 1977.
432. Higgins, J.: Effects of child rearing by schizophrenic mothers: A follow-up. *J. Psychiatr. Res.* **13**(1):1, 1976.
433. Himmelhoch, A.: The effects of chronic illness on children of 0–3 years and their families. *Australas. Nurses J.* **6**(11):14, 1977.
434. Hughes, J.: The emotional impact of chronic disease. The pediatrician's responsibilities. *Am. J. Dis. Child.* **130**(11):1199, 1976.
435. Jones, P.: Malignant disease in childhood: The problems in general practice. *Aust. Fam. Physician.* **6**(3):234, 1977.
436. Kaffman, M.: Kibbutz civilian population under war stress. *Br. J. Psychiatry* **130**:489, 1977.
437. Katz, J.: Psychiatric aspects of accidental poisoning in childhood. *Med. J. Aust.* **2**(2):59, 1976.
438. Kavanaugh, R.: Dealing naturally with the dying. *Nursing (Jenkintown)* **6**(10):23, 1976.
439. Kestenbaum, C.: The effects of fatherless homes upon daughters: Clinical impressions regarding paternal deprivation. *J. Am. Acad. Psychoanal.* **4**(2):171, 1976.
440. Koch, C., Hermann, J., and Donaldson, M.: Supportive care of the child with cancer and his family. *Semin. Oncol.* **1**(1):81, 1974.
441. Lambert, C., and Lambert, V.: Divorce: A psychodynamic development involving grief. *J. Psychiatr. Nurs.* **15**(1):37, 1977.
442. Lansky, S., Lowman, J., Gyulay, J., and Briscoe, K.: A team approach to coping with cancer. 291, in Cullen, J., *et al.* (Eds.), *Cancer: The Behavioral Dimensions.* New York, Raven Press, 1976.
443. Lawson, B.: Chronic illness in the school-aged child: Effects on the total family. *McN.* **2**(1):49, 1977.
444. Legg, C., and Sherick, I.: The replacement child—a developmental tragedy: Some preliminary comments. *Child Psychiatry Hum. Dev.* **7**(2):113, 1976.
445. Leyn, R.: Terminally ill children and their families: A study of the variety of responses to fatal illness. *Matern. Child. Nurs. J.* **5**(3):179, 1976.
446. Miles, M., Mattioli, L., and Diehl, A.: Parent counseling: Psychological support of parents of children with critical heart disease. *J. Kans. Med. Soc.* **78**(3):134, 1977.
447. Parkes, C.: Family reactions to child bereavement letter. *Proc. R. Soc. Med.* **70**(1):54, 1977.

448. Pless, I.: Individual and family needs in the health care of children with developmental disorders. *Birth Defects* **12**(4):91, 1976.
449. Primeaux, M.: Caring for the American Indian patient. *Am. J. Nurs.* **77**(1):91, 1977.
450. Raphael, B.: The Granville train disaster: Psychological needs and their management. *Med. J. Aust.* **1**(9):303, 1977.
451. Rice, N., Satterwhite, B., and Pless, I.: Family counselors in a pediatric specialty clinic setting. *Soc. Work Health Care* **2**(2):193, 1976–1977.
452. Rodriguez, R.: After the child dies: Autopsies and forgotten parents letter. *J. Pediatr.* **89**(5):859, 1976.
453. Rohn, R., Sarles, R., Kenny, T., Reynolds, B., and Heald, F.: Adolescents who attempt suicide. *J. Pediatr.* **90**(4):636, 1977.
454. Sack, W.: Children of imprisoned fathers. *Psychiatry* **40**(2):163, 1977.
455. Samaniego, L., Caldwell, H., Nitschke, R., and Humphrey, G.: Exploring the physically ill child's self-perceptions and the mother's perceptions of her child's needs: Insights gained from the Firo-BC-a behavior test for use with children. *Clin. Pediatr. (Phila.)* **16**(2):154, 1977.
456. Sheer, B.: Help for parents in a difficult job—broaching the subject of death. *Mon.* **2**(5):320, 1977.
457. Shepherd, D., and Barraclough, B.: The aftermath of parental suicide for children. *Br. J. Psychiatry* **129**:267, 1976.
458. Sheridan, M.: Renal disease and the social worker: A review. *Health Soc. Work* **2**(2):122, 1977.
459. Shirkey, H.: Facing the inevitable: The physician's supportive role in the death of a child. *Hosp. Formul.* **11**(3):146, 1976.
460. Shneidman, E.: Some psychological reflections on the death of Malcolm Melville. *Suicide Life Threat. Behav.* **6**(4):231, 1976.
461. Simonds, J.: Psychiatric consultations for 112 pediatric inpatients. *South Med. J.* **70**(8):980, 1977.
462. Steinhausen, H.: Hemophilia: A psychological study in chronic disease in juveniles. *J. Psychosom. Res.* **20**(5):461, 1976.
463. Stevenson, I.: The explanatory value of the idea of reincarnation. *J. Nerv. Ment. Dis.* **164**(5):305, 1977.
464. Sturges, J.: Talking with children about mental illness in the family. *Health Soc. Work* **2**(3):87, 1977.
465. Tew, B., Laurence, K., Payne, H., and Rawnsley, K.: Marital stability following the birth of a child with spina bifida. *Br. J. Psychiatry* **131**:79, 1977.
466. Tietz, W., McSherry, L., and Britt, B.: Family sequelae after a child's death due to cancer. *Am. J. Psychother.* **31**(3):417, 1977.
467. Wass, V., Barratt, T., Howarth, R., Marshall, W., Chantler, C., Ogg, C., Cameron, J., Baillod, R., and Moorhead, J.: Home haemodialysis in children. Report of the London Children's Home Dialysis Group. *Lancet* **1**(8005):2426, 1977.
468. A Recommended 'health care policy relating to children and their families.' *Med. J. Aust.* **2**(2suppl):1, 1975.
469. Editorial: Suicide in children. *Br. Med. J.* **1**(5958):592, 1975.
470. Ethics of selective treatment of spina bifida. Report by a working party. *Lancet* **1**(7898):85, 1975.
471. Aleksandrowicz, M.: The biological strangers: An attempted suicide of a seven-and-a-half-year-old girl. *Bull. Menninger Clin.* **39**(2):163, 1975.
472. Awad, G., and Poznanski, E.: Psychiatric consultations in a pediatric hospital. *Am. J. Psychiatry* **132**(9):915, 1975.

473. Barry, J.: 8 weeks and wheezing. *Pediatr. Nurs.* **1**(5):27, 1975.
474. Battle, C.: Symposium on behavioral pediatrics. Chronic physical disease. Behavioral aspects. *Pediatr. Clin. North Am.* **22**(3):525, 1975.
475. Blank, H.: Crisis Consultation. *Int. J. Soc. Psychiatry* **21**(3):179, 1975.
476. Browne, W., and Palmer, A.: A preliminary study of schizophrenic women who murdered their children. *Hosp. Community Psychiatry* **26**(2):71, 1975.
477. Cassel, J.: The contribution of the social environment to host resistance: The fourth Wade Hampton Frost lecture. *Am. J. Epidemiol.* **104**(2):107, 1976.
478. Crase, D., and Crase, D.: Death and the young child. *Clin. Pediatr. (Phila.)* **14**(8):747, 1975.
479. Crewe, H.: Fears and anxiety in childhood. *Public Health* **87**(5):165, 1973.
480. Cummings, S.: The impact of the child's deficiency on the father: A study of fathers of mentally retarded and of chronically ill children. *Am. J. Orthopsychiatry* **46**(2):246, 1976.
481. D'Heurle, A., and Feimer, J.: Lost children: The role of the child in psychological plays of Henrik Ibsen. *Psychoanal. Rev.* **63**(1):27, 1976.
482. David, C.: Grief, mourning, and pathological mourning. *Primary Care* **2**(1):81, 1975.
483. Degnan, M., Peters, J., Porter, I., and Gottesman, D.: Genetic counseling. *Am. Fam. Physician* **12**(1):111, 1975.
484. Dondlinger, P.: Was Terry dead? *Nursing (Jenkintown)* **5**(6):57, 1975.
485. Draughon, M.: Step-mother's model of identification in relation to mourning in the child. *Psychol. Rep.* **36**(1):183, 1975.
486. Elfert, H.: The nurse and the grieving parent. *Can. Nurse.* **71**(2):30, 1975.
487. Felner, R., Stolberg, A., and Cowen, E.: Crisis events and school mental health referral patterns of young children. *J. Consult. Clin. Psychol.* **43**(3):305, 1975.
488. Fischhoff, J., and O'Brien, N.: After the child dies. *J. Pediatr.* **88**(1):140, 1976.
489. French, A., and Steward, M.: Family dynamics, childhood depression, and attempted suicide in a 7-year-old boy: A case study. *Suicide* **5**(1):29, 1975.
490. Gaffney, K.: Helping grieving parents. *Jen.* **2**(4):42, 1976.
491. Gardner, G.: Childhood, death, and human dignity: Hypnotherapy for David. *Int. J. Clin. Exp. Hypn.* **24**(2):122, 1976.
492. Garmezy, N., and Streitman, S.: Children at risk: The search for the antecedents of schizophrenia. Part I. Conceptual models and research methods. *Schizo. Bull.* **8**:14, 1974.
493. Gayford, J.: Wife battering: A preliminary survey of 100 cases. *Br. Med. J.* **1**(5951):194, 1975.
494. Greene, P.: Acute leukemia in children. *Am. J. Nurs.* **75**(10):1709, 1975.
495. Grossman, M.: The psychosocial approach to the medical management of patients with cystic fibrosis. *Clin. Pediatr. (Phila.)* **14**(9):830, 1975.
496. Gyulay, J.: Care of the dying child. *Nurs. Clin. North Am.* **11**(1):95, 1976.
497. Gyulay, J.: The forgotten grievers. *Am. J. Nurs.* **75**(9):1476, 1975.
498. Hall, A.: Treatment of anorexia nervosa. *N. Z. Med. J.* **82**(543):10, 1975.
499. Hall, W., Hall, C., and Hyde, R.: "Visiting the iniquity of the fathers upon the children." *Am. J. Dis. Child* **129**(8):887, 1975.
500. Heard, D.: Crisis intervention guided by attachment concepts—a case study. *J. Child Psychol. Psychiatry* **15**(2):111, 1974.
501. Heffron, W.: Group therapy in family practice. *Am. Fam. Physician* **10**(5):176, 1974.
502. Herzberg, J., and Wolff, S.: Chronic factitious fever in puberty and adolescence: A diagnostic challenge to the family physician. *Psychiatry Med.* **3**(3):205, 1972.

503. Iga, M., Yamamoto, J., and Noguchi, T.: The vulnerability of young Japanese women and suicide. *Suicide* **5**(4):207, 1975.
504. Jackson, P.: The child's developing concept of death: Implications for nursing care of the terminally ill child. *Nurs. Forum.* **14**(2):204, 1975.
505. Jacobson, S., Fasman, J., and DiMascio, A.: Deprivation in the childhood of depressed women. *J. Nerv. Ment. Dis.* **160**(1):5, 1975.
506. Janken, J.: The nurse in crisis. *Nurs. Clin. North Am.* **9**(1):17, 1974.
507. Kahana, R.: Studies in medical psychology: A brief survey. *Psychiatry Med.* **3**(1):1, 1972.
508. Kanthor, H., Pless, B., Satterwhite, B., and Myers, G.: Areas of responsibility in the health care of multiply handicapped children. *Pediatrics* **54**(6):779, 1974.
509. Kartha, M., and Ertel, I.: Short-term group therapy for mothers of leukemic children. What professional staff can do to help parents cope with chronic illness. *Clin. Pediatr. (Phila.)* **15**(9):803, 1976.
510. Kastenbaum, R.: On death and dying. Should we have mixed feelings about our ambivalence toward the aged. *J. Geriatr. Psychiatry* **7**(1):94, 1974.
511. Kemph, J.: Psychotherapy with donors and recipients of kidney transplants. *Semin. Psychiatry* **3**(1):145, 1971.
512. Kliman, A., and Schutz, M.: Children in crisis. *J. N.Y. State Nurs. Assoc.* **7**(2):21, 1976.
513. Knowelden, J.: 7. Information for community planning and coordination of longterm care services. *Med. Care* **14**(5 suppl):78, 1976.
514. Kraft, D., and Babigian, H.: Suicide by persons with and without psychiatric contacts. *Arch. Gen. Psychiatry* **33**(2):209, 1976.
515. Krant, M., Beiser, M., Adler, G., and Johnston, L.: The role of a hospital-based psychosocial unit in terminal cancer illness and bereavement. *J. Chronic Dis.* **29**(2):115, 1976.
516. Lansky, S.: Childhood leukemia. The child psychiatrist as a member of the oncology team. *J. Am. Acad. Child Psychiatry* **13**(3):499, 1974.
517. Lester, D., and Beck, A.: Early loss as a possible 'sensitizer' to later loss in attempted suicides. *Psychol. Rep.* **39**(1):121, 1976.
518. Levin, S.: The bonds of behaviour. *S. Afr. Med. J.* **50**(19):749, 1976.
519. Lewis, T.: A culturally patterned depression in a mother after loss of a child. *Psychiatry* **38**(1):92, 1975.
520. Lifshitz, M.: Long range effects of father's loss: The cognitive complexity of bereaved children and their school adjustment. *Br. J. Med. Psychol.* **49**(2):189, 1976.
521. Mann, S.: Coping with a child's fatal illness. A parent's dilemma. *Nurs. Clin. North Am.* **9**(1):81, 1974.
522. Marshall, C.: The indigenous nurse as community crisis intervener. *Semin. Psychiatry* **3**(1):264, 1971.
523. Martinson, I.: Why don't we let them die at home. *Rn.* **39**(1):58, 1976.
524. Mazzola, R., and Jacobs, G.: Helping the patient and the family deal with a crisis situation. *J. Neurosurg. Nurs.* **6**(2):85, 1974.
525. McAnarney, E.: Suicidal behavior of children and youth. *Pediatr. Clin. North Am.* **22**(3):595, 1975.
526. Mills, J., Williams, C., Sale, I., Perkin, G., and Henderson, S.: The epidemiology of self-poisoning in Hobart, 1968–1972. *Aust. Nz. J. Psychiatry.* **8**(3):167, 1974.
527. North, A.: When should a child be in the hospital. *Pediatrics* **57**(4):540, 1976.
528. O'Malley, P.: Attempted suicide, suicide and communal violence. *Ir. Med. J.* **68**(5):103, 1975.

529. Orgel, S.: Fusion with the victim and suicide. *Int. J. Psychoanal.* **55**(4):531, 1974.

530. Parnes, E.: Effects of experiences with loss and death among preschool children. *Child Today* **4**(6):2, 1975.

531. Pattison, E.: The fatal myth of death in the family. *Am. J. Psychiatry* **133**(6):674, 1976.

532. Paulley, J.: Cultural influences on the incidence and pattern of disease. *Psychother. Psychosom.* **26**(1):2, 1975.

533. Richardson, P.: A multigravida's use of a living child in the grief and mourning for a lost child. *Matern. Child. Nurs. J.* **3**(3):181, 1974.

534. Rorsman, B.: Mortality among psychiatric patients. *Acta. Psychiatr. Scand.* **50**(3):354, 1974.

535. Rose, M.: Problems families face in home care. *Am. J. Nurs.* **76**(3):416, 1976.

536. Sarles, R.: Symposium on behavioral pediatrics. *Incest. Pediatr. Clin. North Am.* **22**(3):633, 1975.

537. Scheideman, J.: Chronicity: A key to learning. *Am. J. Nurs.* **75**(3):446, 1975.

538. Schowalter, J.: How do children and funerals mix? *J. Pediatr.* **89**(1):139, 1976.

539. Sendi, I., and Blomgren, P.: A comparative study of predictive criteria in the predisposition of homicidal adolescents. *Am. J. Psychiatry* **132**(4):423, 1975.

540. Serban, G.: Parental stress in the development of schizophrenic offspring. *Compr. Psychiatry* **16**(1):23, 1975.

541. Shaffer, D.: Suicide in childhood and early adolescence. *J. Child. Psychol. Psychiatry* **15**(4):275, 1974.

542. Singher, L.: The slowly dying child. *Clin. Pediatr. (Phila.)* **13**(10):861, 1974.

543. Spinetta, J., Rigler, D., and Karon, M.: Personal space as a measure of a dying child's sense of isolation. *J. Consult Clin. Psychol.* **42**(6):751, 1974.

544. Stein, M., Levy, M., and Glasberg, H.: Separations in black and white suicide attempters. *Arch. Gen. Psychiatry* **31**(6):815, 1974.

545. Steinhauer, P., Mushin, D., and Rae-Grant, Q.: Psychological aspects of chronic illness. *Pediatr. Clin. North Am.* **21**(4):825, 1974.

546. Tavormina, J., Kastner, L., Slater, P., and Watt, S.: Chronically ill children. A psychologically and emotionally deviant population. *J. Abnorm. Child. Psychol.* **4**(2):99, 1976.

547. Tessler, R., and Mechanic, D.: Factors affecting the choice between prepaid group practice and alternative insurance programs. *Milbank Mem. Fund Q.* **53**(2):149, 1975.

548. Tiedt, E.: The psychodynamic process of the oncological experience. *Nurs. Forum.* **14**(3):264, 1975.

549. Tietz, W., and Powars, D.: The pediatrician and the dying child. 'Physician, know thyself.' *Clin. Pediatr. (Phila.)* **14**(6):585, 1975.

550. Tietz, W., and Vidmar, J.: The impact of coping styles on the control of juvenile diabetes. *Psychiatry Med.* **3**(1):67, 1972.

551. Tinkelman, D., Brice, J., Yoshida, G., and Sadler, J.: The impact of chronic asthma on the developing child: Observations made in a group setting. *Ann. Allergy* **37**(3):174, 1976.

552. Toolan, J.: Suicide in children and adolescents. *Am. J. Psychother.* **29**(3):339, 1975.

553. Tooley, K.: The choice of a surviving sibling as 'scapegoat' in some cases of maternal bereavement—a case report. *J. Child Psychol. Psychiatry* **16**(4):331, 1975.

554. Walsh, B., Walsh, D., and Whelan, B.: Suicide in Dublin: II. The influence of some social and medical factors on coroners' verdicts. *Br. J. Psychiatry* **126**:309, 1975.

555. Werry, J., and Pedder, J.: Self poisoning in Auckland. *N. Z. Med. J.* **83**(560):183, 1976.

556. Wessel, M.: A death in the family. The impact on children. *JAMA* **234**(8):865, 1975.

557. West, N.: Child's response to death loss. *Nebr. Med. J.* **60**(7):228, 1975.

558. Wilkes, J.: Don't forget the children. *Can. Med. Assoc. J.* **115**(6):528, 1976.
559. Young, J.: A mother's grief work following the death of her deformed child. *Matern. Child. Nurs. J.* **4**(1):57, 1975.
560. London Children's Home Dialysis Group: *Lancet* **1**(8005):2426, 1977.

AUTISM AND BURNS

JOURNAL ARTICLES (1974–1979)

561. Nursing Grand Rounds. Realistic goals don't mean failure. *Nursing (Horsham)* **9**(5):54, 1979.
562. Opinion Exchange: Diane's desperate decision. *RN* **42**(4):95, 1979.
563. National Society for Autistic Children definition of the syndrome of autism. *J. Autism Child. Schizophr.* **8**(2):162, 1978.
564. Risks and benefits in the treatment of autistic children. *J. Autism Child. Schizophr.* **8**(1):99, 1978.
565. Amon, L.: The management of the burned child. *Curationis.* **2**(1):30, 1979.
566. Ando, H., and Yoshimura, I.: Effects of age on communication skill levels and prevalence of maladaptive behaviors in autistic and mentally retarded children. *J. Autism Child. Schizophr.* **9**(1):83, 1979.
567. Applebaum, E., Egel, A., Koegel, R., and Imhoff, B.: Measuring musical abilities of autistic children. *J. Autism Dev. Disord.* **9**(3):279, 1979.
568. Arick, J., and Krug, D.: Autistic children: A study of learning characteristics and programming needs. *Am. J. Ment. Defic.* **83**(2):200, 1978.
569. Attwood, T.: The Croydon workshop for the parents of severely handicapped school age children. *Child Care Health Dev.* **5**(3):177, 1979.
570. Ayoub, C., and Pfeifer, D.: Burns as a manifestation of child abuse and neglect. *Am. J. Dis. Child.* **133**(9):910, 1979.
571. Baker, A.: Cognitive functioning of psychotic children: A reappraisal. *Except. Child.* **45**(5):344, 1979.
572. Baker, G., and Gester, K.: Vestibular stimulation with autistic and schizophrenic children. *Ala. J. Med. Sci.* **14**(4):434, 1977.
573. Barlow, C.: Mental retardation and related disorders. *Contemp. Neurol. Ser.* **17**:1, 1978.
574. Barry, R., and James, A.: Handedness in autistics, retardates, and normals of a wide age range. *J. Autism Child. Schizophr.* **8**(3):315, 1978.
575. Bartolucci, G., and Pierce, S.: A preliminary comparison of phonological development in autistic, normal, and mentally retarded subjects. *Br. J. Disord. Commun.* **12**(2):137, 1977.
576. Bauer, R., and McCarressi, I.: Strategies of therapeutic contact: Working with children with severe object relationship disturbance. *Am. J. Psychother.* **31**(4):605, 1977.
577. Bemporad, J.: Adult recollections of a formerly autistic child. *J. Autism Dev. Disord.* **9**(2):179, 1979.
578. Ben-Aaron, M., and Jarus, A.: Remarks on the communication of psychotic children as seen in group therapy. *Ment. Health Soc.* **5**(3–4):224, 1979.
579. Benaroya, S., Wesley, S., Ogilvie, H., Klein, L., and Clarke, E.: Sign language and multisensory input training of children with communication and related developmental disorders: Phase II. *J. Autism Dev. Disord.* **9**(2):219, 1979.
580. Bhalerao, V., Desai, V., and Pai, D.: Study of socio-psychological aspects of burns in females. *J. Postgrad. Med.* **22**(3):147, 1976.

581. Bharucha, M.: Infantile autism and childhood schizophrenia. *Indian. Pediatr.* **16**(1):65, 1979.
582. Blackstock, E.: Cerebral asymmetry and the development of early infantile autism. *J. Autism Child. Schizphr.* **8**(3):339, 1978.
583. Blades, B., Jones, C., and Munster, A.: Quality of life after major burns. *J. Trauma.* **19**(8):556, 1979.
584. Boucher, J.: Echoic memory capacity in autistic children. *J. Child Psychol. Psychiatry* **19**(2):161, 1978.
585. Brady, D., and Smouse, A.: A simultaneous comparison of three methods for language training with an autistic child: An experimental single case analysis. *J. Autism Child. Schizophr.* **8**(3):271, 1978.
586. Burgdorf, M.: Coping behaviors of a school age child hospitalized with burns. *Matern. Child Nurs. J.* **7**(1):11, 1978.
587. Cahners, S.: Group meetings benefit families of burned children. *Scand. J. Plast. Reconstr. Surg.* **13**(1):169, 1979.
588. Cahners, S.: A strong hospital-school liaison: A necessity for good rehabilitation planning for disfigured children. *Scand. J. Plast. Reconstr. Surg.* **13**(1):167, 1979.
589. Cahners, S.: Group meetings for families of burned children. *Health Soc. Work* **3**(3):165, 1978.
590. Cahners, S., and Bernstein, N.: Rehabilitating families with burned children. *Scan. J. Plast. Reconstr. Surg.* **13**(1):173, 1979.
591. Campbell, M., Anderson, L., and Meier, M.: A comparison of haloperidol, behavior therapy, and their interaction in autistic children proceedings. *Psychopharmacol. Bull.* **15**(2):84, 1979.
592. Campbell, M., Anderson, L., Meier, M., Cohen, I., Small, A., Samit, C., and Sachar, E.: A comparison of haloperidol and behavior therapy and their interaction in autistic children. *J. Am. Acad. Child Psychiatry* **17**(4):640, 1978.
593. Campbell, M., and Cohen, I.: Treatment of infantile autism. *Compr. Ther.* **4**(5):33, 1978.
594. Cantwell, D., Baker, L., and Rutter, M.: A comparative study of infantile autism and specific developmental receptive language disorder—IV. Analysis of syntax and language function. *J. Child. Psychol. Psychiatry* **19**(4):351, 1978.
595. Cantwell, D., and Baker, L.: Imitations and echoes in autistic and dysphasic children. *J. Am. Acad. Child Psychiatry* **17**(4):614, 1978.
596. Cantwell, D., and Baker, L.: The language environment of autistic and dysphasic children. *J. Am. Acad. Child Psychiatry* **17**(4):604, 1978.
597. Cantwell, D., Baker, L., and Rutter, M.: Families of autistic and dysphasic children. I. Family life and interaction patterns. *Arch. Gen. Psychiatry* **36**(6):682, 1979.
598. Cantwell, D., Baker, L., and Rutter, M.: Families of autistic and dysphasic children. II. Mothers' speech to the children. *J. Autism Child. Schizophr.* **7**(4):313, 1977.
599. Caparulo, B., and Cohen, D.: Cognitive structures, language, and emerging social competence in autistic and aphasic children. *J. Am. Acad. Child Psychiatry* **16**(4):620, 1977.
600. Carlson-Leavitt, J.: The importance of choice. *J. Autism Child. Schizophr.* **8**(2):243, 1978.
601. Carr, E., Binkoff, J., Kologinsky, E., and Eddy, M.: Acquisition of sign language by autistic children. I. Expressive labelling. *J. Appl. Behav. Anal.* **11**(4):489, 1978.
602. Casey, L.: Development of communicative behavior in autistic children: A parent program using manual signs. *J. Autism Child. Schizophr.* **8**(1):45, 1978.
603. Chess, S.: Discussion: Language, cognition, and autism by Rutter, Studies of the Autistic Syndromes by Coleman. *Res. Publ. Assoc. Res. Nerv. Ment. Dis.* (57):277, 1979.

604. Christ, A.: Assessment of cognitive organization in emotionally disturbed adolescents: A way of reducing parental perplexity. *Psychiatr. Q.* **50**(1):44, 1978.
605. Clark, P., and Rutter, M.: Task difficulty and task performance in autistic children. *J. Child Psychol. Psychiatry* **20**(4):271, 1979.
606. Cohen, D., Caparulo, B., Gold, J., Waldo, M., Shaywitz, B., Ruttenberg, B., and Rimland, B.: Agreement in diagnosis: Clinical assessment and behavior rating scales for pervasively disturbed children. *J. Amer. Acad. Child Psychiatry* **17**(4):589, 1978.
607. Cohen, I., Anderson, L., and Campbell, M.: Measurement of drug effects in autistic children. *Psychopharmacol. Bull.* **14**(4):68, 1978.
608. Constable, J., Bernstein, N., and Sheehy, E.: Unreasonable expectations of reconstructive patients affecting rehabilitation. *Scand. J. Plast. Reconstr. Surg.* **13**(1):177, 1979.
609. Cooper, N.: Observations on a therapeutic residential setting for autistic children. *Child Care Health Dev.* **3**(6):437, 1977.
610. Copeland, A., and Golden, D.: Assessing and facilitating play in handicapped children. *Child Care Health Dev.* **5**(5):335, 1979.
611. Coss, R.: Perceptual determinants of gaze aversion by normal and psychotic children: The role of two facing eyes. *Behavior* **69**(3–4):228, 1979.
612. Curcio, F.: Sensorimotor functioning and communication in mute autistic children. *J. Autism Child. Schizophr.* **8**(3):281, 1978.
613. Davids, A., and Berenson, J.: Integration of a behavior modification program into a traditionally oriented residential treatment center for children. *J. Autism Child. Schizophr.* **7**(3):269, 1977.
614. Deslauriers, A.: The cognitive-affective dilemma in early infantile autism: The case of Clarence. *J. Autism Child. Schizophr.* **8**(2):219, 1978.
615. Dunlap, G., Koegel, R., and Egel, A.: Autistic children in school. *Except Child.* **45**(7):552, 1979.
616. Ellerstein, N.: The cutaneous manifestations of child abuse and neglect. *Am. J. Dis. Child.* **133**(9):906, 1979.
617. Elmhirst, S.: Time and the Pre-verbal transference. *Int. J. Psychoanal.* **59**(2–3):173, 1978.
618. Eme, R.: Sex differences in childhood psychopathology: A review. *Psychol. Bull.* **86**(3):574, 1979.
619. Fay, W.: Personal pronouns and the autistic child. *J. Autism. Dev. Disord.* **9**(3):247, 1979.
620. Fein, D., Tinder, P., and Waterhouse, L.: Stimulus generalization in autistic and normal children. *J. Child Psychol. Psychiatry* **20**(4):325, 1979.
621. Fenichel, J.: Rights of the handicapped. *J. Autism Child. Schizophr.* **8**(2):243, 1978.
622. Findji, F., Harrison-Covello, A., and Lairy, G.: Long duration EEG studies in the case of a psychotic child. *Electroencephalogr. Clin. Neurophysiol.* **46**(5):592, 1979.
623. Folstein, S., and Rutter, M.: Infantile autism: A genetic study of 21 twin pairs. *J. Child Psychol. Psychiatry.* **18**(4):297, 1977.
624. Fordham, M.: "A possible root of active imagination." *J. Anal. Psychol.* **22**(4):317, 1977.
625. Frank, H., and Green, L.: Successful use of a bulk laxative to control the diarrhea of tube feeding. *Scand. J. Plast. Reconstr. Surg.* **13**(1):193, 1979.
626. Frankel, F., Freeman, B., Ritvo, E., and Pardo, R.: The effect of environmental stimulation upon the stereotyped behavior of autistic children. *J. Autism Child. Schizophr.* **8**(4):389, 1978.
627. Fredricks, R.: Working for all the autistic. *J. Autism Child. Schizophr.* **8**(2):245, 1978.

628. Freedman, D.: The sensory deprivations. An approach to the study of the emergence of affects and the capacity for object relations. *Bull. Menninger. Clin.* **43**(1):29, 1979.

629. Freeman, B., Guthrie, D., Ritvo, E., Schroth, P., Glass, R., and Frankel, F.: Behavior observation scale: Preliminary analysis of the similarities and differences between autistic and mentally retarded children. *Psychol. Rep.* **44**(2):519, 1979.

630. Freeman, B., Ritvo, E., Guthrie, D., Schroth, P., and Ball, J.: The behavior observation scale for autism: Initial methodology, data analysis, and preliminary findings on 89 children. *J. Am. Acad. Child Psychiatry* **17**(4):576, 1978.

631. Fyffe, C., and Prior, M.: Evidence for language recoding in autistic, retarded and normal children: A re-examination. *Br. J. Psychol.* **69**(3):393, 1978.

632. Gordon, N.: Neurological processes concerned with communication and their analysis. *Child Care Health Dev.* **5**(1):29, 1979.

633. Graveling, R., and Brooke, J.: Hormonal and cardiac response of autistic children to changes in environmental stimulation. *J. Autism Child Schizophr.* **8**(4):441, 1978.

634. Green, D.: Aspects of infantile autism. *N. Z. Nurs. J.* **72**(1):18, 1979.

635. Handleman, J.: Generalization by autistic-type children of verbal responses across settings. *J. Appl. Behav. Anal.* **12**(2):273, 1979.

636. Hansen, C.: The need for sharing. *J. Autism Child. Schizophr.* **8**(2):242, 1978.

637. Harris, M.: Understanding the autistic child. *Am. J. Nursing* **78**(10):1682, 1978.

638. Harris, S., and Ersner-Hershfield, R.: Behavioral suppression of seriously disruptive behavior in psychotic and retarded patients: A review of punishment and its alternatives. *Psychol. Bull.* **85**(6):1352, 1978.

639. Harris, S., and Wolchik, S.: Suppression of self-stimulation: Three alternative strategies. *J. Appl. Behav. Anal.* **12**(2):185, 1979.

640. Hayter, J.: Emergency nursing care of the burned patient. *Nurs. Clin. North Am.* **13**(2):223, 1978.

641. Helm, P., Head, M., Pullium, G., O'Brien, M., and Cromes, G.: Burn rehabilitation— a team approach. *Surg. Clin. North Am.* **58**(6):1263, 1978.

642. Hier, D., LeMay, M., and Rosenberger, P.: Autism and unfavorable left-right asymmetries of the brain. *J. Autism Dev. Disord.* **9**(2):153, 1979.

643. Hight, D., Bakalar, H., and Lloyd, J.: Inflicted burns in children. Recognition and treatment. *JAMA* **242**(6):517, 1979.

644. Hightower-Vandamm, M.: Developmental disabilities act: An historic perspective, Part I. *Ajot.* **33**(6):355, 1979.

645. Hung, D.: Using self-stimulation as reinforcement for autistic children. *J. Autism Child. Schizophr.* **8**(3):355, 1978.

646. Hung, D., and Thelander, M.: Summer camp treatment program for autistic children. *Except. Child.* **44**(7):534, 1978.

647. Jayne, D.: The burn survivor's point of view. *J. Trauma* **19**(11 suppl.):920, 1979.

648. Jones, C., and Feller, I.: Burns: The home stretch . . . rehabilitation. *Nursing (Horsham)* **7**(12):54, 1977.

649. Jones, C., and Feller, I.: Burns: Avoiding and coping with complications before and after grafting. *Nursing (Horsham)* **7**(11):72, 1977.

650. Kelley, J., and Samuels, M.: A new look at childhood autism: School–parent collaboration. *J. Sch. Health.* **47**(9):538, 1977.

651. Kessler, J.: Parenting the handicapped child. *Pediatr. Ann.* **6**(10):654, 1977.

652. Koegel, R., and Egel, A.: Motivating autistic children. *J. Abnorm. Psychol.* **88**(4):418, 1979.

653. Koegel, R., and Lovaas, O.: Comments on autism and stimulus overselectivity. *J. Abnorm. Psychol.* **87**(5):563, 1978.

654. Koegel, R., and Schreibman, L.: Teaching autistic children to respond to simultaneous multiple cues. *J. Exp. Child. Psychol.* **24**(2):299, 1977.
655. Kolff, W.: Dialysis of schizophrenics. Weird and novel applications of dialysis, hemofiltration, hemoperfusion and peritoneal dialysis: Witchcraft. *Artif. Organs.* **2**(3):277, 1978.
656. Konstantareas, M., Webster, C., and Oxman, J.: Manual language acquisition and its influence on other areas of functioning in four autistic and autistic-like children. *J. Child Psychol. Psychiatry* **20**(4):337, 1979.
657. Kotsopoulos, S., and Kutty, K.: Histidinemia and infantile autism. *J. Autism Child. Schizophr.* **9**(1):55, 1979.
658. Kramer, D., and McKinney, W.: The overlapping territories of psychiatry and ethology. *J. Nerv. Ment. Dis.* **167**(1):3, 1979.
659. Lennox, C., Callias, M., and Rutter, M.: Cognitive characteristics of parents of autistic children. *J. Autism Child. Schizophr.* **7**(3):243, 1977.
660. Litrownik, A., McInnis, E., Wetzel-Pritchard, A., and Filipelli, D.: Restricted stimulus control and inferred attentional deficits in autistic and retarded children. *J. Abnorm. Psychol.* **87**(5):554, 1978.
661. Lovaas, O., Koegel, R., and Schreibman, L.: Stimulus overselectivity in autism: A review of research. *Psychol. Bull.* **86**(6):1236, 1979.
662. Lubinsky, M.: Behavioral consequences of congenital rubella letter. *J. Pediatr.* **97**(4):678, 1979.
663. Lyell, A.: Cutaneous artifactual disease. A review, amplified by personal experience. *J. Am. Acad. Dermatol.* **1**(5):391, 1979.
664. Mansheim, P.: Tuberous sclerosis and autistic behavior. *J. Clin. Psychiatry* **40**(2):97, 1979.
665. Marcus, L., Lansing, M., Andrews, C., and Schopler, E.: Improvement of teaching effectiveness in parents of autistic children. *J. Am. Acad. Child Psychiatry* **17**(4):625, 1978.
666. Martin, H.: Adolescence with burns anzea, 1977. *Autralas. Nurses J.* **7**(3):8, 1977.
667. Massie, H.: The early natural history of childhood psychosis. Ten cases studied by analysis of family home movies of the infancies of the children. *J. Am. Acad. Child Psychiatry* **17**(1):29, 1978.
668. Massie, H.: Patterns of mother–infant behavior and subsequent childhood psychosis: A research and case report. *Child Psychiatry Hum. Dev.* **7**(4):211, 1977.
669. Mazuryk, G., Barker, P., and Harasym, L.: Behavior therapy for autistic children: A study of acceptability and outcome. *Child Psychiatry Hum. Dev.* **9**(2):119, 1978.
670. McElwee, H., Sirinek, K., and Levine, B.: Cimetidine affords protection equal to antacids in prevention of stress ulceration following thermal injury. *Surgery* **86**(4):620, 1979.
671. McHugh, M., Dimitroff, K., and Davis, N.: Family support group in a burn unit. *Am. J. Nursing* **79**(12):2148, 1979.
672. Miller, S., and Toca, J.: Adapted melodic intonation therapy: A case study of an experimental language program for an autistic child. *J. Clin. Psychiatry* **40**(4):201, 1979.
673. Moersch, M.: Developmental disabilities. *Ajot.* **32**(2):93, 1978.
674. Molinaro, J.: The social fate of children disfigured by burns. *Am. J. Psychiatry* **135**(8):979, 1978.
675. Morris, J., and McFadd, A.: The mental health team on a burn unit: A multidisciplinary approach. *J. Trauma.* **18**(9):658, 1978.
676. Morton-Evans, A., and Hensley, R.: Paired associate learning in early infantile autism and receptive developmental aphasia. *J. Autism Child. Schizophr.* **8**(1):61, 1978.

677. Munster, A.: The early management of thermal burns. *Surgery* **87**(1):29, 1980.
678. Netley, C., and Lockyer, L.: Methodological problems in the study of parental cognitive characteristics. *J. Autism Child. Schizophr.* **8**(1):115, 1978.
679. Neuman, C., and Hill, S.: Self-recognition and stimulus preference in autistic children. *Dev. Psychobiol.* **11**(6):571, 1978.
680. Newson, J., and Newson, E.: The handicapped child: What is an autistic child. *Nurs. Times* **75**(37):4, 1979.
681. Ney, P.: A psychopathogenesis of autism. *Child Psychiatry Hum. Dev.* **9**(4):195, 1979.
682. Noyes, R., Frye, S., Slymen, D., and Canter, A.: Stressful life events and burn injuries. *J. Trauma.* **19**(3):141, 1979.
683. O'Banion, D., Armstrong, B., Cummings, R., and Stange, J.: Disruptive behavior: A dietary approach. *J. Autism Child. Schizophr.* **8**(3):325, 1978.
684. O'Dell, S., Blackwell, L., Larcen, S., and Hogan, J.: Competency-based training for severely behaviorally handicapped children and their parents. *J. Autism Child. Schizophr.* **7**(3):231, 1977.
685. O'Gorman, G.: Childhood autism. *Practitioner* **221**(1323):365, 1978.
686. Ohira, K.: The context theory of autism. *Folia. Psychiatr. Neurol. Jpn.* **33**(1):35, 1979.
687. Palyo, W., Cooke, T., Schuler, A., and Apolloni, T.: Modifying echolalic speech in preschool children: Training and generalization. *Am. J. Ment. Defic.* **83**(5):480, 1979.
688. Pawle, M.: Care of the burned patient. *Aust. Nurses J.* **9**(1):40, 1979.
689. Peeling, B.: One day at a time on a burn unit. *Can. Nurse* **74**(10):38, 1978.
690. Piggott, L.: Overview of selected basic research in autism. *J. Autism Dev. Disord.* **9**(2):199, 1979.
691. Plummer, S., Baer, D.,and LeBlanc, J.: Functional considerations in the use of procedural timeout and in effective alternative. *J. Appl. Behav. Anal.* **10**(4):689, 1977.
692. Prior, M., and Bradshaw, J.: Hemisphere functioning in autistic children. *Cortex* **15**(1):73, 1979.
693. Prior, M., and Hall, L.: Comprehension of transitive and intransitive phrases by autistic, retarded, and normal children. *J. Commun. Disord.* **12**(2):103, 1979.
694. Rathkey, J., Flax, S., Krug, D., and Arick, J.: A microprocessor-based aid for training autistic children. *Isa. Trans.* **18**(2):79, 1979.
695. Rathkey, J., Krug, D., Flax, S., and Arick, J.:A microprocessor-based aid for training autistic children. *Biomed. Sci. Instrum.* **14**:87, 1978.
696. Rincover, A.: Sensory extinction: A procedure form eliminating self-stimulatory behavior in developmentally disabled children. *J. Abnorm. Child. Psychol.* **6**(3):299, 1978.
697. Rincover, A.: Variables affecting stimulus fading and discriminative responding in psychotic children. *J. Abnorm. Psychol.* **87**(5):541, 1978.
698. Rincover, A., Cook, R. Peoples, A., and Packard, D.: Sensory extinction and sensory reinforcement principles for programming multiple adaptive behavior change. *J. Appl. Behav. Anal.* **12**(2):221, 1979.
699. Rincover, A., Newsom, C., Lovaas, O., and Koegel, R.: Some motivational properties of sensory stimulation in psychotic children. *J. Exp. Child Psychol.* **24**(2):312, 1977.
700. Ritvo, E., and Freeman, B.: Current research on the syndrome of autism: Introduction. The National Society for Autistic Children's definition of the syndrome of autism. *J. Am. Acad. Child Psychiatry* **17**(4):565, 1978.
701. Robertson, A.: Malingering, occupational medicine, and the law. *Lancet* **2**(8094):828, 1978.
702. Russo, D., Koegel, R., and Lovaas, O.: A comparison of human and automated instruction of autistic children. *J. Abnorm. Child Psychol.* **6**(2):189, 1978.

703. Rutter, M.: Language, cognition, and autism. *Res. Publ. Assoc. Res. Nerv. Ment. Dis.* **57**:247, 1979.

704. Rutter, M.: Diagnosis and definition of childhood autism. *J. Autism Child. Schizophr.* **8**(2):139, 1978.

705. Schopler, E., Andrews, C., and Strupp, K.: Do autistic children come from upper-middle-class parents. *J. Autism Dev. Disord.* **9**(2):139, 1979.

706. Schroeder, S., Schroeder, C., Smith, B., and Dalldorf, J.: Prevalence of self-injurious behaviors in a large state facility for the retarded: A three-year follow-up study. *J. Autism Child. Schizophr.* **8**(3):261, 1978.

707. Schuler, A.: Echolalia: Issues and clinical applications. *J. Speech Hear. Disord.* **44**(4):411, 1979.

708. Shapiro, T.: The quest for a linguistic model to study the speech of autistic children. Studies on echoing. *J. Am. Acad. Child Psychiatry* **16**(4):608, 1977.

709. Shapiro, T., and Kapit, R.: Linguistic negation in autistic and normal children. *J. Psycholing. Res.* **7**(5):337, 1978.

710. Shapiro, T., and Lucy, P.: Echoing in autistic children: A chronometric study of semantic processing. *J. Child Psychol. Psychiatry* **19**(4):373, 1978.

711. Shoham, S., Weissbrod, L., Markowsky, R., and Stein, Y.: The differential pressures towards schizophrenia and delinquency. *Genet. Psychol. Monogr.* **96**(second half):165, 1977.

712. Silva, P., Buckfield, P., Spears, G., and Williams, S.: Poisoning, burns, and other accidents experienced by a thousand dunedin three year olds: A report from the Dunedin multidisciplinary child development study. *N. Z. Med. J.* **87**(609):242, 1978.

713. Silverman, M.: Beyond the mainstream: The special needs of the chronic child patient. *Am. J. Orthopsychiatry* **49**(1):62, 1979.

714. Simons, R., McFadd, A., Frank, H., Green, L., Malin, R., and Morris, J.: Behavioral contracting in a burn care facility: A strategy for patient participation. *J. Trauma.* **18**(4):257, 1978.

715. Singh, N.: Reprogramming the social environment of an autistic child. *N. Z. Med. J.* **87**(606):135, 1978.

716. Sloan, J.: Differential development of autistic symptoms in a pair of fraternal twins. *J. Autism Child. Schizophr.* **8**(2):191, 1978.

717. Steinberg, D.: Some common psychiatric problems in adolescence. *Ir. Med. J.* **72**(9):366, 1979.

718. Stone, M.: Etiological factors in schizophrenia: A reevaluation in the light of contemporary research. *Psychiatr. Q.* **50**(2):83, 1978.

719. Strain, P., Kerr, M., and Ragland, E.: Effects of peer-mediated social initiations and prompting/reinforcement procedures on the social behavior of autistic children. *J. Autism Child. Schizophr.* **9**(1):41, 1979.

720. Sullivan, R.: Siblings of autistic children. *J. Autism Dev. Disord.* **9**(3):287, 1979.

721. Sullivan, R.: The politics of definitions: How autism got included in the developmental disabilities act. *J. Autism Dev. Disord.* **9**(2):221, 1979.

722. Sullivan, R.: The burn-out syndrome. *J. Autism Child. Schizophr.* **9**(1):112, 1979.

723. Sullivan, R.: Tbe hostage parent. *J. Autism Child. Schizophr.* **8**(2):233, 1978.

724. Sullivan, R.: Poems on autism: Beyond research data. *J. Autism Child. Schizophr.* **7**(4):397, 1977.

725. Susz, E., and Marberg, H.: Autistic withdrawal of a small child under stress. *Acta. Paedopsychiatr. (Basel)* **43**(4):149, 1978.

726. Sverd, J., Kupietz, S., Winsberg, B., Hurwic, M., and Becker, L.: Effects of L-5-hydroxytryptophan in autistic children. *J. Autism Child. Schizophr.* **8**(2):171, 1978.

727. Torisky, C.: Constance Torisky's response to Frank Warren's response. *J. Autism Child. Schizophr.* **8**(2):247, 1978.

728. Torisky, C.: The hostage parent: A life-style or a challenge. *J. Autism Child. Schizophr.* **8**(2):234, 1978.

729. Varni, J. Lovaas, O., Koegel, R., and Everett, N.: An analysis of observational learning in autistic and normal children. *J. Abnorm. Child Psychol.* **7**(1):31, 1979.

730. Vera Buhrmann, M.: Early recognition of infantile autism. *S. Afr. Med. J.* **56**(18):724, 1979.

731. Wakeman, J., and Kaplan, J.: An experimental study of hypnosis in painful burns. *Am. J. Clin. Hypn.* **21**(1):3, 1978.

732. Ward, A.: Early childhood autism and structural therapy: Outcome after 3 years. *J. Consult. Clin. Psychol.* **46**(3):586, 1978.

733. Warlick, S.: Sam—the patient nobody wanted to visit. *Nursing (Horsham)* **8**(7):56, 1978.

734. Warren, F.: The child as victim. *J. Autism Child. Schizophr.* **8**(2):240, 1978.

735. Waserman, M.: Relieving parental anxiety: John Warren's 1792 letter to the father of a burned child. *New Eng. J. Med.* **299**(3):135, 1978.

736. Watson, C., Burke, M., and Plemel, D.: The relationship of personality style to abstract thinking deficits in schizophrenia. *J. Clin. Psychol.* **35**(2):247, 1979.

737. Watson, C., Plemel, D., and Burke, M.: Proverb test deficit in schizophrenic and brain-damaged patients. *J. Nerv. Ment. Dis.* **167**(2):561, 1979.

738. Wells, K., Forehand, R., Hickey, K., and Green, K.: Effects of a procedure derived from the overcorrection principle on manipulated and nonmanipulated behaviors. *J. Appl. Behav. Anal.* **10**(4):679, 1977.

739. Wessels, W., Pompe Van Meerdervoort, M.: Monozygotic twins with early infantile autism. A case report. *S. Afr. Med. J.* **55**(23):955, 1979.

740. West, D., and Shuck, J.: Emotional problems of the severely burned patient. *Surg. Clin. North Amer.* **58**(6):1189, 1978.

741. Wildman, R., and Simon, S.: An Indirect method for increasing the rate of social interaction in an autistic child. *J. Clin. Psychol.* **34**(1):144, 1978.

742. Wing, L.: The current status of childhood autism. *Psychol. Med.* **9**(1):9, 1979.

743. Wolff, S., and Barlow, A.: Schizoid personality in childhood: A comparative study of schizoid, autistic and normal children. *J. Child Psychol. Psychiatry* **20**(1):29, 1979.

744. Zifferblatt, S., Burton, S., Horner, R., and White, T.: Establishing generalization effects among autistic children. *J. Autism Child. Sciizophr.* **7**(4):337, 1977.

745. Akerley, M.: Reactions to employing electric shock with autistic children. *J. Autism Child. Schizophr.* **6**(3):289, 1976.

746. Baltaxe, C., and Simmons, J.: Bedtime soliloquies and linguistic competence in autism. *J. Speech Hear. Disord.* **42**(3):376, 1977.

747. Baltaxe, C., and Simmons, J.: Language in childhood psychosis: A review. *J. Speech Hear. Disord.* **40**(4):439, 1975.

748. Benaroya, S., Wesley, S., Ogilvie, H., Klein, L., and Meaney, M.: Sign language and multisensory input training of children with communication and related developmental disorders. *J. Autism Child. Schizophr.* **7**(1):23, 1977.

749. Boucher, J.: Birmingham University, England. Hand preference in autistic children and their parents. *J. Autism Child. Schizophr.* **7**(2):177, 1977.

750. Bram, S., and Meir, M.: A relationship between motor control and language development in an autistic child. *J. Autism Child. Schizophr.* **7**(1):57, 1977.

751. Cantwell, D., and Baker, L.: Psychiatric disorder in children with speech and language retardation. A critical review. *Arch. Gen. Psychiatry* **34**(5):583, 1977.

752. Cohen, D., Caparulo, B., and Shaywitz, B.: Primary childhood aphasia and childhood autism: Clinical, biological, and conceptual observations. *J. Am. Acad. Child Psychiatry* **15(4):604, 1976.**

753. Cohen, D., Young, J., and Roth, J.: Platelet mondamine oxidase in early childhood autism. *Arch. Gen. Psychiatry* **34(5):534, 1977.**

754. Colletti, G., and Harris, S.: Behavior modification in the home: Siblings as behavior modifiers, parents as observers. *J. Abnorm. Child Psychol.* **5(1):21, 1977.**

755. Cox, M., and Klinge, V.: Treatment and management of a case of self-burning. *Behav. Res. Ther.* **14(5):382, 1976.**

756. Donley, D.: The immune system: Nursing the patient who is immunosuppressed. *Am. J. Nurs.* **76(10):1619, 1976.**

757. Emig, E., and Lloyd, J.: How to get burned children home sooner. *Rn* **40(7):37, 1977.**

758. Koegel, R., and Rincover, A.: Research on the difference between generalization and maintenance in extra-therapy responding. *J. Appl. Behav. Anal.* **10(1):1, 1977.**

759. Koegel, R., Russo, D., and Rincover, A.: Assessing and training teachers in the generalized use of behavior modification with autistic children. *J. Appl. Behav. Anal.* **10(2):197, 1977.**

760. Koh, T.: Cognitive-perceptual treatment of exceptional children. *Child Care Health Dev.* **2(5):251, 1976.**

761. Kopel, H.: The autistic child in dental practice. *J. Dent. Child.* **44(4):302, 1977.**

762. Kotsopoulos, S.: Infantile autism in dizygotic twins. A case report. *J. Autism Child. Schizophr.* **6(2):133, 1976.**

763. Lavigna, G.: Communication training in mute autistic adolescents using the written work. *J. Autism Child. Schizophr.* **7(2):135, 1977.**

764. Lichstein, K., and Schreibman, L.: Employing electric shock with autistic children. A review of the side effects. *J. Autism Child. Schizophr.* **6(2):163, 1976.**

765. Lindley, P., Marks, I., Philpott, R., and Snowden, J.: Treatment of obsessive–compulsive neurosis with history of childhood autism. *Br. J. Psychiatry* **130:592, 1977.**

766. MacLennan, B.: Modifications of activity group therapy for children. *Int. J. Group Psychother.* **27(1):85, 1977.**

767. Marcus, L.: Patterns of coping in families of psychotic children. *Am. J. Orthopsychiatry* **47(3):388, 1977.**

768. Margolies, P.: Behavioral approaches to the treatment of early infantile autism: A review. *Psychol. Bull.* **84(2):249, 1977.**

769. Marsh, R.: The diagnosis, epidemiology, and etiology of childhood schizophrenia. *Genet. Psychol. Monogr.* **95(2):267, 1977.**

770. Miller, W., Gardner, N., and Mlott, S.: Psychosocial support in the treatment of severely burned patients. *J. Trauma.* **16(9):722, 1976.**

771. Ricketts, L.: Music and handicapped children. *J.R. Coll. Gen. Pract.* **26(169):585, 1976.**

772. Rosenbaum, M., and Breiling, J.: The development and functional control of reading–comprehension behavior. *J. Appl. Behav. Anal.* **9(3):323, 1976.**

773. Savedra, M.: Coping with pain: Strategies of severely burned children. *Can. Nurse* **73(8):28, 1977.**

774. Savedra, M.: Coping with pain: Strategies of severely burned children. *Matern. Child Nurs. J.* **5(3):197, 1976.**

775. Savedra, M.: The severely burned child. Moving from hospital to home. *Mon.* **2(4):220, 1977.**

776. Savedra, M.: The severely burned child. The child and his family at home: What then? *Mon.* **2(4):224, 1977.**

777. Simon, C.: The environmental language intervention strategy: A laudatory comment regarding the versatility of its clinical applications letter. *J. Speech Hear. Disord.* **41**(4):557, 1976.
778. Singletary, Y.: More than skin deep. *J. Psychiatr. Nurs.* **15**(2):7, 1977.
779. Swift, W.: Emotional care and support: An overseas tour around burns units and centres. *Aust. Nurses. J.* **6**(5):31, 1976.
780. Upton, G., Vikert, D., and Herkert, J.: A video unit in school—is it worthwhile? *Spec. Educ. Forward Trends* **4**(1):23, 1977.
781. Wenar, C., and Ruttenberg, R.: The use of briac for evaluating therapeutic effectiveness. *J. Autism Child. Schizophr.* **6**(2):175, 1976.
782. Wing, L., Gould, J. Yeates, S., and Brierley, L.: Symbolic play in severely mentally retarded and in autistic children. *J. Child Psychol. Psychiatry* **18**(2):167, 1977.
783. Wing, L., and Ricks, D.: The aetiology of childhood autism: A criticism of the Tinbergen's ethological theory. *Psychol. Med.* **6**(4):533, 1976.
784. Editorial: Reacting to autistic children. *Br. Med. J.* **11**(6023):1425, 1976.
785. The sensory deprivations: An approach to the study of the induction of affects. *J. Am. Psychoanal. Assoc.* **22**(3):626, 1974.
786. Akerley, M.: Springing the tradition trap. *J. Autism Child. Schizophr.* **5**(4):373, 1975.
787. Akerley, M.: The invulnerable parent. *J. Autism Child. Schizophr.* **5**(3):275, 1975.
788. Aleksandrowicz, M.: The little prince: Psychotherapy of a boy with borderline personality structure. *Int. J. Psychoanal. Psychother.* **4**:410, 1975.
789. Anderson, L.: Both burn patient and brother. *Aorn. J.* **20**(5):863, 1974.
790. Ando, H., and Tsuda, K.: Intrafamilial incidence of autism, cerebral palsy, and mongolism. *J. Autism Child. Schizophr.* **5**(3):267, 1975.
791. Bartle, R.: Teaching the handicapped child to learn. *Midwives Chron.* **88**(1051):274, 1975.
792. Bender, L.: The family patterns of 100 schizophrenic children observed at Bellevue, 1935–1952. *J. Autism Child. Schizophr.* **4**(4):279, 1974.
793. Benians, R.: A child psychiatrist looks at burned children and their families. *Guys Hosp. Rep.* **123**(2):149, 1974.
794. Bentley, D.: Incommunicado: A review of childhood autism. *Health Visit.* **47**(1):4, 1974.
795. Bentley, D.: Incommunicado: A review of childhood autism. *Australas. Nurses J.* **2**(35):9, 1974.
796. Berlin, I., Chess, S., Eisenberg, L., Goldfarb, W., and Wing, L.: Springing the tradition trap continued. *J. Autism Child. Schizophr.* **6**(1):93, 1976.
797. Bernstein, J.: The autistic character. *Psychoanal. Rev.* **62**(4):537, 1975–1976.
798. Blank, H.: Reflections on the special senses in relation to the development of affect with special emphasis on blindness. *J. Am. Psychoanal. Assoc.* **23**(1):32, 1975.
799. Breslin, P.: The psychological reactions of children to burn traumata: A review. Part II. *Ill. Med. J.* **148**(6):595, 1975.
800. Breslin, P.: The psychological reactions of children to burn traumata: A review. *Ill. Med. J.* **148**(5):519, 1975.
801. Brodie, C., and Mitchell, B.: Pathophysiology of stress ulcer, in Clearfield, H., and Dinoso, V. (Eds), *Gastrointestinal Emergencies.* New York, Grune & Stratton, 1976.
802. Byassee, J., and Murrell, S.: Interaction patterns in families of autistic, disturbed, and normal children. *Am. J. Orthopsychiatry* **45**(3):473, 1975.
803. Campbell, L.: Special behavioral problems of the burned child. *Am. J. Nurs.* **76**(2):220, 1976.
804. Chang, F., and Herzog, B.: Burn morbidity: A followup study of physical and psychological disability. *Ann. Surg.* **183**(1):34, 1976.

805. Clancy, H., and McBride, G.: The isolation syndrome in childhood. *Dev. Med. Child Neurol.* **17**(2):198, 1975.
806. Cline, J.: Movement therapy with an autistic boy. *Perspect. Psychiatr. Care.* **13**(1):19, 1975.
807. Cohen, D., and Caparulo, B.: Childhood autism. *Child Today* **4**(4):2, 1975.
808. Colby, K., and Kraemer, H.: An objective measurement of nonspeaking children's performance with a computer-controlled program for the stimulation of language behavior. *J. Autism Child. Schizophr.* **5**(2):139, 1975.
809. Condon, W.: Multiple response to sound in dysfunctional children. *J. Autism Child. Schizophr.* **5**(1):37, 1975.
810. Cox, A., Rutter, M., Newman, S., and Bartak, L.: A comparative study of infantile autism and specific developmental receptive language disorder. II. Parental Characteristics. *Br. J. Psychiatry* **126**:146, 1975.
811. Davids, A.: Childhood psychosis. The problem of differential diagnosis. *J. Autism Child. Schizophr.* **5**(2):129, 1975.
812. Davids, A.: Effects of human and nonhuman stimuli on attention and learning in psychotic children. *Child Psychiatry Hum. Dev.* **5**(2):108, 1974.
813. Demyer, M.: Research in infantile autism: A strategy and its results. *Biol. Psychiatry* **10**(4):433, 1975.
814. Dewey, M., and Everard, M.: Autism, early infantile. Parents speak. *J. Autism Child. Schizophr.* **4**(4):347, 1974.
815. Drabman, R.: An integrated approach to treating low-functioning children. *Curr. Psychiatr. Ther.* **15**:45, 1975.
816. Dubner, H.: Listening: A goal of therapy for the autistic child. *Rehabil. Lit.* **36**(10):306, 1975.
817. Dyer, C., and Hadden, A.: A multi-axial classification for the education of autistic children. *Child Care Health Dev.* **2**(3):155, 1976.
818. Elgar, S.: First year at Somerset court. *Spec. Educ. Forward Trends* **2**(2):14, 1975.
819. Fagerhaugh, S.: Pain expression and control on a burn care unit. *Nurs. Outlook* **22**(10):645, 1974.
820. Ferster, C.: Clinical reinforcement. *Semin. Psychiatry* **4**(2):101, 1972.
821. Fish, B.: Biologic antecendents of psychosis in children. *Res. Publ. Assoc. Nerv. Ment. Dis.* **54**:49, 1975.
822. Fisher, S.: On the development of the capacity to use transitional objects. A case study of an autistic child. *J. Am. Acad. Child Psychiatry* **14**(1):114, 1975.
823. Forness, S.: Educational approaches to autism. *Train. Sch. Bull. (Vinel)* **71**(3):167, 1974.
824. Forrest, A.: Psychological medicine. Mental handicap and syndromes of brain damage in children. *Br. Med. J.* **2**(5962):71, 1975.
825. Frank, S., Allen, D., Stein, L., and Myers, B.: Linguistic performance in vulnerable and autistic children and their mothers. *Am. J. Psychiatry* **133**(8):909, 1976.
826. Frankel, F., and Graham, V.: Systematic observation of classroom behavior of retarded and autistic preschool children. *Am. J. Ment. Defic.* **81**(1):73, 1976.
827. Freeman, B., Ritvo, E., and Miller, R.: An operant procedure to teach an echolaic, autistic child to answer questions appropriately. *J. Autism Child. Schizophr.* **5**(2):169, 1975.
828. Freeman, B., Somerset, T., and Ritvo, E.: Effect of duration of time out in suppressing disruptive behavior of a severely autistic child. *Psychol. Rep.* **38**(1):124, 1976.
829. Friedman, S., and Morse, C.: Child abuse: A five-year follow-up of early case finding in the emergency department. *Pediatrics* **54**(4):404, 1974.
830. Gallagher, J., and Wiegerink, R.: Educational strategies for the autistic child. *J. Autism Child. Schizophr.* **6**(1):15, 1976.

831. Grossman, M.: Early child development in the context of mothering experiences. *Child Psychiatry Hum. Dev.* **5**(4):216, 1975.

832. Gruen, A.: The discontinuity in the ontogeny of self: Possibilities for integration or destructiveness. *Psychoanal. Rev.* **61**(4):557, 1974–1975.

833. Hamburg, D.: Coping behavior in life-threatening circumstances. *Psychother. Psychosom.* **23**(1–6):13, 1974.

834. Hargrave, E., and Swisher, L.: Modifying the verbal expression of a child with autistic behaviors. *J. Autism Child. Schizophr.* **5**(2):147, 1975.

835. Harper, J.: Age and type of onset as critical variables in early infantile autism. *J. Autism Child. Schizophr.* **5**(1):25, 1975.

836. Harris, S.: Teaching language to nonverbal children—with emphasis on problems of generalization. *Psychol. Bull.* **82**(4):565, 1975.

837. Harris, S.: Letter: Contingent and noncontingent reinforcement. *J. Autism Child. Schizophr.* **4**(2):94, 1974.

838. Helm, D.: Psychodynamic and behavior modification approaches to the treatment of infantile autism empirical similarities. *J. Autism Child. Schizophr.* **6**(1):27, 1976.

839. Holroyd, J., and McArthur, D.: Mental retardation and stress on the parents: A contrast between Down's syndrome and childhood autism. *Am. J. Ment. Defic.* **80**(4):431, 1976.

840. Jellis, T., and Grainger, S.: The back projection of kaleidoscopic patterns as a technique for eliciting verbalizations in an autistic child. A final note. *Br. J. Disord. Commun.* **9**(1):65, 1974.

841. Kean, J.: The development of social skills in autistic twins. *N. Z. Med. J.* **81**(534):204, 1975.

842. King, P.: Early infantile autism. Relation to schizophrenia. *J. Am. Acad. Child Psychiatry* **14**(4):666, 1975.

843. Kirman, B.: The clinical assessment of mental handicap. *Br. J. Psychiatry* **Spec no.** 9:337, 1975.

844. Koegel, R., Firestone, P., Kramme, K., and Dunlap, G.: Increasing spontaneous play by suppressing self-stimulation in autistic children. *J. Appl. Behav. Anal.* **7**(4):521, 1974.

845. Koegel, R., and Rincover, A.: Treatment of psychotic children in a classroom environment: I. Learning in a large group. *J. Appl. Behav. Anal.* **7**(1):45, 1974.

846. Kueffner, M.: Passage through hospitalization of severely burned, isolated school-age children. *Commun. Nurs. Res.* **7**:181, 1976.

847. Lotter, V.: Factors related to outcome in autistic children. *J. Autism Child. Schizophr.* **4**(3):263, 1974.

848. Lovaas, O., Schreibman, L., and Koegel, R.: A behavior modification approach to the treatment of autistic children. *J. Autism Child. Schizophr.* **4**(2):111, 1974.

849. MacDonald, R., and Allan, J.: The use of fantasy enactment in the treatment of an emerging autistic child. *J. Anal. Psychol.* **20**(1):57, 1975.

850. Mahoney, G.: Ethological approach to delayed language acquisition. *Am. J. Ment. Defic.* **80**(2):139, 1975.

851. Marchant, R., Howlin, P., Yule, W., and Rutter, M.: Graded change in the treatment of the behavior of autistic children. *J. Child Psychol. Psychiatry* **15**(3):221, 1974.

852. Massie, H.: The early natural history of childhood psychosis. *J. Am. Acad. Child Psychiatry* **14**(4):683, 1975.

853. McQuaid, P.: Infantile autism in twins. *Br. J. Psychiatry* **127**:530, 1975.

854. Meltzer, D.: Mutism in infantile autism, schizophrenia and manic-depressive states: The correlation of clinical psychopathology and linguistics. *Int. J. Psychoanal.* **55**(3):397, 1974.

855. Menyuk, P.: The bases of language acquisition: Some questions. *J. Autism Child. Schizophr.* **4**(4):325, 1974.

856. Moss, S., and Moss, M.: Surrogate mother–child relationships. *Am. J. Orthopsychiatry* **45**(3):382, 1975.

857. Narasimhachari, N., and Himwich, H.: Biochemical studies in early infantile autism. *Biol. Psychiatry* **10**(4):425, 1975.

858. Noyes, R., Andreasen, N., and Hartford, C.: The psychological reaction to severe burns. *Psychosomatics* **12**(6):416, 1971.

859. Olin, R.: Differentiating the psychotic child from the mentally retarded child. *Minn. Med.* **58**(6):489, 1975.

860. Ornitz, E., and Ritvo, E.: The syndrome of autism: A critical review. *Am. J. Psychiatry* **133**(6):609, 1976.

861. Park, D.: Operant conditioning of a speaking autistic child. *J. Autism Child. Schizophr.* **4**(2):189, 1974.

862. Piggott, L., and Simson, C.: Changing diagnosis of childhood psychosis. *J. Autism Child. Schizophr.* **5**(3):239, 1975.

863. Ploog, D.: Psychobiology of partnership behavior. *Psychol. Med.* **5**(4):327, 1975.

864. Prior, M., and Chen, C.: Learning set acquisition in autistic children. *J. Abnorm. Psychol.* **84**(6):701, 1975.

865. Reardon, J.: Occupational therapy treatment of the patient with thermally injured upper extremity. *Major Probl. Clin. Surg.* **19**:127, 1976.

866. Rincover, A., and Koegel, R.: Setting generality and stimulus control in autistic children. *J. Appl. Behav. Anal.* **8**(3):235, 1975.

867. Robson, K.: Development of object relations during the first year of life. *Semin. Psychiatry* **4**(4):301, 1972.

868. Romanczyk, R., Diament, C., Goren, E., Trunell, G., and Harris, S.: Increasing isolate and social play in severely disturbed children: Intervention and postintervention effectiveness. *J. Autism Child Schizophr.* **5**(1):57, 1975.

869. Rutter, M., and Sussenwein, F.: A developmental and behavioral approach to the treatment of preschool autistic children. *J. Autism Child. Schizophr.* **1**(4):376, 1971.

870. Rutter, M.: Psychiatric disorder and intellectual impairment in childhood. *Br. J. Psychiatry* **Spec No. 9**:344, 1975.

871. Sakamoto, M.: A case of acute childhood psychosis. *Folia Psychiatr. Neurol. Jpn.* **28**(4):307, 1974.

872. Schmitt, B., and Kempe, C.: The pediatrician's role in child abuse and neglect. *Curr. Probl. Pediatr.* **5**(5):3, 1975.

873. Schopler, E.: Toward reducing behavior problems in autistic children. *J. Autism Child. Schizophr.* **6**(1):1, 1976.

874. Schopler, E.: The stress of autism as ethology. *J. Autism Child. Schizophr.* **4**(3):193, 1974.

875. Schreibman, L.: Effects of within-stimulus and extra-stimulus prompting on discrimination learning in autistic children. *J. Appl. Behav. Anal.* **8**(1):91, 1975.

876. Seligman, R., Carroll, S., MacMillan, B.: Emotional responses of burned children in a pediatric intensive care unit. *Psychiatry Med.* **3**(1):59, 1972.

877. Sherman, T., and Webster, C.: The effects of stimulus-fading on acquisition of a visual position discrimination in autistic, retarded, and normal children. *J. Autism Child. Schizophr.* **4**(4):301, 1974.

878. Simon, N.: Echolaic speech in childhood autism. Consideration of possible underlying loci of brain damage. *Arch. Gen. Psychiatry* **32**(11):1439, 1975.

879. Smith, R., and Lau, M.: Developmental lines of achievement. *Child Psychiatry Hum. Dev.* **5**(2):117, 1974.

880. Smith, S., and Hanson, R.: 134 battered children: A medical and psychological study. *Br. Med. J.* **3**(5932):666, 1974.

881. Solnit, A., and Priel, B.: Psychological reactions to facial and hand burns in young men. Can I see myself through your eyes. *Psychoanal. Study Child.* **30**:549, 1975.

882. Solnit, A., and Priel, B.: Scared and scarred—psychological aspects in the treatment of soldiers with burns. *Isr. Ann. Psychiatry* **13**(3):213, 1975.

883. Stevens-Long, J., and Rasmussen, M.: The acquisition of simple and compound sentence structure in an autistic child. *J. Appl. Behav. Anal.* **7**(3):473, 1974.

884. Talabere, L., and Graves, P.: A tool for assessing families of burned children. *Am. J. Nurs.* **76**(2):225, 1976.

885. Watt, J.: The injuries of four centuries of naval warfare. *Ann. R. Coll. Surg. Engl.* **57**(1):3, 1975.

886. Wilhelm, H., and Lovaas, O.: Stimulus overselectivity: A common feature in autism and mental retardation. *Am. J. Ment. Defic.* **81**(1):26, 1976.

887. Wilmore, D.: Nutrition and metabolism following thermal injury. *Clin. Plast. Surg.* **1**(4):603, 1974.

888. Wing, L.: The syndrome of early childhood autism. *Br. J. Psychiatry* **Spec No. 9**:349, 1975.

889. Wood, E.: The wild boy of Aveyron (Itard's syndrome?). *Nurs. Mirror.* **140**(18):61, 1975.

890. Wright, L., and Fulwiler, R.: Long range emotional sequelae of burns: Effects on children and their mothers. *Pediatr. Res.* **8**(12):931, 1974.

891. Wulbert, M., Barach, R., Perry, M., Straughan, J., Sulzbacher, S., Turner, I., and Wiltz, N.: The generalization of newly acquired behaviors by parents and child across three different settings. A study of an autistic child. *J. Abnorm. Child Psychol.* **2**(2):87, 1974.

892. Zisman, A.: Letter: Environmental autism. *Pediatrics* **58**(2):297, 1976.

Index